...thematics
Revision and Practice

General Mathematics:
Revision and Practice

D. Rayner

Second edition
with investigations and answers

Oxford University Press

Oxford University Press, Walton Street, Oxford OX2 6DP
Oxford New York Toronto
Delhi Bombay Calcutta Madras Karachi
Petaling Jaya Singapore Hong Kong Tokyo
Nairobi Dar es Salaam Cape Town
Melbourne Auckland

and associated companies in
Berlin Ibadan

Oxford is a trade mark of Oxford University Press

© D Rayner 1988
ISBN 0 19 914278 5

First edition published in 1984
Second edition publishcd 1988
Reprinted 1990 (twice), 1991, 1992

Acknowledgements
The publishers would like to thank the following for
permission to use photographs:
Bettmann Archive
Popperfoto
Ann Ronan Picture Library

Printed and bound in Great Britain by
Butler & Tanner Ltd, Frome and London

Preface

This book is for candidates working towards GCSE in mathematics: it covers the requirements of all the major examination boards. The book contains teaching notes, worked examples and carefully graded exercises. These can be used selectively for classwork, for homework and for later revision.

The work is divided into short topics to suit any required order of treatment. The wide choice of questions provides plenty of practice in the basic skills and leads on to work of a more demanding nature. Each major part of the book concludes with two exercises: one of short revision questions and another of actual past examination questions. All numerical answers are given at the end of the book.

The author is indebted to the many pupils and colleagues who have assisted him in this work. He is particularly grateful to Reg Moxom, Philip Cutts, Michael Day and Micheline Dubois for their invaluable work of correction and checking.

For the second edition the whole text has been revised to reflect the new GCSE syllabuses. Redundant material has been removed and in some cases new work has been added. Many exercises have been rewritten to add extra interest and relevance to everyday life.

In the new chapter eleven there are numerous ideas for investigations, practical problems and puzzles, many of which can be extended at the discretion of the teacher and used as a basis for assessment. Also new is chapter twelve which consists of mental arithmetic tests and multiple choice revision tests.

The author would like to thank Lawrence Campbell and all his colleagues at school for all their help and suggestions.

Thanks are also due to the following examination boards for kindly allowing the use of questions from their past mathematics papers.

Associated Examining Board
East Anglian Examination Board
Joint Matriculation Board
Northern Examining Association for Joint Examinations
Oxford and Cambridge Schools Examination Board
Oxford Delegacy for Local Examinations
Southern Universities' Joint Board
University of Cambridge Local Examinations Syndicate
University of London School Examinations Department
Welsh Joint Education Committee.

Contents

Contents

1 Number

Karl Friedrich Gauss (1777–1855) was the son of a German labourer and is thought by many to have been the greatest all-round mathematician of all time. He considered that his finest discovery was the method for constructing a regular seventeen-sided polygon. This was not of the slightest use outside the world of mathematics, but was a great achievement of the human mind. Gauss would not have understood the modern view held by many that mathematics must somehow be 'useful' to be worthy of study.

1.1 ARITHMETIC

Decimals

Example 1

Evaluate: (a) $7·6 + 19$
        (b) $3·4 − 0·24$
        (c) $7·2 × 0·21$
        (d) $0·84 ÷ 0·2$
        (e) $3·6 ÷ 0·004$

(a)
$$\begin{array}{r} 7·6 \\ + 19.0 \\ \hline 26·6 \\ \hline \end{array}$$

(b)
$$\begin{array}{r} 3·\overset{31}{4}0 \\ - 0·24 \\ \hline 3·16 \\ \hline \end{array}$$

(c)
$$\begin{array}{r} 7·2 \\ × 0·21 \\ \hline 72 \\ 1440 \\ \hline 1.512 \\ \hline \end{array}$$
No decimal points in the working, '3 figures after the points in the question *and* in the answer'.

(d) $0·84 : 0·2 = 8·4 : 2$

$$\begin{array}{r} 4·2 \\ 2\overline{)8·4} \end{array}$$
Multiply both numbers by 10 so that we can divide by a whole number

(e) $3·6 ÷ 0·004 = 3600 ÷ 4$
                 $= 900$

Exercise 1

Evaluate the following without a calculator:

1. $7·6 + 0·31$
2. $15 + 7·22$
3. $7·004 + 0·368$
4. $0·06 + 0·006$
5. $4·2 + 42 + 420$
6. $3·84 − 2·62$
7. $11·4 − 9·73$
8. $4·61 − 3$
9. $17 − 0·37$
10. $8·7 + 19·2 − 3·8$
11. $25 − 7·8 + 9·5$
12. $3·6 − 8·74 + 9$
13. $20·4 − 20·399$
14. $2·6 × 0·6$
15. $0·72 × 0·04$
16. $27·2 × 0·08$

17. $0{\cdot}1 \times 0{\cdot}2$
18. $(0{\cdot}01)^2$
19. $2{\cdot}1 \times 3{\cdot}6$
20. $2{\cdot}31 \times 0{\cdot}34$
21. $0{\cdot}36 \times 1000$
22. $0{\cdot}34 \times 100\,000$
23. $3{\cdot}6 \div 0{\cdot}2$
24. $0{\cdot}592 \div 0{\cdot}8$
25. $0{\cdot}1404 \div 0{\cdot}06$
26. $3{\cdot}24 \div 0{\cdot}002$
27. $0{\cdot}968 \div 0{\cdot}11$
28. $600 \div 0{\cdot}5$
29. $0{\cdot}007 \div 4$
30. $2640 \div 200$
31. $1100 \div 5{\cdot}5$
32. $(11 + 2{\cdot}4) \times 0{\cdot}06$
33. $(0{\cdot}4)^2 \div 0{\cdot}2$
34. $77 \div 1000$
35. $(0{\cdot}3)^2 \div 100$
36. $(0{\cdot}1)^4 \div 0{\cdot}01$
37. $\dfrac{92 \times 4{\cdot}6}{2{\cdot}3}$
38. $\dfrac{180 \times 4}{36}$
39. $\dfrac{0{\cdot}55 \times 0{\cdot}81}{4{\cdot}5}$
40. $\dfrac{63 \times 600 \times 0{\cdot}2}{360 \times 7}$

Exercise 2

1. A maths teacher bought 40 calculators at £8·20 each and a number of other calculators costing £2·95 each. In all she spent £387. How many of the cheaper calculators did she buy?

2. At a temperature of 20°C the common amoeba reproduces by splitting in half every 24 hours. If we start with a single amoeba how many will there be after (a) 8 days, (b) 16 days?

3. Copy and complete.
$3^2 + 4^2 + 12^2 = 13^2$
$5^2 + 6^2 + 30^2 = 31^2$
$6^2 + 7^2 +\ \ \ =$
$x^2 +\ \ +\ \ =$

4. Find all the missing digits in these multiplications.

(a)
```
   5 *
   9 ×
  ────
  0 * * 6
```
(b)
```
   * 7
   * ×
  ────
  4 * 6
```
(c)
```
   5 *
   * ×
  ────
  1 * 4
```

5. Pages 6 and 27 are on the same (double) sheet of a newspaper. What are the page numbers on the opposite side of the sheet? How many pages are there in the newspaper altogether?

6. Use the numbers 1, 2, 3, 4, 5, 6, 7, 8, 9 once each and in their natural order to obtain an answer of 100. You may use only the operations $+$, $-$, $\times$, $\div$.

7. The ruler below has eleven marks and can be used to measure lengths from one unit to twelve units.

Design a ruler which can be used to measure all the lengths from one unit to twelve units but this time put the minimum possible number of marks on the ruler.

8. Each packet of washing powder carries a token and four tokens can be exchanged for a free packet. How many free packets will I receive if I buy 64 packets?

9. Put three different numbers in the circles so that when you add the numbers at the end of each line you always get a square number.

10. Put four different numbers in the circles so that when you add the numbers at the end of each line you always get a square number.

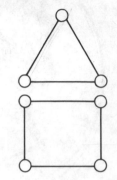

11. A group of friends share a bill for £13·69 equally between them. How many were in the group?

Fractions

Common fractions are added or subtracted from one another directly only when they have a common denominator.

Example 1

Evaluate: (a) $\frac{3}{4} + \frac{2}{5}$ (b) $2\frac{3}{8} - 1\frac{5}{12}$

(c) $\frac{2}{5} \times \frac{6}{7}$ (d) $2\frac{2}{5} \div 6$

(a) $\frac{3}{4} + \frac{2}{5} = \frac{15}{20} + \frac{8}{20}$

$= \frac{23}{20}$

$= 1\frac{3}{20}$

(b) $2\frac{3}{8} - 1\frac{5}{12} = \frac{19}{8} - \frac{17}{12}$

$= \frac{57}{24} - \frac{34}{24}$

$= \frac{23}{24}$

(c) $\frac{2}{5} \times \frac{6}{7} = \frac{12}{35}$

(d) $2\frac{2}{5} \div 6 = \frac{12}{5} \div \frac{6}{1}$

$= \frac{12}{5} \times \frac{1}{6} = \frac{2}{5}$

Exercise 3

Evaluate and simplify your answer.

1. $\frac{3}{4} + \frac{4}{5}$ 2. $\frac{1}{3} + \frac{1}{8}$ 3. $\frac{5}{6} + \frac{6}{9}$

4. $\frac{3}{4} - \frac{1}{3}$ 5. $\frac{3}{5} - \frac{1}{3}$ 6. $\frac{1}{2} - \frac{2}{5}$

7. $\frac{2}{3} \times \frac{4}{5}$ 8. $\frac{1}{7} \times \frac{5}{6}$ 9. $\frac{5}{8} \times \frac{12}{13}$

10. $\frac{1}{3} \div \frac{4}{5}$ 11. $\frac{3}{4} \div \frac{1}{6}$ 12. $\frac{5}{6} \div \frac{1}{2}$

13. $\frac{3}{8} + \frac{1}{5}$ 14. $\frac{3}{8} \times \frac{1}{5}$ 15. $\frac{3}{8} \div \frac{1}{5}$

16. $1\frac{3}{4} - \frac{2}{3}$ 17. $1\frac{3}{4} \times \frac{2}{3}$ 18. $1\frac{3}{4} \div \frac{2}{3}$

19. $3\frac{1}{2} + 2\frac{3}{5}$ 20. $3\frac{1}{2} \times 2\frac{3}{5}$ 21. $3\frac{1}{2} \div 2\frac{3}{5}$

22. $(\frac{3}{4} - \frac{2}{3}) \div \frac{3}{4}$ 23. $(\frac{3}{5} + \frac{1}{3}) \times \frac{5}{7}$

24. $\dfrac{\frac{3}{8} - \frac{1}{5}}{\frac{7}{10} - \frac{2}{3}}$ 25. $\dfrac{\frac{2}{3} + \frac{1}{5}}{\frac{3}{4} - \frac{1}{3}}$

26. Arrange the fractions in order of size:
    (a) $\frac{7}{12}, \frac{1}{2}, \frac{2}{3}$ (b) $\frac{3}{4}, \frac{2}{3}, \frac{5}{6}$
    (c) $\frac{1}{3}, \frac{17}{24}, \frac{5}{8}, \frac{3}{4}$ (d) $\frac{5}{6}, \frac{8}{9}, \frac{11}{12}$

27. Find the fraction which is mid-way between the two fractions given:
    (a) $\frac{2}{5}, \frac{3}{5}$ (b) $\frac{5}{8}, \frac{7}{8}$
    (c) $\frac{3}{3}, \frac{3}{4}$ (d) $\frac{1}{3}, \frac{4}{9}$
    (e) $\frac{4}{15}, \frac{1}{3}$ (f) $\frac{3}{8}, \frac{11}{24}$

28. In the equation below all the asterisks stand for the same number. What is the number?
    $$\left[\frac{*}{*} - \frac{*}{6} = \frac{*}{30}\right]$$

29. When it hatches from its egg, the shell of a certain crab is 1 cm across. When fully grown the shell is approximately 10 cm across. Each new shell is one-third bigger than the previous one.
    How many shells does a fully grown crab have during its life?

30. Glass A contains 100 ml of water and glass B contains 100 ml of wine.

A 10 ml spoonful of wine is taken from glass B and mixed thoroughly with the water in glass A. A 10 ml spoonful of the mixture from A is returned to B. Is there now more wine in the water or more water in the wine?

Fractions and decimals

A decimal fraction is simply a fraction expressed in tenths, hundredths etc.

Example 1

(a) Change $\frac{7}{8}$ to a decimal fraction.
(b) Change 0·35 to a vulgar fraction.
(c) Change $\frac{1}{3}$ to a decimal fraction.

(a) $\frac{7}{8}$, divide 8 into 7

$$\frac{7}{8} = 0·875 \qquad \begin{array}{r} 0·875 \\ 8)\overline{7·000} \end{array}$$

(b) $0·35 = \frac{35}{100} = \frac{7}{20}$

(c) $\frac{1}{3}$, divide 3 into 1

$$\frac{1}{3} = 0·\dot{3} \ (0·3 \text{ recurring}) \qquad \begin{array}{r} 0·3\,3\,3\,3 \\ 3)\overline{1·0^10^10^10^1000} \end{array}$$

Exercise 4

In questions **1** to **24**, change the fractions to decimals.

1. $\frac{1}{4}$ 2. $\frac{2}{5}$ 3. $\frac{4}{5}$ 4. $\frac{3}{4}$

5. $\frac{1}{2}$ 6. $\frac{3}{8}$ 7. $\frac{9}{10}$ 8. $\frac{5}{8}$

9. $\frac{5}{12}$ 10. $\frac{1}{6}$ 11. $\frac{2}{3}$ 12. $\frac{5}{6}$

13. $\frac{2}{7}$ 14. $\frac{3}{7}$ 15. $\frac{4}{9}$ 16. $\frac{5}{11}$

17. $1\frac{1}{5}$ 18. $2\frac{5}{8}$ 19. $2\frac{1}{3}$ 20. $1\frac{7}{10}$

21. $2\frac{3}{16}$ 22. $2\frac{2}{7}$ 23. $2\frac{6}{7}$ 24. $3\frac{19}{100}$

In questions **25** to **40**, change the decimals to vulgar fractions and simplify.

25. 0·2   26. 0·7   27. 0·25   28. 0·45
29. 0·36   30. 0·52   31. 0·125   32. 0·625
33. 0·84   34. 2·35   35. 3·95   36. 1·05
37. 3·2   38. 0·27   39. 0·007   40. 0·00011

Evaluate, giving the answer to 2 decimal places:

41. $\frac{1}{4} + \frac{1}{3}$ 42. $\frac{2}{3} + 0·75$

43. $\frac{8}{9} - 0·24$ 44. $\frac{7}{8} + \frac{5}{9} + \frac{2}{11}$

45. $\frac{1}{3} \times 0·2$ 46. $\frac{5}{8} \times \frac{1}{4}$

47. $\frac{8}{11} \div 0·2$ 48. $(\frac{4}{7} - \frac{1}{3}) \div 0·4$

Arrange the numbers in order of size (smallest first)

49. $\frac{1}{3}, 0·33, \frac{4}{15}$ 50. $\frac{2}{7}, 0·3, \frac{4}{9}$

51. $0·71, \frac{7}{11}, 0·705$ 52. $\frac{4}{13}, 0·3, \frac{5}{18}$

1.2 NUMBER FACTS AND SEQUENCES

Number facts

(a) A *prime* number is divisible only by itself and by one.
e.g. 2, 3, 5, 7, 11, 13 . . .

(b) The *multiples* of 12 are 12, 24, 36, 48 . . .

(c) The *factors* of 12 are 1, 2, 3, 4, 6, 12.

(d) Rational and irrational numbers.
The *exact* value of a *rational* number can be written down as the ratio of two whole numbers.
e.g. 3, $2\frac{1}{2}$, 5·72, $-3\frac{1}{4}$.
The exact value of an *irrational* number *cannot* be written down.
e.g. π, $\sqrt{2}$, $\sqrt{3}$, $\sqrt{5}$.

Exercise 5

1. Which of the following are prime numbers?
3, 11, 15, 19, 21, 23, 27, 29, 31, 37, 39, 47, 51, 59, 61, 67, 72, 73, 87, 99.

2. Write down the first five multiples of the following numbers:
(a) 4   (b) 6   (c) 10
(d) 11   (e) 20.

3. Write down the first six multiples of 4 and of 6. What are the first two *common* multiples of 4 and 6? [i.e. multiples of both 4 and 6]

4. Write down the first six multiples of 3 and of 5. What is the lowest common multiple of 3 and 5?

5. Write down all the factors of the following:
(a) 6   (b) 9   (c) 10
(d) 15   (e) 24   (f) 32

6. Decide which of the following are rational numbers and which are irrational:
(a) 3·5   (b) 3·153   (c) $\sqrt{7}$
(d) $\frac{1}{3}$ (e) 0·072   (f) $\sqrt{2}$
(g) $\sqrt{4}$ (h) π (i) $\dfrac{\sqrt{3}}{2}$
(j) $\sqrt{100}$ (k) $-2\frac{3}{7}$ (l) $\sqrt{5}$

7. (a) Is 263 a prime number? By how many numbers do you need to divide 263 so that you can find out?
(b) Is 527 a prime number?
(c) Suppose you used a computer to find out if 1147 was a prime number. Which numbers would you tell the computer to divide by?

8. Make six prime numbers using the digits 1, 2, 3, 4, 5, 6, 7, 8, 9 once each.

Sequences

Exercise 6

Write down each sequence and find the next two numbers.

1. 2, 6, 10, 14   2. 2, 9, 16, 23
3. 95, 87, 79, 71   4. 13, 8, 3, −2
5. 7, 9, 12, 16   6. 20, 17, 13, 8
7. 1, 2, 4, 7, 11   8. 1, 2, 4, 8
9. 55, 49, 42, 34   10. 10, 8, 5, 1
11. −18, −13, −9, −6   12. 120, 60, 30, 15
13. 27, 9, 3, 1   14. 162, 54, 18, 6
15. 2, 5, 11, 20   16. 1, 4, 20, 120
17. 2, 3, 1, 4, 0   18. 720, 120, 24, 6

We can describe a sequence using symbols u_n to stand for the *n*th term, and u_{n+1} to stand for the $(n + 1)$th term.

For the sequence 3, 8, 13, 18 . . .

First term, $u_1 = 3$
$$u_{n+1} = u_n + 5$$
So $u_2 = u_1 + 5$
$$u_3 = u_2 + 5 \quad \text{and so on.}$$

By inspection, the *n*th term, $u_n = 3 + (n - 1)5$

Exercise 7

In questions **1** to **10** write down the first five terms.

1. $u_1 = 3$, $u_{n+1} = u_n + 2$ 2. $u_1 = 11$, $u_{n+1} = u_n + 5$
3. $u_1 = 12$, $u_{n+1} = u_n - 2$ 4. $u_1 = 5$, $u_{n+1} = u_n + 1\frac{1}{2}$
5. $u_1 = 3$, $u_{n+1} = 2u_n$ 6. $u_1 = 64$, $u_{n+1} = \frac{1}{2}u_n$
7. $u_1 = 2$, $u_{n+1} = (u_n)^2$ 8. $u_1 = 0$, $u_{n+1} = (u_n)^2 + 1$
9. $u_1 = 3$, $u_{n+1} = 7 - u_n$ 10. $u_1 = 2$, $u_{n+1} = \dfrac{1}{u_n}$

In questions **11** to **18** write down the first four terms. Start with $n = 1$.

11. $u_n = n^2$ 12. $u_n = n^2 - 1$
13. $u_n = 2n^2$ 14. $u_n = 2n + 1$
15. $u_n = n(n + 1)$ 16. $u_n = 2^n$
17. $u_n = (n + 1)(n + 2)$ 18. $u_n = 3 \times 2^n$

In questions **19** to **24** find a formula for the *n*th term.

19. 3, 6, 9, 12   20. 5, 10, 15, 20
21. 1, 4, 9, 16   22. $\frac{1}{2}$, 1, 2, 4, 8
23. 3, 9, 27, 81   24. 1×3, 2×4, 3×5, 4×6

1.3 APPROXIMATIONS

Example 1

(a) $7·8126 = 8$ to the nearest whole number
↑ This figure is '5 or more'.

(b) $7·8126 = 7·81$ to three significant figures
↑ This figure is not '5 or more'.

(c) $7·8126 = 7·813$ to three decimal places
↑ This figure is '5 or more'.

Exercise 8

Write the following numbers correct to
(a) the nearest whole number
(b) three significant figures
(c) two decimal places

1. 8·174	2. 19·617	3. 20·041
4. 0·81452	5. 311·14	6. 0·275
7. 0·00747	8. 15·62	9. 900·12
10. 3·555	11. 5·454	12. 20·961
13. 0·0851	14. 0·5151	15. 3·071

Write the following numbers correct to one decimal place.

16. 5·71	17. 0·7614	18. 11·241
19. 0·0614	20. 0·0081	21. 11·12

22. If $a = 3·1$ and $b = 7·3$ correct to 1 decimal place, find the largest possible value of
(i) $a + b$ (ii) $b - a$

23. If $x = 5$ and $y = 7$ to one significant figure, find the largest and smallest possible values of
(i) $x + y$ (ii) $y - x$ (iii) $\dfrac{x}{y}$

24. In the diagram, ABCD and EFGH are rectangles with AB = 10 cm, BC = 7 cm, EF = 7 cm and FG = 4 cm, all figures accurate to 1 significant figure.
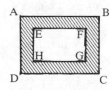
Find (a) the largest and (b) the smallest possible values of the shaded area.

25. The velocity v of a body is calculated from the formula $v = \dfrac{2s}{t} - u$ where u, s and t are measured correct to 1 decimal place. Find the largest possible value for v when $u = 2·1$, $s = 5·7$ and $t = 2·2$. Find also the smallest possible value for v consistent with these figures.

Estimation

It is always sensible to check that the answer to a calculation is 'about the right size'.

Example 2

Estimate the value of $\dfrac{57·2 \times 110}{2·146 \times 46·9}$, correct to one significant figure.

We have approximately, $\dfrac{50 \times 100}{2 \times 50} \approx 50$

On a calculator the value is $62·52$ (to 4 significant figures).

Exercise 9

In this exercise there are 25 questions, each followed by three possible answers. In each case only one answer is correct.
Write down each question and decide (by estimating) which answer is correct.

1. $7·2 \times 9·8$	[52 16, 98 36, 70·56]
2. $2·03 \times 58·6$	[118·958, 87·848, 141·116]
3. $23·4 \times 19·3$	[213·32, 301·52, 451·62]
4. $313 \times 107·6$	[3642·8, 4281·8, 33678·8]
5. $6·3 \times 0·098$	[0·6174, 0·0622, 5·98]
6. $1200 \times 0·89$	[722, 1068, 131]
7. $0·21 \times 93$	[41·23, 9·03, 19·53]
8. $88·8 \times 213$	[18914·4, 1693·4, 1965·4]
9. $0·04 \times 968$	[38·72, 18·52, 95·12]
10. $0·11 \times 0·089$	[0·1069, 0·0959, 0·00979]
11. $13·92 \div 5·8$	[0·52, 4·2, 2·4]
12. $105·6 \div 9·6$	[8·9, 11, 15]
13. $8405 \div 205$	[4·6, 402, 41]
14. $881·1 \div 99$	[4·5, 8·9, 88]
15. $4·183 \div 0·89$	[4·7, 48, 51]
16. $6·72 \div 0·12$	[6·32, 21·2, 56]
17. $20·301 \div 1010$	[0·0201, 0·211, 0·0021]
18. $0·28896 \div 0·0096$	[312, 102·1, 30·1]
19. $0·143 \div 0·11$	[2·3, 1·3, 11·4]
20. $159·65 \div 515$	[0·11, 3·61, 0·31]
21. $(5·6 - 0·21) \times 39$	[389·21, 210·21, 20·51]
22. $\dfrac{17·5 \times 42}{2·5}$	[294, 504m 86]
23. $(906 + 4·1) \times 0·31$	[473·21, 282·131, 29·561]
24. $\dfrac{543 + 472}{18·1 + 10·9}$	[65, 35, 85]
25. $\dfrac{112·2 \times 75·9}{6·9 \times 5·1}$	[242, 20·4, 25·2]

1.4   STANDARD FORM

When dealing with either very large or very small numbers, it is not convenient to write them out in full in the normal way. It is better to use standard form. Most calculators represent large and small numbers in this way.

The number $a \times 10^n$ is in standard form when $1 \leqslant a < 10$ and n is a positive or negative integer.

Example 1

Write the following numbers in standard form:

(a) $2000 = 2 \times 1000 = 2 \times 10^3$

(b) $150 = 1 \cdot 5 \times 100 = 1 \cdot 5 \times 10^2$

(c) $0 \cdot 0004 = 4 \times \dfrac{1}{10\,000} = 4 \times 10^{-4}$

Exercise 10

Write the following numbers in standard form:

1. 4000	**2.** 500	**3.** 70 000
4. 60	**5.** 2400	**6.** 380
7. 46 000	**8.** 46	**9.** 900 000
10. 2560	**11.** 0·007	**12.** 0·0004
13. 0·0035	**14.** 0·421	**15.** 0·000 055
16. 0·01	**17.** 564 000	**18.** 19 million

19. The population of China is estimated at 1 100 000 000. Write this in standard form.
20. A hydrogen atom weighs 0.000 000 000 000 000 000 000 001 67 grams. Write this weight in standard form.
21. The area of the surface of the Earth is about 510 000 000 km². Express this in standard form.
22. A certain crius is 0.000 000 000 25 cm in diameter. Write this in standard form.
23. Avogadro's number is 602 300 000 000 000 000 000 000. Express this in standard form.
24. The speed of light is 300 000 km/s. Express this speed in cm/s in standard form.
25. A very rich oil sheikh leaves his fortune of £$3 \cdot 6 \times 10^8$ to be divided between his 100 children. How much does each child receive? Give the answer in standard form.

Example 2

Work out $1500 \times 8\,000\,000$

$$1500 \times 8\,000\,000 = (1 \cdot 5 \times 10^3) \times (8 \times 10^6)$$
$$= 12 \times 10^9$$
$$= 1 \cdot 2 \times 10^{10}$$

Notice that we multiply the numbers and the powers of 10 separately.

Exercise 11

In questions **1** to **12** give the answer in standard form.

1. $5\,000 \times 3\,000$	**2.** $60\,000 \times 5\,000$
3. $0 \cdot 000\,07 \times 400$	**4.** $0 \cdot 0007 \times 0 \cdot 000\,01$
5. $8\,000 \div 0 \cdot 004$	**6.** $(0 \cdot 002)^2$
7. $150 \times 0 \cdot 0006$	**8.** $0 \cdot 000\,033 \div 500$
9. $0 \cdot 007 \div 20\,000$	**10.** $(0 \cdot 0001)^4$
11. $(2\,000)^3$	**12.** $0 \cdot 00592 \div 8\,000$

13. If $a = 512 \times 10^2$
    $b = 0 \cdot 478 \times 10^6$
    $c = 0 \cdot 0049 \times 10^7$
    arrange a, b and c in order of size (smallest first).
14. If the number $2 \cdot 74 \times 10^{15}$ is written out in full, how many zeros follow the 4?
15. If the number $7 \cdot 31 \times 10^{-17}$ is written out in full, how many zeros would there be between the decimal point and the first significant figure?
16. If $x = 2 \times 10^5$ and $y = 3 \times 10^{-3}$ correct to one significant figure, find the greatest and least possible values of

    (i) xy \qquad\qquad (ii) $\dfrac{x}{y}$

17. Oil flows through a pipe at a rate of 40 m³/s. How long will it take to fill a tank of volume $1 \cdot 2 \times 10^5$ m³?
18. Given that $L = 2\sqrt{\dfrac{a}{k}}$, find the value of L in standard form when $a = 4 \cdot 5 \times 10^{12}$ and $k = 5 \times 10^7$.
19. (a) The number 10 to the power 100 (10 000 sexdecillion) is called a 'Googol'! If it takes $\frac{1}{5}$ second to write a zero and $\frac{1}{10}$ second to write a 'one', how long would it take to write the number 100 'Googols' in full?

    (b) The number 10 to the power of a 'Googol' is called a 'Googolplex'. Using the same speed of writing, how long in years would it take to write 1 'Googolplex' in full? You may assume that your pen has enough ink.

1.5 RATIO AND PROPORTION

The word 'ratio' is used to describe a fraction. If the *ratio* of a boy's height to his father's height is $4:5$, then he is $\frac{4}{5}$ as tall as his father.

Example 1

Change the ratio $2:5$ into the form
(a) $1:n$ (b) $m:1$

(a) $2:5 = 1:\frac{5}{2}$ (b) $2:5 = \frac{2}{5}:1$
$\quad\quad = 1:2\cdot5$ $\quad\quad = 0\cdot4:1$

Example 2

Divide £60 between two people A and B in the ratio $5:7$.

Consider £60 as 12 equal parts (i.e. $5+7$). Then A receives 5 parts and B receives 7 parts.

∴   A receives $\frac{5}{12}$ of £60 = £25
    B receives $\frac{7}{12}$ of £60 = £35

Example 3

Divide 200 kg in the ratio $1:3:4$.
The parts are $\frac{1}{8}$, $\frac{3}{8}$ and $\frac{4}{8}$ (of 200 kg). i.e. 25 kg, 75 kg and 100 kg.

Exercise 12

Express the following ratios in the form $1:n$:

1. $2:6$ **2.** $5:30$ **3.** $2:100$
4. $5:8$ **5.** $4:3$ **6.** $8:3$
7. $22:550$ **8.** $45:360$

Express the following ratios in the form $n:1$:

9. $12:5$ **10.** $5:2$ **11.** $4:5$
12. $2:100$

In questions **13** to **18** divide the quantity in the ratio given.

13. £40; $(3:5)$ **14.** £120; $(3:7)$
15. 250 m; $(14:11)$ **16.** £117; $(2:3:8)$
17. 180 kg; $(1:5:6)$ **18.** 184 minutes; $(2:3:3)$

19. When £143 is divided in the ratio $2:4:5$, what is the difference between the largest share and the smallest share?
20. Divide 180 kg in the ratio $1:2:3:4$.
21. Divide £4000 in the ratio $2:5:5:8$.

22. If $\frac{5}{8}$ of the children in a school are boys, what is the ratio of boys to girls?
23. A man and a woman share a bingo prize of £1000 between them in the ratio $1:4$. The woman shares her part between herself, her mother and her daughter in the ratio $2:1:1$. How much does her daughter receive?
24. A man and his wife share a sum of money in the ratio $3:2$. If the sum of money is doubled, in what ratio should they divide it so that the man still receives the same amount?
25. In a herd of x cattle, the ratio of the number of bulls to cows is $1:6$. Find the number of bulls in the herd in terms of x.
26. If $x:3 = 12:x$, calculate the positive value of x.
27. If $y:18 = 8:y$, calculate the positive value of y.
28. £400 is divided between Ann, Brian and Carol so that Ann has twice as much as Brian and Brian has three times as much as Carol. How much does Brian receive?
29. A cake weighing 550 g has three ingredients: flour, sugar and raisins. There is twice as much flour as sugar and one and a half times as much sugar as raisins. How much flour is there?
30. A brother and sister share out their collection of 5000 stamps in the ratio $5:3$. The brother then shares his stamps with two friends in the ratio $3:1:1$, keeping most for himself. How many stamps do each of his friends receive?

Proportion

The majority of problems where proportion is involved are usually solved by finding the value of a unit quantity.

Example 4

If a wire of length 2 metres costs £10, find the cost of a wire of length 35 cm.

200 cm costs 1000 pence

∴   1 cm costs $\frac{1000}{200}$ pence = 5 pence

∴   35 cm costs 5×35 pence = 175 pence
$\quad\quad\quad\quad\quad\quad\quad\quad\quad = £1\cdot75$

Example 5

Eight men can dig a trench in 4 hours. How long will it take five men to dig the same size trench?

8 men take 4 hours

1 man would take 32 hours

5 men would take $\frac{32}{5}$ hours = 6 hours 24 minutes.

Exercise 13

1. Five cans of beer cost £1.20. Find the cost of seven cans.
2. A man earns £140 in a 5-day week. What is his pay for 3 days?
3. Three men build a wall in 10 days. How long would it take five men?
4. Nine milk bottles contain $4\frac{1}{2}$ litres of milk between them. How much do five bottles hold?
5. A car uses 10 litres of petrol in 75 km. How far will it go on 8 litres?
6. A wire 11 cm long has a mass of 187 g. What is the mass of 7 cm of this wire?
7. A shopkeeper can buy 36 toys for £20·52. What will he pay for 120 toys?
8. A ship has sufficient food to supply 600 passengers for 3 weeks. How long would the food last for 800 people?
9. The cost of a phone call lasting 3 minutes 30 seconds was 52·5 p. At this rate, what was the cost of a call lasting 5 minutes 20 seconds?
10. 80 machines can produce 4800 identical pens in 5 hours. At this rate
    (a) how many pens would one machine produce in one hour?
    (b) how many pens would 25 machines produce in 7 hours?
11. Three men can build a wall in 10 hours. How many men would be needed to build the wall in $7\frac{1}{2}$ hours?
12. If it takes 6 men 4 days to dig a hole 3 feet deep, how long will it take 10 men to dig a hole 7 feet deep?
13. Find the cost of 1 km of pipe at 7p for every 40 cm.
14. A wheel turns through 90 revolutions per minute. How many degrees does it turn through in 1 second?
15. Find the cost of 20 grams of lead at £60 per kilogram.
16. The height of an office building is 623 feet. Express this height to the nearest metre using 1 m = 3·281 feet.

17. A floor is covered by 800 tiles measuring 10 cm square. How many square tiles of side 8 cm would be needed to cover the same floor?
18. Jackie can drive to work in 18 minutes if she travels at an average speed of 35 m.p.h. How long will the journey take if she drives at an average speed of 45 m.p.h.?
19. A battery has enough energy to operate 8 toy bears for 21 hours. For how long could the battery operate 15 toy bears?
20. An aircraft flying at 500 km/h completes a journey in 8·4 h. How long would the journey take if it flew at a speed of 420 km/h?
21. An engine has enough fuel to operate at full power for 20 minutes. For how long could the engine operate at 35% of full power?
22. A large drum, when full, contains 260 kg of oil of density 0·9 g/cm^3. What weight of petrol, of density 0·84 g/cm^3, can be contained in the drum?
23. A wall can be built by 6 men working 8 hours per day in 5 days. How many days will it take 4 men to build the wall if they work only 5 hours per day?

Foreign exchange

Money is changed from one currency into another using the method of proportion.

Exchange rates

Country	Rate of exchange
Belgium (franc)	BF 87·0 = £1
France (franc)	Fr 10·9 = £1
Germany (mark)	DM 4·2 = £1
Italy (lire)	lire 2280 = £1
Spain (peseta)	Ptas 182 = £1
United States (dollar)	$1·74 = £1

Example 6

If a bottle of wine costs 8 francs in France, what is the cost in British money?

10·9 francs = £1

$\therefore$ 1 franc = £$\frac{1}{10·9}$

8 francs = £$\frac{1}{10·9} \times 8$ = £0·73

(to the nearest penny)

The bottle costs approximately 73p in British money.

Example 7

Convert $500 into German marks.

$$\$1\cdot74 = £1$$
$$DM\ 4\cdot2 = £1$$
$$\therefore\quad \$1\cdot74 = DM\ 4\cdot2$$
$$\$1 = DM\left(\frac{4\cdot2}{1\cdot74}\right)$$
$$\therefore\quad \$500 = DM\left(\frac{4\cdot2}{1\cdot74}\times 500\right)$$
$$= DM\ 1206$$

Exercise 14

Give your answers correct to two decimal places. Use the exchange rates given in the table.

1. Change the amount of British money into the foreign currency stated.
   (a) £20 [French francs]
   (b) £70 [dollars]
   (c) £200 [pesetas]
   (d) £1·50 [marks]
   (e) £2·30 [lire]
   (f) 90p [dollars]

2. Change the amount of foreign currency into British money.
   (a) Fr. 500   (b) $2500
   (c) DM 7·5   (d) BF 900
   (e) Lire 500,000   (f) Ptas 950

3. An L.P. costs £4·50 in Britain and $4·70 in the United States. How much cheaper, in British money, is the record when bought in the USA?

4. A bottle of Cointreau costs 582 pesetas in Spain and Fr 48 in France. Which is the cheaper in British money, and by how much?

5. The EEC 'Butter Mountain' was estimated in 1988 to be costing Fr 218 000 per day to maintain the storage facilities. How much is this in pounds?

6. A Jaguar XJS is sold in several countries at the prices given below.

Britain	£15,000
Belgium	BF 1 496 400
France	Fr 194 020
Germany	DM 52 080
USA	$24 882

   Write out in order a list of the prices converted into pounds.

7. A traveller in Switzerland exchanges 1300 Swiss francs for £400. What is the exchange rate?

8. An Irish gentleman on holiday in Germany finds that his wallet contains $700. If he changes the money at a bank how many marks will he receive?

9. An English soccer fan is arrested in France and has to pay a fine of 2 000 francs. He has 10 000 German marks in his wallet. How much has he left in British money, after paying the fine?

10. In Britain, a pint of beer costs 65p. In France a third of a litre of the same beer costs 4 francs. If 1 pint is approximately 0·568 litre, calculate the cost in pence of a pint of the beer bought in France.

Map scales

Example 8

A map is drawn to a scale of 1 to 50 000.
Calculate:
(a) the length of a road which appears as 3 cm long on the map.
(b) the length on the map of a lake which is 10 km long.

(a) 1 cm on the map is equivalent to 50 000 cm on the Earth
$$\therefore\quad 3\ cm \equiv 3\times 50\,000 = 150\,000\ cm$$
$$= 1500\ m$$
$$= 1\cdot5\ km$$

The road is 1·5 km long.

(b) $50\,000\ cm \equiv 1\ cm$ on the map
$$1\ cm \equiv \tfrac{1}{50\,000}\ cm \text{ on the map}$$
$$10\ km = 1\,000\,000\ cm$$
$$\therefore\quad 10\ km \equiv \left(1\,000\,000\times\frac{1}{50\,000}\right)\ cm \text{ on the map}$$
$$\equiv 20\ cm \text{ on the map}$$

The lake appears 20 cm long on the map.

Exercise 15

1. Find the actual length represented on a drawing by
   (a) 14 cm   (b) 3·2 cm
   (c) 0·71 cm   (d) 21·7 cm
   when the scale is 1 cm to 5 m.

2. Find the length on a drawing that represents
   (a) 50 m   (b) 35 m
   (c) 7·2 m   (d) 28·6 m
   when the scale is 1 cm to 10 m.

3. If the scale is 1 : 10 000, what length will 45 cm on the map represent:
   (a) in cm;   (b) in m;   (c) in km?
4. On a map of scale 1 : 100 000, the distance between Tower Bridge and Hammersmith Bridge is 12·3 cm. What is the actual distance in km?
5. On a map of scale 1 : 15 000, the distance between Buckingham Palace and Brixton Underground Station is 31·4 cm. What is the actual distance in km?
6. If the scale of a map is 1 : 10 000, what will be the length on this map of a road which is 5 km long?
7. The distance from Hertford to St Albans is 32 km. How far apart will they be on a map of scale 1 : 5000?
8. The 17th hole at the famous St Andrews golf course is 420 m in length. How long will it appear on a plan of the course of scale 1 : 8000?

An area involves two dimensions multiplied together and hence the scale is multiplied *twice*.

For example, if the linear scale is $\frac{1}{100}$, then the area scale is $\frac{1}{100} \times \frac{1}{100} = \frac{1}{10\,000}$.

Exercise 16

1. The scale of a map is 1 : 1000. What are the actual dimensions of a rectangle which appears as 4 cm by 3 cm on the map? What is the area on the map in cm^2? What is the actual area in m^2?
2. The scale of a map is 1 : 100. What area does 1 cm^2 on the map represent? What area does 6 cm^2 represent?
3. The scale of a map is 1 : 20 000. What area does 8 cm^2 represent?
4. The scale of a map is 1 : 1000. What is the area on the map of a lake of area 5 km^2?
5. The scale of a map is 1 cm to 5 km. A farm is represented by a rectangle measuring 1·5 cm by 4 cm. What is the actual area of the farm?
6. On a map of scale 1 cm to 2 km the area of a car park is 3 cm^2. What is the actual area of the car park in hectares? (1 hectare = 10 000 m^2)
7. The area of the playing surface at Wembley Stadium is $\frac{3}{5}$ of a hectare. What area will it occupy on a plan drawn to a scale of 1 : 500?
8. On a map of scale 1 : 20 000 the area of a forest is 50 cm^2. On another map the area of the forest is 8 cm^2. Find the scale of the second map.

1.6   PERCENTAGES

Percentages are simply a convenient way of expressing fractions or decimals. '50% of £60' means $\frac{50}{100}$ of £60, or more simply $\frac{1}{2}$ of £60. Percentages are used very frequently in everyday life and are misunderstood by a large number of people. What are the implications if 'inflation falls from 10% to 8%'? Does this mean prices will fall?

Example 1

(a) change 80% to a fraction
(b) change $\frac{3}{8}$ to a percentage
(c) change 8% to a decimal.

(a) $80\% = \frac{80}{100} = \frac{4}{5}$

(b) $\frac{3}{8} = \left(\frac{3}{8} \times \frac{100}{1}\right)\% = 37\frac{1}{2}\%$

(c) $8\% = \frac{8}{100} = 0 \cdot 08$

Exercise 17

1. Change to fractions
   (a) 60%   (b) 24%
   (c) 35%   (d) 2%
2. Change to percentages
   (a) $\frac{1}{4}$ (b) $\frac{1}{10}$
   (c) $\frac{7}{8}$ (d) $\frac{1}{3}$
   (e) 0·72   (f) 0·31
3. Change to decimals
   (a) 36%   (b) 28%
   (c) 7%   (d) 13·4%
   (e) $\frac{3}{5}$ (f) $\frac{7}{8}$
4. Arrange in order of size (smallest first)
   (a) $\frac{1}{2}$; 45%; 0.6
   (b) 0·38; $\frac{6}{16}$; 4%
   (c) 0·111; 11%; $\frac{1}{9}$
   (d) 32%; 0·3; $\frac{1}{3}$

5. The following are marks obtained in various tests.
    Convert them to percentages.
    (a) 17 out of 20
    (b) 31 out of 40
    (c) 19 out of 80
    (d) 112 out of 200
    (e) $2\frac{1}{2}$ out of 25
    (f) $7\frac{1}{2}$ out of 20

Example 2

A car costing £2400 is reduced in price by 10%.
Find the new price.

$$10\% \text{ of } £2400 = \frac{10}{100} \times \frac{2400}{1}$$
$$= £240$$

New price of car = £(2400 − 240)
                 = £2160

Example 3

After a price increase of 10% a television set
costs £286. What was the price before the
increase?

The price before the increase is 100%

∴   110% of old price = £286

∴   1% of old price = £$\frac{286}{110}$

∴   100% of old price = £$\frac{286}{110} \times \frac{100}{1}$

     Old price of TV = £260

Exercise 18

1. Calculate
    (a) 30% of £50   (b) 45% of 2000 kg
    (c) 4% of $70   (d) 2·5% of 5000 people

2. In a sale, a jacket costing £40 is reduced by
    20%. What is the sale price?

3. The charge for a telephone call costing 12p is
    increased by 10%. What is the new
    charge?

4. In peeling potatoes 4% of the mass of the
    potatoes is lost as 'peel'. How much is *left* for use
    from a bag containing 55 kg?

5. Work out, to the nearest penny
    (a) 6·4% of £15·95   (b) 11·2% of £192·66
    (c) 8·6% of £25·84   (b) 2·9% of £18·18

6. Find the total bill:
    5 golf clubs at £18·65 each
    60 golf balls at £16·50 per dozen
    1 bag at £35·80
    V.A.T. at 15% is added to the total cost.

7. In 1987 a club has 250 members who each pay £95
    annual subscription. In 1988 the membership
    increases by 4% and the annual subscription is
    increased by 6%. What is the total income from
    subscriptions in 1988?

8. In 1987 the prison population was 48700 men and
    1600 women. What percentage of the total prison
    population were men?

9. In 1988 there were 21 280 000 licensed vehicles on
    the road. Of these, 16 486 000 were private cars.
    What percentage of the licensed vehicles were
    private cars?

10. A quarterly telephone bill consists of £19·15 rental
    plus 4·7 p for each dialled unit. V.A.T. is added at
    15%. What is the total bill for Mrs Jones who
    used 915 dialled units?

11. Johnny thinks his goldfish got chickenpox. He lost
    70% of his collection of goldfish. If he has 60
    survivors, how many did he have originally?

12. The average attendance at Everton football club
    fell by 7% in 1982. If 2030 fewer people went to
    matches in 1982, how many went in 1981?

13. When heated an iron bar expands by 0·2%. If the
    increase in length is 1 cm, what is the original
    length of the bar?

14. In the last two weeks of a sale, prices are reduced
    first by 30% and then by a *further* 40% of the new
    price. What is the final sale price of a shirt which
    originally cost £15?

15. During a Grand Prix car race, the tyres on a car
    are reduced in weight by 3%. If they weigh 388 kg
    at the end of the race, how much did they weigh
    at the start?

16. Over a period of 6 months, a colony of rabbits
    increases in number by 25% and then by a further
    30%. If there were originally 200 rabbits in the
    colony how many were there at the end?

17. A television costs £270·25 including 15% V.A.T.
    How much of the cost is tax?

18. The cash price for a car was £7640. Mr Elder
    bought the car on the following hire purchase
    terms: 'A deposit of 20% of the cash price and 36
    monthly payments of £191·60'
    Calculate the total amount Mr Elder paid.

Exercise 19

British Gas

Share Offer

In 1986 the Government sold shares in British Gas to members of the public. Because the shares were oversubscribed, most people did not receive all the shares they applied for. A simple formula was devised to work out the allocation of shares. People received 10% of the number they applied for plus another 300 shares. So if someone applied for 4000 shares, she received (10% of 4000) + 300 shares (ie 700).

People had to pay 50p for each share they received. When dealing in the shares opened on the Stock Market the price of shares went up to a value between 60p and 75p (share prices vary from day to day).

Answer the following questions.

1. Mr Jones applied for 500 shares. How much profit did he make when he sold his entire allocation at 64p per share?

2. Mrs Miller applied for 5000 shares and sold her entire allocation at a price of 67p per share. How much profit did she make?

3. Mr Gandee sold his allocation when the price was 65p. He received a total of £1105. How many shares did he apply for?

4. Ms Ludwell sold her allocation when the price was 69p. She received a total of £586·50. How many shares did she apply for?

5. Mr Wyatt sold his allocation when the price was 66p and he made a *profit* of £400. How many shares did he apply for?

6. Mrs Morgan sold her allocation when the price was 68p and she made a profit of £270. How many shares did she apply for?

7. Miss Green applied for 2600 shares and then sold her allocation to make a profit of £67·20. At what price did she sell the shares?

8. Mr Singh applied for 60 000 shares and then sold his allocation to make a profit of £1386. At what price did he sell the shares?

In the next exercise use the formulae:

$$\text{Percentage profit} = \frac{\text{Actual profit}}{\text{Original price}} \times \frac{100}{1}$$

$$\text{Percentage loss} = \frac{\text{Actual loss}}{\text{Original price}} \times \frac{100}{1}$$

Example 4

A radio is bought for £16 and sold for £20. What is the percentage profit?

$$\text{Actual profit} = \text{£4}$$

$$\therefore \quad \text{Percentage profit} = \frac{4}{16} \times \frac{100}{1} = 25\%$$

The radio is sold at a 25% profit.

Example 5

A car is sold for £2280, at a loss of 5% on the cost price. Find the cost price.

Do *not* calculate 5% of £2280!
The loss is 5% of the cost price.

$$\therefore \quad 95\% \text{ of cost price} = \text{£2280}$$

$$1\% \text{ of cost price} = \text{£} \frac{2280}{95}$$

$$\therefore \quad 100\% \text{ of cost price} = \text{£} \frac{2280}{95} \times \frac{100}{1}$$

$$\text{Cost price} = \text{£2400.}$$

Exercise 20

1. The first figure is the cost price and the second figure is the selling price. Calculate the percentage profit or loss in each case.
   (a) £20, £25   (b) £400, £500
   (c) £60, £54   (d) £9000, $10 800
   (e) £460, £598   (f) £512, £550·40
   (g) £45, £39·60   (h) 50p, 23p

2. A car dealer buys a car for £500, gives it a clean, and then sells it for £640. What is the percentage profit?

3. A damaged carpet which cost £180 when new, is sold for £100. What is the percentage loss?

4. During the first four weeks of her life, a baby girl increases her weight from 3·2 kg to 4·7 kg. What percentage increase does this represent? (give your answer to 3 sig. fig.)

5. When V.A.T. is added to the cost of a car tyre, its price increases from £16·50 to £18·48. What is the rate at which V.A.T. is charged?

6. In order to increase sales, the price of a Concorde airliner is reduced from £30 000 000 to £28 400 000. What percentage reduction is this?

7. Find the *cost* price of the following:
   (a) selling price £55, profit 10%
   (b) selling price £558, profit 24%
   (c) selling price £680, loss 15%
   (d) selling price £11.78, loss 5%

8. An oven is sold for £600, thereby making a profit of 20%, on the cost price. What was the cost price?

9. A pair of jeans is sold for £15, thereby making a profit of 25% on the cost price. What was the cost price?

10. A book is sold for £5·40, at a profit of 8% on the cost price. What was the cost price?

11. An obsolete can of worms is sold for 48p, incurring a loss of 20%. What was the cost price?

12. A car, which failed its MOT test, was sold for £143, thereby making a loss of 35% on the cost price. What was the cost price?

13. If an employer reduces the working week from 40 hours to 35 hours, with no loss of weekly pay, calculate the percentage increase in the hourly rate of pay.

14. The rental for a television set changed from £80 per year to £8 per month. What is the percentage increase in the yearly rental?

15. A greengrocer sells a cabbage at a profit of $37\frac{1}{2}\%$ on the price he pays for it. What is the ratio of the cost price to the selling price?

16. Given that $G = ab$, find the percentage increase in G when both a and b increase by 10%.

17. Given that $T = \dfrac{kx}{y}$, find the percentage increase in T when k, x and y all increase by 20%.

Percentage error

Suppose you work out π by measuring cylinders and get an answer of 3·27. The value of π correct to 4 s.f. is 3·142.

The percentage error is

$$\frac{3\cdot27 - 3\cdot142}{3\cdot142} \times \frac{100}{1} = 4\cdot1\% \ (2 \text{ sig. fig.})$$

In general, % error $= \dfrac{\text{actual error}}{\text{exact value}} \times \dfrac{100}{1}$

Exercise 21

Give answers correct to 3 significant figures.

1. A man's salary is estimated to be £20 000. His actual salary is £21,540. Find the percentage error.

2. The speedometer of a car showed a speed of 68 m.p.h. The actual speed was 71 m.p.h. Find the percentage error.

3. The rainfall during a storm was estimated at 3·5 cm. The exact value was 3·1 cm. Find the percentage error.

4. The dimensions of a rectangle were taken as 80×30. The exact dimensions were $81\cdot5 \times 29\cdot5$. Find the percentage error in the area.

5. The answer to the calculation $3\cdot91 \times 21\cdot8$ is estimated to be 80. Find the percentage error.

6. An approximate value for π is $\frac{22}{7}$. A more accurate value is given on most calculators. Find the percentage error.

7. A wheel has a diameter of 58·2 cm. Its circumference is estimated at 3×60. Find the percentage error.

8. To the calculation $56\cdot2 \times 9\cdot1 - 4\cdot53$, Katy and Louise gave estimates of $50 \times 10 - 4$ and $60 \times 9 - 5$ respectively. Who made the more accurate estimate?

9. A speed of 60 km/h is taken to be roughly 18 m/s. Find the percentage error.

Compound interest

Suppose a bank pays a fixed interest of 10% on money in deposit accounts. A man puts £500 in the bank.

After one year he has
   500 + 10% of 500 = £550

After two years he has
   550 + 10% of 550 = £605

   [Check that this is $1\cdot10^2 \times 500$]

After three years he has
   605 + 10% of 605 = £665·50

   [Check that this is $1\cdot10^3 \times 500$]

In general after n years the money in the bank will be £$(1\cdot10^n \times 500)$

Exercise 22

These questions are easier if you use a calculator with a $\boxed{y^x}$ button.

1. A bank pays interest of 9% on money in deposit accounts. Mrs Wells puts £2000 in the bank. How much has she after (a) one year, (b) two years, (c) three years?
2. A bank pays interest of 11%. Mr Olsen puts £5000 in the bank. How much has he after (a) one year, (b) three years, (c) five years.
3. A computer operator is paid £10 000 a year. Assuming her pay is increased by 7% each year, what will her salary be in four years time?
4. A new car is valued at £15 000. At the end of each year its value is reduced by 15% of its value at the start of the year. What will it be worth after 6 years?
5. Twenty years ago a bus driver was paid £50 a week. He is now paid £185 a week. Assuming an average rate of inflation of 7%, has his pay kept up with inflation?
6. Assuming an average inflation rate of 8%, work out the probable cost of the following items in 10 years:
   (a) car £6500
   (b) T.V. £340
   (c) house £50 000
7. The population of an island increases by 10% each year. After how many years will the original population be doubled?
8. A bank pays interest of 11% on money in deposit accounts. After how many years will a sum of money have trebled?
9. (a) Draw the graph of $y = 1 \cdot 08^x$ for values of x from 0 to 10.
   (b) Solve approximately the equation $1 \cdot 08^x = 2$.
   (c) Money is invested at 8% interest. After how many years will the money have doubled?

Income tax

The tax which an employee pays on her income depends on (a) how much she is paid
　　　　　　　　　　　(b) her allowances
　　　　　　　　　　　(c) the rate of taxation.

Tax is paid only on the 'taxable income'.

Total income = Allowances + Taxable income.

Example 6

Calculate the tax paid each month by a man whose annual salary is £8500, with allowances of £2100, where the rate of taxation is 30% of the first £10 000 of taxable income.

Total income = Allowances + Taxable income

$\therefore$ 　　Taxable income = 8500 − 2100
　　　　　　　　　　= £6400

Tax paid in one year = 30% of £6400
　　　　　　　　　= £1920

Tax paid per month = 1920 ÷ 12
　　　　　　　　= £160

Example 7

Calculate the annual salary of a woman who has allowances of £2350 and who pays £1404 in tax. The rate of taxation is 30%.

Let her taxable income be £x

$$\frac{30}{100} \times x = 1404$$

$\therefore$ 　　　　$x = 1404 \times \dfrac{100}{30}$

　　　　　　　= 4680

$\therefore$ 　Annual salary = Taxable income + Allowances
　　　　　　　　= £4680 + £2350
　　　　　　　　= £7030

Exercise 23

In questions **1** to **3**, the rates of taxation are as follows:

taxable income	rate
£1 → 11 250	30%
£11 251 → 13 250	40%
£13 251 → 16 750	45%
over £16 751	50%

1. Calculate the tax paid by someone with income and allowances as follows:
   (a) income £6000, allowances £1400
   (b) income £2750, allowances £1375
   (c) income £12 500, allowances £1375
   (d) income £14 500, allowances £2050
   (e) income £17 600, allowances £2760
   (f) income £35 000, allowances £2000

2. Calculate the income of a person whose amount of tax paid and allowances are as follows:
   (a) tax paid £2400, allowances £2000
   (b) tax paid £960, allowances £2000
   (c) tax paid £3360, allowances £1950
   (d) tax paid £4200, allowances £2660
   (e) tax paid £4000, allowances £3400
   (f) tax paid £5600, allowances £4055

3. In the budget the Chancellor raises allowances by 8% and reduces the rate of taxation from 32% to 30%.
   (a) How much less tax is payable on an income of £8800 with original allowances of £1800?
   (b) What is the saving for someone with an income of £13 000 and allowances of £2000?

4. A man with an income of £7500 and allowances of £1800, paid £1824 in tax. What is the rate of taxation?

5. A man with an income of £10 850 and allowances of £2140, paid £2525·90 in tax. What is the rate of taxation?

6. Mr Green earns £95 per week and has allowances of £2450. Mrs Patel earns £19 400 per annum and has allowances of £3680. They both pay all their tax at basic rate. How much does each person *save* per year when the basic rate of tax is reduced from 29% to 25%?

1.7   SPEED, DISTANCE AND TIME

Calculations involving these three quantities are simpler when the speed is *constant*. The formulae connecting the quantities are as follows:
(a) distance = speed × time

(b) speed = $\dfrac{\text{distance}}{\text{time}}$

(c) time = $\dfrac{\text{distance}}{\text{speed}}$

A helpful way of remembering these formulae is to write the letters D, S and T in a triangle,

thus:

to find D, cover D and we have ST

to find S, cover S and we have $\dfrac{D}{T}$

to find T, cover T and we have $\dfrac{D}{S}$

Great care must be taken with the units in these questions.

Example 1

A man is running at a speed of 8 km/h for a distance of 5200 metres. Find the time taken in minutes.

$$5200 \text{ metres} = 5\cdot2 \text{ km}$$

$$\text{time taken in hours} = \left(\frac{D}{S}\right) = \frac{5\cdot2}{8}$$
$$= 0\cdot65 \text{ hours}$$

time taken in minutes = $0\cdot65 \times 60$
$$= 39 \text{ minutes.}$$

Example 2

Change the units of a speed of 54 km/h into metres per second.

54 km/hour = 54 000 metres/hour

$$= \frac{54\,000}{60} \text{ metres/minute}$$

$$= \frac{54\,000}{60 \times 60} \text{ metres/second}$$

$$= 15 \text{ m/s.}$$

Exercise 24

1. Find the time taken for the following journeys:
   (a) 100 km at a speed of 40 km/h
   (b) 250 miles at a speed of 80 miles per hour
   (c) 15 metres at a speed of 20 cm/s. (answer in seconds)
   (d) 10^4 metres at a speed of 2·5 km/h

2. Change the units of the following speeds as indicated:
   (a) 72 km/h into m/s
   (b) 108 km/h into m/s
   (c) 300 km/h into m/s
   (d) 30 m/s into km/h
   (e) 22 m/s into km/h
   (f) 0·012 m/s into cm/s
   (g) 9000 cm/s into m/s
   (h) 600 miles/day into miles per hour
   (i) 2592 miles/day into miles per second

3. Find the speeds of the bodies which move as follows:
   (a) a distance of 600 km in 8 hours
   (b) a distance of 31·64 km in 7 hours
   (c) a distance of 136·8 m in 18 seconds
   (d) a distance of 4×10^4 m in 10^{-2} seconds
   (e) a distance of 5×10^5 cm in 2×10^{-3} seconds
   (f) a distance of 10^8 mm in 30 minutes (in km/h)
   (g) a distance of 500 m in 10 minutes (in km/h)

4. Find the distance travelled (in metres) in the following:
   (a) at a speed of 55 km/h for 2 hours
   (b) at a speed of 40 km/h for $\frac{1}{4}$ hour
   (c) at a speed of 338.4 km/h for 10 minutes
   (d) at a speed of 15 m/s for 5 minutes
   (e) at a speed of 14 m/s for 1 hour
   (f) at a speed of 4×10^3 m/s for 2×10^{-2} seconds
   (g) at a speed of 8×10^5 cm/s for 2 minutes

5. A car travels 60 km at 30 km/h and then a further 180 km at 160 km/h. Find
   (a) the total time taken
   (b) the average speed for the whole journey

6. A cyclist travels 25 kilometres at 20 km/h and then a further 80 kilometres at 25 km/h. Find
   (a) the total time taken
   (b) the average speed for the whole journey

7. A swallow flies at a speed of 50 km/h for 3 hours and then at a speed of 40 km/h for a further 2 hours. Find the average speed for the whole journey.

8. Sebastian Coe ran two laps around a 400 m track. He completed the first lap in 50 seconds and then decreased his speed by 5% for the second lap. Find
   (a) his speed on the first lap
   (b) his speed on the second lap
   (c) his total time for the two laps
   (d) his average speed for the two laps.

9. The airliner Concorde flies 2000 km at a speed of 1600 km/h and then returns due to bad weather at a speed of 1000 km/h. Find the average speed for the whole trip.

10. A train travels from A to B, a distance of 100 km, at a speed of 20 km/h. If it had gone two and a half times as fast, how much earlier would it have arrived at B?

11. Two men running towards each other at 4 m/s and 6 m/s respectively are one kilometre apart. How long will it take before they meet?

12. A car travelling at 90 km/h is 500 m behind another car travelling at 70 km/h in the same direction. How long will it take the first car to catch the second?

13. How long is a train which passes a signal in twenty seconds at a speed of 108 km/h?

14. A train of length 180 m approaches a tunnel of length 620 m. How long will it take the train to pass completely through the tunnel at a speed of 54 km/h?

15. An earthworm of length 15 cm is crawling along at 2 cm/s. An ant overtakes the worm in 5 seconds. How fast is the ant walking?

16. A train of length 100 m is moving at a speed of 50 km/h. A horse is running alongside the train at a speed of 56 km/h. How long will it take the horse to overtake the train?

17. A car completes a journey at an average speed of 40 m.p.h. At what speed must it travel on the return journey if the average speed for the complete journey (out and back) is 60 m.p.h.?

1.8 CALCULATOR

In this book, the keys are described thus:

$+$	add,
$-$	subtract,
$\times$	multiply,
$\div$	divide,
$=$	equals,

$\sqrt{}$	square root
x^2	square
$1/x$	reciprocal
y^x	raise number y to the power x
M+	add number to memory
MR	recall number from memory

Example 1

Evaluate the following to 4 significant figures:

(a) $\dfrac{2 \cdot 3}{4 \cdot 7 + 3 \cdot 61}$ (b) $\left(\dfrac{1}{0 \cdot 084}\right)^4$

(c) $\sqrt[3]{[3 \cdot 2 \times (1 \cdot 7 - 1 \cdot 64)]}$

(a) $\dfrac{2 \cdot 3}{4 \cdot 7 + 3 \cdot 61}$ Find the denominator first

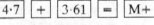

| 4·7 | + | 3·61 | = | M+ |

| 2·3 | ÷ | MR | = |

Answer 0·2768 (to four sig. fig.)

Alternatively:

| 4·7 | + | 3·61 | = | ÷ | 2·3 | = | 1/x |

(b) $\left(\dfrac{1}{0 \cdot 084}\right)^4$

| 0·084 | 1/x | y^x | 4 | = |

Answer 20 090 to (four sig. fig.)

(c) $\sqrt[3]{[3 \cdot 2(1 \cdot 7 - 1 \cdot 64)]}$

| 1·7 | − | 1·64 | = | × | 3·2 | = |

| y^x | 0·333333 | = |

Answer 0·5769 (to four sig. fig.)

Note: To find a cube root, raise to the power $\frac{1}{3}$, or as a decimal 0.333 . . .

Exercise 25

Use a calculator to evaluate the following, giving the answers to 4 significant figures:

1. $\dfrac{7 \cdot 351 \times 0 \cdot 764}{1 \cdot 847}$

2. $\dfrac{0 \cdot 0741 \times 14700}{0 \cdot 746}$

3. $\dfrac{0 \cdot 0741 \times 9 \cdot 61}{23 \cdot 1}$

4. $\dfrac{417 \cdot 8 \times 0 \cdot 00841}{0 \cdot 07324}$

5. $\dfrac{8 \cdot 41}{7 \cdot 601 \times 0 \cdot 00847}$

6. $\dfrac{4 \cdot 22}{1 \cdot 701 \times 5 \cdot 2}$

7. $\dfrac{9 \cdot 61}{17 \cdot 4 \times 1 \cdot 51}$

8. $\dfrac{8 \cdot 71 \times 3 \cdot 62}{0 \cdot 84}$

9. $\dfrac{0 \cdot 76}{0 \cdot 412 - 0 \cdot 317}$

10. $\dfrac{81 \cdot 4}{72 \cdot 6 + 51 \cdot 92}$

11. $\dfrac{111}{27 \cdot 4 + 2960}$

12. $\dfrac{27 \cdot 4 + 11 \cdot 61}{5 \cdot 9 - 4 \cdot 763}$

13. $\dfrac{6 \cdot 51 - 0 \cdot 1114}{7 \cdot 24 + 1 \cdot 653}$

14. $\dfrac{5 \cdot 71 + 6 \cdot 093}{9 \cdot 05 - 5 \cdot 77}$

15. $\dfrac{0 \cdot 943 - 0 \cdot 788}{1 \cdot 4 - 0 \cdot 766}$

16. $\dfrac{2 \cdot 6}{1 \cdot 7} + \dfrac{1 \cdot 9}{3 \cdot 7}$

17. $\dfrac{8 \cdot 06}{5 \cdot 91} - \dfrac{1 \cdot 594}{1 \cdot 62}$

18. $\dfrac{4 \cdot 7}{11 \cdot 4 - 3 \cdot 61} + \dfrac{1 \cdot 6}{9 \cdot 7}$

19. $\dfrac{3 \cdot 74}{1 \cdot 6 \times 2 \cdot 89} - \dfrac{1}{0 \cdot 741}$

20. $\dfrac{1}{7 \cdot 2} - \dfrac{1}{14 \cdot 6}$

21. $\dfrac{1}{0 \cdot 961} \times \dfrac{1}{0 \cdot 412}$

22. $\dfrac{1}{7} + \dfrac{1}{13} - \dfrac{1}{8}$

23. $4 \cdot 2 \left(\dfrac{1}{5 \cdot 5} - \dfrac{1}{7 \cdot 6}\right)$

24. $\sqrt{(9 \cdot 61 + 0 \cdot 1412)}$

25. $\sqrt{\left(\dfrac{8 \cdot 007}{1 \cdot 61}\right)}$

26. $(1 \cdot 74 + 9 \cdot 611)^2$

27. $\left(\dfrac{1 \cdot 63}{1 \cdot 7 - 0 \cdot 911}\right)^2$

28. $\left(\dfrac{9 \cdot 6}{2 \cdot 4} - \dfrac{1 \cdot 5}{0 \cdot 74}\right)^2$

29. $\sqrt{\left(\dfrac{4 \cdot 2 \times 1 \cdot 611}{9 \cdot 83 \times 1 \cdot 74}\right)}$

30. $(0 \cdot 741)^3$

31. $(1 \cdot 562)^5$

32. $(0 \cdot 32)^3 + (0 \cdot 511)^4$

33. $(1 \cdot 71 - 0 \cdot 863)^6$

34. $\left(\dfrac{1}{0 \cdot 971}\right)^4$

35. $\sqrt[3]{(4 \cdot 714)}$

36. $\sqrt[3]{(0 \cdot 9316)}$

37. $\sqrt[3]{\left(\dfrac{4 \cdot 114}{7 \cdot 93}\right)}$

38. $\sqrt[4]{(0 \cdot 8145 - 0 \cdot 799)}$

39. $\sqrt[5]{(8 \cdot 6 \times 9 \cdot 71)}$

40. $\sqrt[3]{\left(\dfrac{1 \cdot 91}{4 \cdot 2 - 3 \cdot 766}\right)}$

41. $\left(\dfrac{1}{7\cdot6}-\dfrac{1}{18\cdot5}\right)^3$

42. $\dfrac{\sqrt{(4\cdot79)}+1\cdot6}{9\cdot63}$

43. $\dfrac{(0\cdot761)^2-\sqrt{(4\cdot22)}}{1\cdot96}$

44. $\sqrt[3]{\left(\dfrac{1\cdot74\times0\cdot761}{0\cdot0896}\right)}$

45. $\left(\dfrac{8\cdot6\times1\cdot71}{0\cdot43}\right)^3$

46. $\dfrac{9\cdot61-\sqrt{(9\cdot61)}}{9\cdot61^2}$

47. $\dfrac{9\cdot6\times10^4\times3\cdot75\times10^7}{8\cdot88\times10^6}$

48. $\dfrac{8\cdot06\times10^{-4}}{1\cdot71\times10^{-6}}$

49. $\dfrac{3\cdot92\times10^{-7}}{1\cdot884\times10^{-11}}$

50. $\left(\dfrac{1\cdot31\times2\cdot71\times10^5}{1\cdot91\times10^4}\right)^5$

51. $\left(\dfrac{1}{9\cdot6}-\dfrac{1}{9\cdot99}\right)^{10}$

52. $\dfrac{\sqrt[3]{(86\cdot6)}}{\sqrt[4]{(4\cdot71)}}$

53. $\dfrac{23\cdot7\times0\cdot0042}{12\cdot48-9\cdot7}$

54. $\dfrac{0\cdot482+1\cdot6}{0\cdot024\times1\cdot83}$

55. $\dfrac{8\cdot52-1\cdot004}{0\cdot004-0\cdot0083}$

56. $\dfrac{1\cdot6-0\cdot476}{2\cdot398\times41\cdot2}$

57. $\left(\dfrac{2\cdot3}{0\cdot791}\right)^7$

58. $\left(\dfrac{8\cdot4}{28\cdot7-0\cdot47}\right)^3$

59. $\left(\dfrac{5\cdot114}{7\cdot332}\right)^5$

60. $\left(\dfrac{4\cdot2}{2\cdot3}+\dfrac{8\cdot2}{0\cdot52}\right)^3$

61. $\dfrac{1}{8\cdot2^2}-\dfrac{3}{19^2}$

62. $\dfrac{100}{11^3}+\dfrac{100}{12^3}$

63. $\dfrac{7\cdot3-4\cdot291}{2\cdot6^2}$

64. $\dfrac{9\cdot001-8\cdot97}{0\cdot95^3}$

65. $\dfrac{10\cdot1^2+9\cdot4^2}{9\cdot8}$

66. $(3\cdot6\times10^{-8})^2$

67. $(8\cdot24\times10^4)^3$

68. $(2\cdot17\times10^{-3})^3$

69. $(7\cdot095\times10^{-6})^{\frac13}$

70. $\sqrt[3]{\left(\dfrac{4\cdot7}{2.3^2}\right)}$

REVISION EXERCISE 1A

1. Evaluate, without a calculator:

(a) $148\div0\cdot8$

(b) $0\cdot024\div0\cdot00016$

(c) $(0\cdot2)^2\div(0\cdot1)^3$

(d) $2-\frac12-\frac13-\frac14$

(e) $1\frac34\times1\frac35$

(f) $\dfrac{1\frac16}{1\frac13+1\frac14}$

2. On each bounce, a ball rises to $\frac45$ of its previous height. To what height will it rise after the third bounce, if dropped from a height of 250 cm?

3. A man spends $\frac13$ of his salary on accommodation and $\frac25$ of the remainder on food. What fraction is left for other purposes?

4. $a=\frac12$, $b=\frac14$. Which one of the following has the greatest value?

(i) ab (ii) $a+b$ (iii) $\dfrac{a}{b}$ (iv) $\dfrac{b}{a}$ (v) $(ab)^2$

5. Express $0\cdot05473$

(a) correct to three significant figures

(b) correct to three decimal places

(c) in standard form.

6. Evaluate $\frac23+\frac47$, correct to three decimal places.

7. Evaluate the following and give the answer in standard form:

(a) $3600\div0\cdot00012$ (b) $\dfrac{3\cdot33\times10^4}{9\times10^{-1}}$

(c) $(30\,000)^3$

8. (a) £143 is divided in the ratio $2:3:6$; calculate the smallest share.

(b) A prize is divided between three people X, Y and Z. If the ratio of X's share to Y's share is $3:1$ and Y's share to Z's share is $2:5$, calculate the ratio of X's share to Z's share.

(c) If $a:3=12:a$, calculate the positive value of a.

9. Labour costs, totalling £47·25, account for 63% of a car repair bill. Calculate the total bill.

10. (a) Convert to percentages

(i) $0\cdot572$ (ii) $\frac78$

(b) Express 2·6 kg as a percentage of 6·5 kg.

(c) In selling a red herring for 92p, a fishmonger makes a profit of 15%. Find the cost price of the fish.

11. The length of a rectangle is decreased by 25% and the breadth is increased by 40%. Calculate the percentage change in the area of the rectangle.

12. (a) What sum of money, invested at 9% interest per year, is needed to provide an income of £45 per year?

(b) A particle increases its speed from 8×10^5 m/s to $1\cdot1\times10^6$ m/s. What is the percentage increase?

13. A family on holiday in France exchanged £450 for francs when the exchange rate was 10·70 francs to the pound. They spent 3700 francs and then changed the rest back into pounds, by which time the exchange rate had become 11·15 francs to the pound. How much did the holiday cost? (Answer in pounds.)

14. A welder has an income of £8400 and allowances which total £1950. How much tax does he pay each month if the rate of taxation is 28% of taxable income? He receives a pay rise and subsequently pays £162·40 in tax each month. Assuming the same allowances and rate of taxation, calculate his new salary.

15. A map is drawn to a scale of 1 : 10 000. Find:
    (a) the distance between two railway stations which appear on the map 24 cm apart.
    (b) the area, in square kilometres, of a lake which has an area of 100 cm^2 on the map.

16. A map is drawn to a scale of 1 : 2000. Find:
    (a) the actual distance between two points, which appear 15 cm apart on the map.
    (b) the length on the map of a road, which is 1·2 km in length.
    (c) the area on the map of a field, with an actual area of 60 000 m^2.

17. (a) On a map, the distance between two points is 16 cm. Calculate the scale of the map if the actual distance between the points is 8 km.
    (b) On another map, two points appear 1·5 cm apart and are in fact 60 km apart. Calculate the scale of the map.

18. (a) A house is bought for £20 000 and sold for £24 400. What is the percentage profit?
    (b) A piece of meat, initially weighing 2·4 kg, is cooked and subsequently weighs 1·9 kg. What is the percentage loss in weight?
    (c) An article is sold at a 6% loss for £225·60. What was the cost price?

19. (a) Convert into metres per second:
    (i) 700 cm/s
    (ii) 720 km/h
    (iii) 18 km/h
    (b) Convert into kilometres per hour:
    (i) 40 m/s   (ii) 0·6 m/s

20. (a) Calculate the speed (in metres per second) of a slug which moves a distance of 30 cm in 1 minute.
    (b) Calculate the time taken for a bullet to travel 8 km at a speed of 5000 m/s.
    (c) Calculate the distance flown, in a time of four hours, by a pigeon which flies at a speed of 12 m/s.

21. A motorist travelled 200 miles in five hours. Her average speed for the first 100 miles was 50 m.p.h. What was her average speed for the second 100 miles?

22. (a) Without using tables, write down the value of:
    (i) $\sqrt{0·0025}$ (ii) $\sqrt{2\frac{1}{4}}$
    (b) Given $\sqrt{15} = 3·87$ and $\sqrt{1·5} = 1·22$, write down:
    (i) $\sqrt{150}$ (ii) $\sqrt{1500}$ (iii) $\sqrt{0·0015}$
    (c) Given $\sqrt{19} = 4·36$ and $\sqrt{1·9} = 1·38$, write down:
    (i) $\sqrt{19\,000}$ (ii) $\sqrt{190}$ (iii) $\sqrt{0·00019}$

23. Write down the first four terms of the following sequences.
    (a) $u_1 = 5$, $u_{n+1} = u_n + 3$
    (b) $u_1 = 2$, $u_{n+1} = 3u_n$
    (c) $u_1 = 60$, $u_{n+1} = u_n - 7$
    (d) $u_1 = 3$, $u_{n+1} = (u_n)^2$

24. Throughout his life Mr Cram's heart has beat at an average rate of 72 beats per minute. Mr Cram is sixty years old. How many times has his heart beat during his life? Give the answer in standard form correct to two significant figures.

25. Estimate the answer correct to one significant figure. Do not use a calculator.
    (a) $(612 \times 52) \div 49·2$ (b) $(11·7 + 997·1) \times 9·2$
    (c) $\sqrt{\left(\dfrac{91·3}{10·1}\right)}$ (d) $\pi\sqrt{(5·2^2 + 18·2^2)}$

26. Evaluate the following using a calculator: (answers to 4 sig. fig.)
    (a) $\dfrac{0·74}{0·81 \times 1·631}$ (b) $\sqrt{\left(\dfrac{9·61}{8·34 - 7·41}\right)}$
    (c) $\left(\dfrac{0·741}{0·8364}\right)^4$ (d) $\dfrac{8·4 - 7·642}{3·333 - 1·735}$

27. Evaluate the following and give the answers to 3 significant figures:
    (a) $\sqrt[3]{(9·61 \times 0·0041)}$ (b) $\left(\dfrac{1}{9·5} - \dfrac{1}{11·2}\right)^3$
    (c) $\dfrac{15·6 \times 0·714}{0·0143 \times 12}$ (d) $\sqrt[4]{\left(\dfrac{1}{5 \times 10^3}\right)}$

28. The edges of a cube are all increased by 10%. What is the percentage increase in the volume?

EXAMINATION EXERCISE 1B

1. (a) Given that
$$t = 2\pi \sqrt{\left(\frac{\ell}{g}\right)},$$

find the value of t, to 3 sig. fig., when $\ell = 2\cdot31$ and $g = 9\cdot81$

(b) Using another formula, the value of t is given by

$$t = 2\pi \sqrt{\left(\frac{2\cdot31^2 + 0\cdot9^2}{2\cdot31 \times 9\cdot81}\right)}.$$

Without using a calculator, and using suitable approximate values for the numbers in the formula, find an estimate for the value of t. (To earn the marks in this part of the question you must show the various stages of your working.) [L]

2.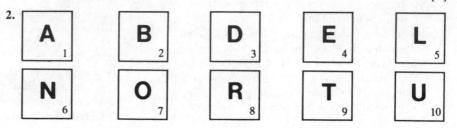

The ten tiles shown above are used in a game to make words and the scores shown on the tiles are added together. For example, BONE scores $2 + 7 + 6 + 4 = 19$.

(a) Calculate the score for DOUBT.
(b) A player makes a five-letter word and scores a total of 17. Four of the letters in the word are A, B, E, O. What is the fifth letter?
(c) Another player makes a five-letter word using the letters L, N, O and two others. The score for the word is 23. What could the other two letters be?
(d) What is the highest possible score for four letters which are all different?
(e) A player makes a four-letter word with a score of 33. All the letters in the word are different. What are the four letters? [M]

3. 1 3 8 9 10

From these numbers, write down:
(a) the prime number, (Note: 1 is NOT a prime number)
(b) a multiple of 5,
(c) two square numbers,
(d) two factors of 32.
(e) Find two numbers m and n from the list such that
$m = \sqrt{n}$ and $n = \sqrt{81}$.
(f) If each of the numbers in the list can be used once, find p, q, r, s, t such that
$(p + q)r = 2(s + t) = 36$. [S]

4. Violet intends to make herself an outfit to wear at her brother's wedding in October. She decides to make a pair of trousers, a blouse and a tunic. The table below shows the quantity of fabric required for her size and the cost per metre of the various materials she has chosen.

Item	Quantity in metres	Cost per metre
Trousers	2·40	£4·90
Blouse	2·90	£3·10
Tunic	1·50	£6·40
Lining	1·50	£1·40

(a) Calculate the cost of each of these four items.
  (i) Trousers
  (ii) Blouse
  (iii) Tunic
  (iv) Lining
(b) Calculate the cost of
  (i) 1 zip (18 cm) at 5p per cm,
  (ii) 6 buttons at 9p each,
  (iii) 3 reels of thread at 57p each,
  (iv) 1·50 m of tape at 12p per metre.
(c) Calculate the TOTAL cost of the items listed in (a) and (b).
(d) Violet is entitled to a discount of 8% on the total cost.
   Calculate, to the nearest penny, how much she actually pays.   [L]

5.

★ ★ ★ ★ per litre

38.6p

Four star petrol costs 38·6p per litre. One gallon is the same as 4·55 litres. What is the cost, to the nearest penny, of one gallon of petrol?   [NI]

6. The scale of a map is 1 : 25000. A forest is represented on the map as a shape whose perimeter is 12 cm and whose area is 10 cm^2.
   (a) Find the perimeter of the forest in kilometres.
   (b) Find the area of the forest in square metres.   [M]

7. A householder is having a garden path made out of concrete. The table shows various mixes of concrete for different uses.

Purpose	Composition by volume		
Mix 1 Foundation, drives, floor slabs	Cement: 1	Sand: $2\frac{1}{2}$	Aggregate: 4
Mix 2 Paths and sections less than 100 mm	Cement: 1	Sand: 2	Aggregate: 3
Mix 3 Paving less than 50 mm thick	Cement: 1	Coarse sand: 3	

If he requires 1·2 m^3 of concrete, how much sand will be required?   [NI]

8. National Westminster Bank issues this advice to people considering buying a house:

   'You can calculate what your maximum potential mortgage will be at the time you apply by multiplying your anticipated gross annual income by 2¾. In appropriate cases this can be increased by up to 1½ times your partner's income.

   N.B. Remember, the actual mortgage available will always be limited to a maximum of 95% of the lower of valuation or purchase price of the property.'

   Joy and Harry Black have seen a house they would like to buy. The purchase price is £41 000 but it is valued by the surveyor at only £38 000.
   (i) What is the maximum mortgage that the bank would give on this house?
   (ii) Henry expects his gross annual income to be £10 200. The bank has agreed to consider Joy's income for mortgage purposes and will make the maximum allowance on it. What is the minimum gross income that Joy must earn in order for the Blacks to be able to ask for the maximum mortage on this house?   [N]

9. Travel Protector Insurance issued the following table of premiums for holiday insurance in 1985.

   Premiums
   per injured person

Period of Travel	Area 'A' UK†	Area 'B' Europe	Area 'C' Worldwide
1– 4 days	£3·60	£ 5·40	£16·90
5– 8 days	£4·50	£ 7·80	£16·90
9–17 days	£5·40	£ 9·95	£21·45
18–23 days	£6·30	£12·30	£27·95
24–31 days	£7·20	£15·25	£32·50
32–62 days	—	£24·10	£44·20
63–90 days	—	£34·50	£53·95

   Winter Sports
   Cover is available at 3 times these premium rates.

   (Source: *Travel Protector Insurance*, published by National Westminster Bank PLC.)

   †Excluding Channel Islands
   Discount for Children
   Under 14 years at date of Application —
   20% reduction
   Under 3 years at date of Application —
   Free of charge

   Find the premiums paid by
   (a) Mr. Jones holidaying in Blackpool with his wife and three children, aged 2, 9 and 15, from 2 August to 16 August,
   (b) Carole Smith going to Switzerland to ski from 19 December to 8 January.   [N]

10. (a) A houseowner sells his house for £32 000. He has to pay Stamp Duty at 2% of the selling price, and agent's fees at $2\frac{1}{4}$% of the selling price. Solicitor's fees amount to £260, and he must repay the balance of a loan from the Building Society amounting to £17 250. How much cash will he have remaining after the sale is complete?

(b) £250 is invested at 12% per annum compound interest in a deposit account.
    (i) Calculate the total amount of money invested in the deposit account after one, two and three years.
    (ii) Calculate the percentage increase on the original £250 over the three years. Give your answer to the nearest 0.1%   [MEI]

11. Parvinda's dad runs Tali's Take-away. He buys his rice from Yaul's Cash and Carry. A 45 kg bag costs £28·50. He uses 150 g rice for a portion.
(a) How many complete portions will he get from one bag?
(b) How much does the rice for one portion cost to the nearest penny?
The foil trays he puts the rice in when he sells it cost £17·60 + VAT for 2000. VAT is 15%.
(c) How much does it cost Parvinda's dad to sell 1 portion of rice?
He charges 35p for a portion of rice.
(d) What is the cost to Parvinda's dad of the rice and container as a percentage of the selling price?
Parvinda's dad fixes all his prices so that the basic cost of the food and container is the same percentage of the selling price as the rice. His weekly expenses are:
Rent £34·00   Water Rates £12·50   Rates £23·00   Gas £49·00   Electricity £19·60
(e) One week he takes £407·50. How much profit does he make?   [S]

12. For a particular make of street lamp, the length of the pavement lit by the lamp is equal to two thirds of the height of the lamp.

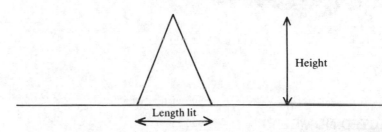

Height

Length lit

(a) Copy and complete the following table.

Lamp	Height (metres)	Length lit (metres)
A	6	
B		6

(b) Two A lamps are positioned 5 metres apart. What length of pavement is unlit between the lamps?
(c) Two B lamps are positioned 5 metres apart. Draw a diagram showing clearly the section of pavement lit by *both* lamps. How long is this section?
(d) The cost of a lamp (in pounds) is given by
Cost = 500 + 125 × Height
where the height is measured in metres.
Find the cost of lamps A and B.
(e) A planning officer is trying to decide what lamps to buy for a new street of total length 100 metres. She has to make sure that the whole length of the pavement will be lit and that the total cost is as low as possible.
Should she buy lamps A or lamps B? Explain your answer fully.   [W]

2 Algebra 1

Isaac Newton (1642–1727) is thought by many to have been one of the greatest intellects of all time. He went to Trinity College Cambridge in 1661 and by the age of 23 he had made three major discoveries: the nature of colours, the calculus and the law of gravitation. He used his version of the calculus to give the first satisfactory explanation of the motion of the Sun, the Moon and the stars. Because he was extremely sensitive to criticism, Newton was always very secretive, but he was eventually persuaded to publish his discoveries in 1687.

2.1 DIRECTED NUMBERS

To add two directed numbers with the same sign, find the sum of the numbers and give the answer the same sign.

e.g. $+3 + (+5) = +3 + 5 = +8$
$-7 + (-3) = -7 - 3 = -10$
$-9\cdot1 + (-3\cdot1) = -9\cdot1 - 3\cdot1 = -12\cdot2$
$-2 + (-1) + (-5) = (-2 - 1) - 5$
$\qquad\qquad\qquad = -3 - 5$
$\qquad\qquad\qquad = -8$

To add two directed numbers with different signs, find the difference between the numbers and give the answer the sign of the larger number.

e.g. $+7 + (-3) = +7 - 3 = +4$
$+9 + (-12) = +9 - 12 = -3$
$-8 + (+4) = -8 + 4 = -4$

To subtract a directed number, change its sign and add.

e.g. $+7 - (+5) = +7 - 5 = +2$
$+7 - (-5) = +7 + 5 = +12$
$-8 - (+4) = -8 - 4 = -12$
$-9 - (-11) = -9 + 11 = +2$

Exercise 1

1. $+7 + (+6)$
2. $+11 + (+200)$
3. $-3 + (-9)$
4. $-7 + (-24)$
5. $-5 + (-61)$
6. $+0\cdot2 + (+5\cdot9)$
7. $+5 + (+4\cdot1)$
8. $-8 + (-27)$
9. $+17 + (+1\cdot7)$
10. $-2 + (-3) + (-4)$
11. $-7 + (+4)$
12. $+7 + (-4)$
13. $-9 + (+7)$
14. $+16 + (-30)$
15. $+14 + (-21)$
16. $-7 + (+10)$
17. $-19 + (+200)$
18. $+7\cdot6 + (-9\cdot8)$
19. $-1\cdot8 + (+10)$
20. $-7 + (+24)$

21. $+7-(+5)$ **22.** $+9-(+15)$
23. $-6-(+9)$ **24.** $-9-(+5)$
25. $+8-(+10)$ **26.** $-19-(-7)$
27. $-10-(+70)$ **28.** $-5\cdot1-(+8)$
29. $-0\cdot2-(+4)$ **30.** $+5\cdot2-(-7\cdot2)$

31. $-4+(-3)$ **32.** $+6-(-2)$
33. $+8+(-4)$ **34.** $-4-(+6)$
35. $+7-(-4)$ **36.** $+6+(-2)$
37. $+10-(+30)$ **38.** $+19-(+11)$
39. $+4+(-7)+(-2)$ **40.** $-3-(+2)+(-5)$

41. $-17-(-1)+(-10)$ **42.** $-5+(-7)-(+9)$
43. $+9+(-7)-(-6)$ **44.** $-7-(-8)$
45. $-10\cdot1+(-10\cdot1)$ **46.** $-75-(-25)$
47. $-204-(+304)$ **48.** $-7+(-11)-(+11)$
49. $+17-(+17)$ **50.** $-6+(-7)-(+8)$

51. $+7+(-7\cdot1)$ **52.** $-11-(-4)+(+3)$
53. $-2-(-8\cdot7)$ **54.** $+7+(-11)+(+5)$
55. $-610+(-240)$ **56.** $-7-(-3)-(-8)$
57. $+9-(-6)+(-9)$ **58.** $-1-(-5)+(-8)$
59. $-2\cdot1+(-9\cdot9)$ **60.** $-47-(-16)$

When two directed numbers with the same sign are multiplied together, the answer is positive.
(a) $+7\times(+3)=+21$
(b) $-6\times(-4)=+24$

When two directed numbers with different signs are multiplied together, the answer is negative.
(a) $-8\times(+4)=-32$
(b) $+7\times(-5)=-35$
(c) $-3\times(+2)\times(+5)=-6\times(+5)=-30$

When dividing directed numbers, the rules are the same as in multiplication.
(a) $-70\div(-2)=+35$
(b) $+12\div(-3)=-4$
(c) $-20\div(+4)=-5$

Exercise 2

1. $+2\times(-4)$ **2.** $+7\times(+4)$
3. $-4\times(-3)$ **4.** $-6\times(-4)$
5. $-6\times(-3)$ **6.** $+5\times(-7)$
7. $-7\times(-7)$ **8.** $-4\times(+3)$
9. $+0\cdot5\times(-4)$ **10.** $-1\frac{1}{2}\times(-6)$
11. $-8\div(+2)$ **12.** $+12\div(+3)$
13. $+36\div(-9)$ **14.** $-40\div(-5)$
15. $-70\div(-1)$ **16.** $-56\div(+8)$
17. $-\frac{1}{2}\div(-2)$ **18.** $-3\div(+5)$
19. $+0\cdot1\div(-10)$ **20.** $-0\cdot02\div(-100)$

21. $-11\times(-11)$ **22.** $-6\times(-1)$
23. $+12\times(-50)$ **24.** $-\frac{1}{2}\div(+\frac{1}{2})$
25. $-600\div(+30)$ **26.** $-5\cdot2\div(+2)$
27. $+7\times(-100)$ **28.** $-6\div(-\frac{1}{3})$
29. $100\div(-0\cdot1)$ **30.** -8×-80
31. $-3\times(-2)\times(-1)$ **32.** $+3\times(-7)\times(+2)$
33. $+0\cdot4\div(-1)$ **34.** $-16\div(+40)$
35. $+0\cdot2\times(-1000)$ **36.** $-7\times(-5)\times(-1)$
37. $-14\div(+7)$ **38.** $-7\div(-14)$
39. $+1\frac{1}{4}\div(-5)$ **40.** $-6\times(-\frac{1}{2})\times(-30)$

Exercise 3

1. $-7+(-3)$ **2.** $-6-(-7)$
3. $-4\times(-3)$ **4.** $-4\times(+7)$
5. $4-(+6)$ **6.** $-4\times(-4)$
7. $+6\div(-2)$ **8.** $+8-(-6)$
9. $-7\times(+4)$ **10.** $-8\div(-2)$
11. $+10\div(-60)$ **12.** $(-3)^2$
13. $40-(+70)$ **14.** $-6\times(-4)$
15. $(-1)^5$ **16.** $-8\div(+4)$
17. $+10\times(-3)$ **18.** $-7\times(-1)$
19. $+10+(-7)$ **20.** $+12-(-4)$

21. $+100+(-7)$ **22.** $-60\times(-40)$
23. $-20\div(-2)$ **24.** $(-1)^{20}$
25. $6-(+10)$ **26.** $-6\times(+4)\times(-2)$
27. $+8\div(-8)$ **28.** $0\times(-6)$
29. $(-2)^3$ **30.** $+100-(-70)$
31. $+18\div(-6)$ **32.** $(-1)^{12}$
33. $-6-(-7)$ **34.** $(-2)^2+(-4)$
35. $+8-(-7)$ **36.** $+7+(-2)$
37. $-6\times(+0\cdot4)$ **38.** $-3\times(-6)\times(-10)$
39. $(-2)^2+(+1)$ **40.** $+6-(+1000)$

41. $(-3)^2-7$ **42.** $-12\div\frac{1}{4}$
43. $-30\div-\frac{1}{2}$ **44.** $5-(+7)+(-0\cdot5)$
45. $(-2)^5$ **46.** $0\div(-\frac{1}{5})$
47. $(-0\cdot1)^2\times(-10)$ **48.** $3-(+19)$
49. $2\cdot1+(-6\cdot4)$ **50.** $(-\frac{1}{2})^2\div(-4)$

2.2 FORMULAE

When a calculation is repeated many times it is often helpful to use a formula. When a building society offers a mortgage it may use a formula like '$2\frac{1}{2}$ times the main salary plus the second salary'. Publishers use a formula to work out the selling price of a book based on the production costs and the expected sales of the book.

Exercise 4

1. The final speed v of a car is given by the formula $v = u + at$. [u = initial speed, a = acceleration, t = time taken]. Find v when $u = 15$, $a = 0.2$, $t = 30$.

2. The time period T of a simple pendulum is given by the formula $T = 2\pi \sqrt{\left(\dfrac{\ell}{g}\right)}$, where ℓ is the length of the pendulum and g is the gravitational acceleration. Find T when $\ell = 0.65$, $g = 9.81$ and $\pi = 3.142$.

3. The total surface area A of a cone is related to the radius r and the slant height ℓ by the formula $A = \pi r(r + \ell)$. Find A when $r = 7$ and $\ell = 11$.

4. The sum S of the squares of the integers from 1 to n is given by $S = \frac{1}{6}n(n + 1)(2n + 1)$. Find S when $n = 12$.

5. The acceleration a of a train is found using the formula $a = \dfrac{v^2 - u^2}{2s}$. Find a when $v = 20$, $u = 9$ and $s = 2.5$.

6. Einstein's famous equation relating energy, mass and the speed of light is $E = mc^2$. Find E when $m = 0.0001$ and $c = 3 \times 10^8$

7. The area A of a parallelogram with sides a and b is given by $A = ab \sin \theta$, where θ is the angle between the sides. Find A when $a = 7$, $b = 3$ and $\theta = 30°$.

8. The distance s travelled by an accelerating rocket is given by $s = ut + \frac{1}{2}at^2$. Find s when $u = 3$, $t = 100$ and $a = 0.1$.

9. Find a formula for the area of the shape below, in terms of a, b and c.

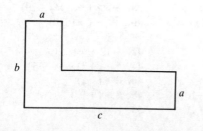

10. Find a formula for the shaded part below, in terms of p, q and r.

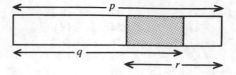

11. An intelligent fish lays brown eggs or white eggs and it likes to lay them in a certain pattern. Each brown egg is surrounded by six white eggs.

Here there are 3 brown eggs and 14 white eggs.
   (a) How many eggs does it lay altogether if it lays 200 brown eggs?
   (b) How many eggs does it lay altogether if it lays n brown eggs?

12. In the diagrams below the rows of black tiles are surrounded by white tiles.

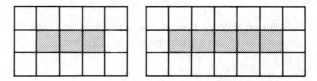

Find a formula for the number of white tiles which would be needed to surround a row of n black tiles.

Example 1

When $a = 3$, $b = -2$, $c = 5$, find the value of:
(a) $3a + b$ (b) $ac + b^2$
(c) $\dfrac{a + c}{b}$ (d) $a(c - b)$

(a) $3a + b = (3 \times 3) + (-2)$
$\qquad\quad = 9 - 2$
$\qquad\quad = 7$

(b) $ac + b^2 = (3 \times 5) + (-2)^2$
$\qquad\qquad = 15 + 4$
$\qquad\qquad = 19$

(c) $\dfrac{a+c}{b}=\dfrac{3+5}{-2}$

$\quad=\dfrac{8}{-2}$

$\quad=-4$

(d) $a(c-b)$

$\quad=3[5-(-2)]$
$\quad=3[7]$
$\quad=21$

Notice that working *down* the page is often easier to follow.

Exercise 5

Evaluate the following:
For questions **1** to **12** $a=3,\ c=2,\ e=5$.

1. $3a-2$ 2. $4c+e$ 3. $2c+3a$
4. $5e-a$ 5. $e-2c$ 6. $e-2a$
7. $4c+2e$ 8. $7a-5e$ 9. $c-e$
10. $10a+c+e$ 11. $a+c-e$ 12. $a-c-e$

For questions **13** to **24** $h=3,\ m=-2,\ t=-3$.

13. $2m-3$ 14. $4t+10$ 15. $3h-12$
16. $6m+4$ 17. $9t-3$ 18. $4h+4$
19. $2m-6$ 20. $m+2$ 21. $3h+m$
22. $t-h$ 23. $4m+2h$ 24. $3t-m$

For questions **25** to **36**, $x=-2,\ y=-1,\ k=0$

25. $3x+1$ 26. $2y+5$ 27. $6k+4$
28. $3x+2y$ 29. $2k+x$ 30. xy
31. xk 32. $2xy$ 33. $2(x+k)$
34. $3(k+y)$ 35. $5x-y$ 36. $3k-2x$

$2x^2$ means $2(x^2)$
$(2x^2)$ means 'work out $2x$ and *then* square it'
$-7x$ means $-7(x)$
$-x^2$ means $-1(x^2)$

Example 2

When $x=-2$, find the value of
(a) $2x^2-5x$ (b) $(3x)^2-x^2$

(a) $2x^2-5x=2(-2)^2-5(-2)$
$\qquad\qquad=2(4)+10$
$\qquad\qquad=18$

(b) $(3x)^2-x^2=(3\times-2)^2-1(-2)^2$
$\qquad\qquad=(-6)^2-1(4)$
$\qquad\qquad=36-4$
$\qquad\qquad=32.$

Exercise 6

If $x=-3$ and $y=2$, evaluate the following:

1. x^2 2. $3x^2$ 3. y^2
4. $4y^2$ 5. $(2x)^2$ 6. $2x^2$
7. $10-x^2$ 8. $10-y^2$ 9. $20-2x^2$
10. $20-3y^2$ 11. $5+4x$ 12. x^2-2x
13. y^2-3x^2 14. x^2-3y 15. $(2x)^2-y^2$
16. $4x^2$ 17. $(4x)^2$ 18. $1-x^2$
19. $y-x^2$ 20. x^2+y^2 21. x^2-y^2
22. $2-2x^2$ 23. $(3x)^2+3$ 24. $11-xy$
25. $12+xy$ 26. $(2x)^2-(3y)^2$ 27. $2-3x^2$

28. y^2-x^2 29. x^2+y^3 30. $\dfrac{x}{y}$

31. $10-3x$ 32. $2y^2$ 33. $25-3y$
34. $(2y)^2$ 35. $-7+3x$ 36. $-8+10y$
37. $(xy)^2$ 38. xy^2 39. $-7+x^2$
40. $17+xy$ 41. $-5-2x^2$ 42. $10-(2x)^2$

43. x^2+3x+5 44. $2x^2-4x+1$ 45. $\dfrac{x^2}{y}$

Example 3

When $a=-2,\ b=3,\ c=-3$, evaluate

(a) $\dfrac{2a(b^2-a)}{c}$ (b) $\sqrt{(a^2+b^2)}$

(a) $(b^2-a)=9-(-2)$
$\qquad\qquad=11$

$\therefore\ \dfrac{2a(b^2-a)}{c}=\dfrac{2\times(-2)\times(11)}{-3}$

$\qquad\qquad\qquad=-14\tfrac{2}{3}$

(b) $a^2+b^2=(-2)^2+(3)^2$
$\qquad\qquad=4+9$
$\qquad\qquad=13$

$\therefore\ \sqrt{(a^2+b^2)}=\sqrt{13}$

Exercise 7

Evaluate the following:
In questions **1** to **16**, $a=4,\ b=-2,\ c=-3$.

1. $a(b+c)$ 2. $a^2(b-c)$
3. $2c(a-c)$ 4. $b^2(2a+3c)$
5. $c^2(b-2a)$ 6. $2a^2(b+c)$
7. $2(a+b+c)$ 8. $3c(a-b-c)$
9. b^2+2b+a 10. c^2-3c+a
11. $2b^2-3b$ 12. $\sqrt{(a^2+c^2)}$
13. $\sqrt{(ab+c^2)}$ 14. $\sqrt{(c^2-b^2)}$

15. $\dfrac{b^2}{a}+\dfrac{2c}{b}$ 16. $\dfrac{c^2}{b}+\dfrac{4b}{a}$

In questions **17** to **32**, $k = -3$, $m = 1$, $n = -4$.

17. $k^2(2m - n)$

18. $5m \sqrt{(k^2 + n^2)}$

19. $\sqrt{(kn + 4m)}$

20. $kmn(k^2 + m^2 + n^2)$

21. $k^2m^2(m - n)$

22. $k^2 - 3k + 4$

23. $m^3 + m^2 + n^2 + n$

24. $k^3 + 3k$

25. $m(k^2 - n^2)$

26. $m\sqrt{(k - n)}$

27. $100k^2 + m$

28. $m^2(2k^2 - 3n^2)$

29. $\dfrac{2k + m}{k - n}$

30. $\dfrac{kn - k}{2m}$

31. $\dfrac{3k + 2m}{2n - 3k}$

32. $\dfrac{k + m + n}{k^2 + m^2 + n^2}$

In questions **33** to **48**, $w = -2$, $x = 3$, $y = 0$, $z = -\frac{1}{2}$

33. $\dfrac{w}{z} + x$

34. $\dfrac{w + x}{z}$

35. $y\left(\dfrac{x + z}{w}\right)$

36. $x^2(z + wy)$

37. $x\sqrt{(x + wz)}$

38. $w^2\sqrt{(z^2 + y^2)}$

39. $2(w^2 + x^2 + y^2)$

40. $2x(w - z)$

41. $\dfrac{z}{w} + x$

42. $\dfrac{z + w}{x}$

43. $\dfrac{x + w}{z^2}$

44. $\dfrac{y^2 - w^2}{xz}$

45. $z^2 + 4z + 5$

46. $\dfrac{1}{w} + \dfrac{1}{z} + \dfrac{1}{x}$

47. $\dfrac{4}{z} + \dfrac{10}{w}$

48. $\dfrac{yz - xw}{xz - w}$

49. Find $K = \sqrt{\left(\dfrac{a^2 + b^2 + c^2 - 2c}{a^2 + b^2 + 4c}\right)}$ if $a = 3$, $b = -2$, $c = -1$.

50. Find $W = \dfrac{kmn(k + m + n)}{(k + m)(k + n)}$ if $k = \frac{1}{2}$, $m = -\frac{1}{3}$, $n = \frac{1}{4}$.

2.3   BRACKETS AND SIMPLIFYING

A term outside a bracket multiplies each of the terms inside the bracket. This is the *distributive law*.

Example 1

$3(x - 2y) = 3x - 6y$

Example 2

$2x(x - 2y + z) = 2x^2 - 4xy + 2xz$

Example 3

$7y - 4(2x - 3) = 7y - 8x + 12$

In general,
 numbers can be added to numbers
  x's can be added to x's
  y's can be added to y's
  x^2's can be added to x^2's

But they must not be mixed.

Example 4

$2x + 3y + 3x^2 + 2y - x = x + 5y + 3x^2$

Example 5

$$7x + 3x(2x - 3) = 7x + 6x^2 - 9x$$
$$= 6x^2 - 2x$$

Exercise 8

Simplify as far as possible:

1. $3x + 4y + 7y$

2. $4a + 7b - 2a + b$

3. $3x - 2y + 4y$

4. $2x + 3x + 5$

5. $7 - 3x + 2 + 4x$

6. $5 - 3y - 6y - 2$

7. $5x + 2y - 4y - x^2$

8. $2x^2 + 3x + 5$

9. $2x - 7y - 2x - 3y$

10. $4a + 3a^2 - 2a$

11. $7a - 7a^2 + 7$

12. $x^2 + 3x^2 - 4x^2 + 5x$

13. $\dfrac{3}{a} + b + \dfrac{7}{a} - 2b$

14. $\dfrac{4}{x} - \dfrac{7}{y} + \dfrac{1}{x} + \dfrac{2}{y}$

15. $\dfrac{m}{x} + \dfrac{2m}{x}$

16. $\dfrac{5}{x} - \dfrac{7}{x} + \dfrac{1}{2}$

17. $\dfrac{3}{a} + b + \dfrac{2}{a} + 2b$

18. $\dfrac{n}{4} - \dfrac{m}{3} - \dfrac{n}{2} + \dfrac{m}{3}$

19. $x^3 + 7x^2 - 2x^3$

20. $(2x)^2 - 2x^2$

21. $(3y)^2 + x^2 - (2y)^2$

22. $(2x)^2 - (2y)^2 - (4x)^2$

23. $5x - 7x^2 - (2x)^2$

24. $\dfrac{3}{x^2} + \dfrac{5}{x^2}$

Remove the brackets and collect like terms:

25. $3x + 2(x + 1)$ **26.** $5x + 7(x - 1)$
27. $7 + 3(x - 1)$ **28.** $9 - 2(3x - 1)$
29. $3x - 4(2x + 5)$ **30.** $5x - 2x(x - 1)$
31. $7x + 3x(x - 4)$ **32.** $4(x - 1) - 3x$
33. $5x(x + 2) + 4x$ **34.** $3x(x - 1) - 7x^2$
35. $3a + 2(a + 4)$ **36.** $4a - 3(a - 3)$
37. $3ab - 2a(b - 2)$ **38.** $3y - y(2 - y)$
39. $3x - (x + 2)$ **40.** $7x - (x - 3)$
41. $5x - 2(2x + 2)$ **42.** $3(x - y) + 4(x + 2y)$
43. $x(x - 2) + 3x(x - 3)$ **44.** $3x(x + 4) - x(x - 2)$
45. $y(3y - 1) - (3y - 1)$ **46.** $7(2x + 2) - (2x + 2)$
47. $7b(a + 2) - a(3b + 3)$ **48.** $3(x - 2) - (x - 2)$

Two brackets

Example 6

$$(x + 5)(x + 3) = x(x + 3) + 5(x + 3)$$
$$= x^2 + 3x + 5x + 15$$
$$= x^2 + 8x + 15$$

Example 7

$$(2x - 3)(4y + 3) = 2x(4y + 3) - 3(4y + 3)$$
$$= 8xy + 6x - 12y - 9$$

Example 8

$$3(x + 1)(x - 2) = 3[x(x - 2) + 1(x - 2)]$$
$$= 3[x^2 - 2x + x - 2]$$
$$= 3x^2 - 3x - 6$$

Exercise 9

Remove the brackets and simplify.

1. $(x + 1)(x + 3)$ **2.** $(x + 3)(x + 2)$
3. $(y + 4)(y + 5)$ **4.** $(x - 3)(x + 4)$
5. $(x + 5)(x - 2)$ **6.** $(x - 3)(x - 2)$
7. $(a - 7)(a + 5)$ **8.** $(z + 9)(z - 2)$
9. $(x - 3)(x + 3)$ **10.** $(k - 11)(k + 11)$

11. $(2x + 1)(x - 3)$ **12.** $(3x + 4)(x - 2)$
13. $(2y - 3)(y + 1)$ **14.** $(7y - 1)(7y + 1)$
15. $(3x - 2)(3x + 2)$ **16.** $(3a + b)(2a + b)$
17. $(3x + y)(x + 2y)$ **18.** $(2b + c)(3b - c)$
19. $(5x - y)(3y - x)$ **20.** $(3b - a)(2a + 5b)$

21. $2(x - 1)(x + 2)$ **22.** $3(x - 1)(2x + 3)$
23. $4(2y - 1)(3y + 2)$ **24.** $2(3x + 1)(x - 2)$
25. $4(a + 2b)(a - 2b)$ **26.** $x(x - 1)(x - 2)$
27. $2x(2x - 1)(2x + 1)$ **28.** $3y(y - 2)(y + 3)$
29. $x(x + y)(x + z)$ **30.** $3z(a + 2m)(a - m)$

Be careful with an expression like $(x - 3)^2$. It is not $x^2 - 9$ or even $x^2 + 9$

$$(x - 3)^2 = (x - 3)(x - 3)$$
$$= x(x - 3) - 3(x - 3)$$
$$= x^2 - 6x + 9$$

Another common mistake occurs with an expression like $4 - (x - 1)^2$

$$4 - (x - 1)^2 = 4 - 1(x - 1)(x - 1)$$
$$= 4 - 1(x^2 - 2x + 1)$$
$$= 4 - x^2 + 2x - 1$$
$$= 3 + 2x - x^2$$

Exercise 10

Remove the brackets and simplify:

1. $(x + 4)^2$ **2.** $(x + 2)^2$
3. $(x - 2)^2$ **4.** $(2x + 1)^2$
5. $(y - 5)^2$ **6.** $(3y + 1)^2$
7. $(x + y)^2$ **8.** $(2x + y)^2$
9. $(a - b)^2$ **10.** $(2a - 3b)^2$
11. $3(x + 2)^2$ **12.** $(3 - x)^2$

13. $(3x + 2)^2$ **14.** $(a - 2b)^2$
15. $(x + 1)^2 + (x + 2)^2$ **16.** $(x - 2)^2 + (x + 3)^2$
17. $(x + 2)^2 + (2x + 1)^2$ **18.** $(y - 3)^2 + (y - 4)^2$
19. $(x + 2)^2 - (x - 3)^2$ **20.** $(x - 3)^2 - (x + 1)^2$
21. $(y - 3)^2 - (y + 2)^2$ **22.** $(2x + 1)^2 - (x + 3)^2$
23. $3(x + 2)^2 - (x + 4)^2$ **24.** $2(x - 3)^2 - 3(x + 1)^2$

2.4 LINEAR EQUATIONS

(a) If the x term is negative, take it to the other side, where it becomes positive.

Example 1

$$4 - 3x = 2$$
$$4 = 2 + 3x$$
$$2 = 3x$$
$$\frac{2}{3} = x$$

(b) If there are x terms on both sides, collect them on one side.

Example 2

$$2x - 7 = 5 - 3x$$
$$2x + 3x = 5 + 7$$
$$5x = 12$$
$$x = \frac{12}{5} = 2\frac{2}{5}$$

(c) If there is a fraction in the x term, multiply out to simplify the equation.

Example 3

$$\frac{2x}{3} = 10$$
$$2x = 30$$
$$x = \frac{30}{2} = 15$$

Exercise 11

Solve the following equations:

1. $2x - 5 = 11$
2. $3x - 7 = 20$
3. $2x + 6 = 20$
4. $5x + 10 = 60$
5. $8 = 7 + 3x$
6. $12 = 2x - 8$
7. $-7 = 2x - 10$
8. $3x - 7 = -10$
9. $12 = 15 + 2x$
10. $5 + 6x = 7$
11. $\frac{x}{5} = 7$
12. $\frac{x}{10} = 13$
13. $7 = \frac{x}{2}$
14. $\frac{x}{2} = \frac{1}{3}$
15. $\frac{3x}{2} = 5$
16. $\frac{4x}{5} = -2$
17. $7 = \frac{7x}{3}$
18. $\frac{3}{4} = \frac{2x}{3}$

19. $\frac{5x}{6} = \frac{1}{4}$
20. $-\frac{3}{4} = \frac{3x}{5}$
21. $\frac{x}{2} + 7 = 12$
22. $\frac{x}{3} - 7 = 2$
23. $\frac{x}{5} - 6 = -2$
24. $4 = \frac{x}{2} - 5$
25. $10 = 3 + \frac{x}{4}$
26. $\frac{a}{5} - 1 = -4$
27. $100x - 1 = 98$
28. $7 = 7 + 7x$
29. $\frac{x}{100} + 10 = 20$
30. $1000x - 5 = -6$
31. $-4 = -7 + 3x$
32. $2x + 4 = x - 3$
33. $x - 3 = 3x + 7$
34. $5x - 4 = 3 - x$
35. $4 - 3x = 1$
36. $5 - 4x = -3$
37. $7 = 2 - x$
38. $3 - 2x = x + 12$
39. $6 + 2a = 3$
40. $a - 3 = 3a - 7$
41. $2y - 1 = 4 - 3y$
42. $7 - 2x = 2x - 7$
43. $7 - 3x = 5 - 2x$
44. $8 - 2y = 5 - 5y$
45. $x - 16 = 16 - 2x$
46. $x + 2 = 3 \cdot 1$
47. $-x - 4 = -3$
48. $-3 - x = -5$
49. $-\frac{x}{2} + 1 = -\frac{1}{4}$
50. $-\frac{3}{5} + \frac{x}{10} = -\frac{1}{5} - \frac{x}{5}$

Example 4

$$x - 2(x - 1) = 1 - 4(x + 1)$$
$$x - 2x + 2 = 1 - 4x - 4$$
$$x - 2x + 4x = 1 - 4 - 2$$
$$3x = -5$$
$$x = -\frac{5}{3}$$

Exercise 12

Solve the following equations:

1. $x + 3(x + 1) = 2x$
2. $1 + 3(x - 1) = 4$
3. $2x - 2(x + 1) = 5x$
4. $2(3x - 1) = 3(x - 1)$
5. $4(x - 1) = 2(3 - x)$
6. $4(x - 1) - 2 = 3x$
7. $4(1 - 2x) = 3(2 - x)$
8. $3 - 2(2x + 1) = x + 17$
9. $4x = x - (x - 2)$
10. $7x = 3x - (x + 20)$
11. $5x - 3(x - 1) = 39$
12. $3x + 2(x - 5) = 15$
13. $7 - (x + 1) = 9 - (2x - 1)$
14. $10x - (2x + 3) = 21$

15. $3(2x + 1) + 2(x - 1) = 23$
16. $5(1 - 2x) - 3(4 + 4x) = 0$
17. $7x - (2 - x) = 0$
18. $3(x + 1) = 4 - (x - 3)$
19. $3y + 7 + 3(y - 1) = 2(2y + 6)$
20. $4(y - 1) + 3(y + 2) = 5(y - 4)$
21. $4x - 2(x + 1) = 5(x + 3) + 5$
22. $7 - 2(x - 1) = 3(2x - 1) + 2$

23. $10(2x + 3) - 8(3x - 5) + 5(2x - 8) = 0$
24. $2(x + 4) + 3(x - 10) = 8$
25. $7(2x - 4) + 3(5 - 3x) = 2$
26. $10(x + 4) - 9(x - 3) - 1 = 8(x + 3)$
27. $5(2x - 1) - 2(x - 2) = 7 + 4x$
28. $6(3x - 4) - 10(x - 3) = 10(2x - 3)$
29. $3(x - 3) - 7(2x - 8) - (x - 1) = 0$
30. $5 + 2(x + 5) = 10 - (4 - 5x)$

31. $6x + 30(x - 12) = 2(x - 1\frac{1}{2})$
32. $3(2x - \frac{2}{3}) - 7(x - 1) = 0$
33. $5(x - 1) + 17(x - 2) = 2x + 1$
34. $6(2x - 1) + 9(x + 1) = 8(x - 1\frac{1}{4})$
35. $7(x + 4) - 5(x + 3) + (4 - x) = 0$
36. $0 = 9(3x + 7) - 5(x + 2) - (2x - 5)$
37. $10(2\cdot3 - x) - 0\cdot1(5x - 30) = 0$
38. $8(2\frac{1}{2}x - \frac{3}{4}) - \frac{1}{4}(1 - x) = \frac{1}{2}$

39. $(6 - x) - (x - 5) - (4 - x) = -\dfrac{x}{2}$

40. $10\left(1 - \dfrac{x}{10}\right) - (10 - x) - \dfrac{1}{100}(10 - x) = 0\cdot05$

Example 5

$$(x + 3)^2 = (x + 2)^2 + 3^2$$
$$(x + 3)(x + 3) = (x + 2)(x + 2) + 9$$
$$x^2 + 6x + 9 = x^2 + 4x + 4 + 9$$
$$6x + 9 = 4x + 13$$
$$2x = 4$$
$$x = 2$$

Exercise 13

Solve the following equations:

1. $x^2 + 4 = (x + 1)(x + 3)$
2. $x^2 + 3x = (x + 3)(x + 1)$
3. $(x + 3)(x - 1) = x^2 + 5$
4. $(x + 1)(x + 4) = (x - 7)(x + 6)$
5. $(x - 2)(x + 3) = (x - 7)(x + 7)$
6. $(x - 5)(x + 4) = (x + 7)(x - 6)$
7. $2x^2 + 3x = (2x - 1)(x + 1)$
8. $(2x - 1)(x - 3) = (2x - 3)(x - 1)$
9. $x^2 + (x + 1)^2 = (2x - 1)(x + 4)$
10. $x(2x + 6) = 2(x^2 - 5)$

11. $(x + 1)(x - 3) + (x + 1)^2 = 2x(x - 4)$
12. $(2x + 1)(x - 4) + (x - 2)^2 = 3x(x + 2)$
13. $(x + 2)^2 - (x - 3)^2 = 3x - 11$
14. $x(x - 1) = 2(x - 1)(x + 5) - (x - 4)^2$
15. $(2x + 1)^2 - 4(x - 3)^2 = 5x + 10$
16. $2(x + 1)^2 - (x - 2)^2 = x(x - 3)$

In questions **17** to **22**, form an equation in x by means of Pythagoras' Theorem, and hence find the length of each side of the triangle. (All the lengths are in cm.)

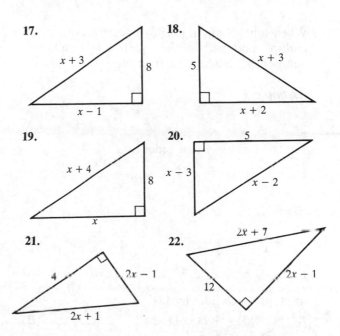

17. $x + 3$, 8, $x - 1$

18. 5, $x + 3$, $x + 2$

19. $x + 4$, 8, x

20. 5, $x - 3$, $x - 2$

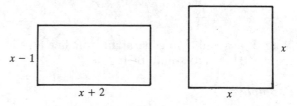

21. 4, $2x - 1$, $2x + 1$

22. $2x + 7$, $2x - 1$, 12

23. The area of the rectangle shown exceeds the area of the square by 2 cm². Find x.

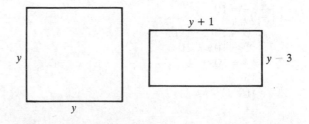

$x - 1$, $x + 2$, x, x

24. The area of the square exceeds the area of the rectangle by 13 m². Find y.

y, y, $y + 1$, $y - 3$

25. The area of the square is half the area of the rectangle. Find x.

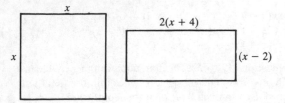

When solving equations involving fractions, multiply both sides of the equation by a suitable number to eliminate the fractions.

Example 6

$$\frac{5}{x} = 2$$

$5 = 2x$ (multiply both sides by x)

$$\frac{5}{2} = x$$

Example 7

$$\frac{x+3}{4} = \frac{2x-1}{3} \qquad \dots(A)$$

$$12\frac{(x+3)}{4} = 12\frac{(2x-1)}{3}$$

(multiply both sides by 12)

$\therefore\ \ 3(x+3) = 4(2x-1) \qquad \dots(B)$

$\ \ \ \ \ \ 3x+9 = 8x-4$

$\ \ \ \ \ \ \ \ \ \ \ \ 13 = 5x$

$\ \ \ \ \ \ \ \ \ \ \ \ \dfrac{13}{5} = x$

$\ \ \ \ \ \ \ \ \ \ \ \ x = 2\frac{3}{5}$

Note: It is possible to go straight from line (A) to line (B) by 'cross-multiplying'.

Example 8

$$\frac{5}{(x-1)} + 2 = 12$$

$$\frac{5}{(x-1)} = 10$$

$\ \ \ \ \ \ 5 = 10(x-1)$

$\ \ \ \ \ \ 5 = 10x - 10$

$\ \ \ \ 15 = 10x$

$\ \ \ \ \dfrac{15}{10} = x$

$\ \ \ \ \ \ x = 1\frac{1}{2}$

Exercise 14

Solve the following equations:

1. $\dfrac{7}{x} = 21$

2. $30 = \dfrac{6}{x}$

3. $\dfrac{5}{x} = 3$

4. $\dfrac{9}{x} = -3$

5. $11 = \dfrac{5}{x}$

6. $-2 = \dfrac{4}{x}$

7. $\dfrac{x}{4} = \dfrac{3}{2}$

8. $\dfrac{x}{3} = 1\frac{1}{4}$

9. $\dfrac{x+1}{3} = \dfrac{x-1}{4}$

10. $\dfrac{x+3}{2} = \dfrac{x-4}{5}$

11. $\dfrac{2x-1}{3} = \dfrac{x}{2}$

12. $\dfrac{3x+1}{5} = \dfrac{2x}{3}$

13. $\dfrac{8-x}{2} = \dfrac{2x+2}{5}$

14. $\dfrac{x+2}{7} = \dfrac{3x+6}{5}$

15. $\dfrac{1-x}{2} = \dfrac{3-x}{3}$

16. $\dfrac{2}{x-1} = 1$

17. $\dfrac{x}{3} + \dfrac{x}{4} = 1$

18. $\dfrac{x}{3} + \dfrac{x}{2} = 4$

19. $\dfrac{x}{2} - \dfrac{x}{5} = 3$

20. $\dfrac{x}{3} = 2 + \dfrac{x}{4}$

21. $\dfrac{5}{x-1} = \dfrac{10}{x}$

22. $\dfrac{12}{2x-3} = 4$

23. $2 = \dfrac{18}{x+4}$

24. $\dfrac{5}{x+5} = \dfrac{15}{x+7}$

25. $\dfrac{9}{x} = \dfrac{5}{x-3}$

26. $\dfrac{4}{x-1} = \dfrac{10}{3x-1}$

27. $\dfrac{-7}{x-1} = \dfrac{14}{5x+2}$

28. $\dfrac{4}{x+1} = \dfrac{7}{3x-2}$

29. $\dfrac{x+1}{2} + \dfrac{x-1}{3} = \dfrac{1}{6}$

30. $\dfrac{1}{3}(x+2) = \dfrac{1}{5}(3x+2)$

31. $\dfrac{1}{2}(x-1) - \dfrac{1}{6}(x+1) = 0$

32. $\dfrac{1}{4}(x+5) - \dfrac{2x}{3} = 0$

33. $\dfrac{4}{x} + 2 = 3$

34. $\dfrac{6}{x} - 3 = 7$

35. $\dfrac{9}{x} - 7 = 1$

36. $-2 = 1 + \dfrac{3}{x}$

37. $4 - \dfrac{4}{x} = 0$

38. $5 - \dfrac{6}{x} = -1$

39. $7 - \dfrac{3}{2x} = 1$

40. $4 + \dfrac{5}{3x} = -1$

41. $\dfrac{9}{2x} - 5 = 0$

42. $\dfrac{x-1}{5} - \dfrac{x-1}{3} = 0$

43. $\dfrac{x-1}{4} - \dfrac{2x-3}{5} = \dfrac{1}{20}$

44. $\dfrac{4}{1-x} = \dfrac{3}{1+x}$

45. $\dfrac{x+1}{4} - \dfrac{x}{3} = \dfrac{1}{12}$

46. $\dfrac{2x+1}{8} - \dfrac{x-1}{3} = \dfrac{5}{24}$

2.5 PROBLEMS SOLVED BY LINEAR EQUATIONS

(a) Let the unknown quantity be x (or any other letter) and state the units (where appropriate).

(b) Express the given statement in the form of an equation.

(c) Solve the equation for x and give the answer in *words*. (Do not finish by writing '$x = 3$'.)

(d) Check your solution using the problem (not your equation).

Example 1

The sum of three consecutive whole numbers is 78. Find the numbers.

(a) Let the smallest number be x; then the other numbers are $(x + 1)$ and $(x + 2)$.

(b) Form an equation:
$x + (x + 1) + (x + 2) = 78$

(c) Solve. $3x = 75$
$\qquad x = 25$
In words:
The three numbers are 25, 26 and 27.

(d) Check. $25 + 26 + 27 = 78$

Example 2

The length of a rectangle is three times the width. If the perimeter is 36 cm, find the width.

(a) Let the width of the rectangle be x cm.
Then the length of the rectangle is $3x$ cm.

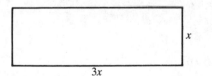

(b) Form an equation.
$x + 3x + x + 3x = 36$

(c) Solve $8x = 36$
$\qquad x = \dfrac{36}{8}$
$\qquad x = 4 \cdot 5$
In words:
The width of the rectangle is $4 \cdot 5$ cm.

(d) Check.   If width $= 4 \cdot 5$ cm
$\qquad\qquad$ length $= 13 \cdot 5$ cm
$\qquad\qquad$ perimeter $= 36$ cm

Exercise 15

Solve each problem by forming an equation. The first questions are easy but should still be solved using an equation, in order to practise the method.

1. The sum of three consecutive numbers is 276. Find the numbers.
2. The sum of four consecutive numbers is 90. Find the numbers.
3. The sum of three consecutive odd numbers is 177. Find the numbers.
4. Find three consecutive even numbers which add up to 1524.
5. When a number is doubled and then added to 13, the result is 38. Find the number.
6. When a number is doubled and then added to 24, the result is 49. Find the number.
7. When 7 is subtracted from three times a certain number, the result is 28. What is the number?
8. The sum of two numbers is 50. The second number is five times the first. Find the numbers.
9. Two numbers are in the ratio 1 : 11 and their sum is 15. Find the numbers.
10. The length of a rectangle is twice the width. If the perimeter is 20 cm, find the width.
11. The width of a rectangle is one third of the length. If the perimeter is 96 cm, find the width.
12. If AB is a straight line, find x.

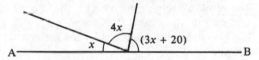

13. If the perimeter of the triangle is 22 cm, find the length of the shortest side.

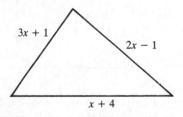

14. If the perimeter of the rectangle is 34 cm, find x.

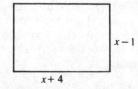

15. The difference between two numbers is 9. Find the numbers, if their sum is 46.

16. The three angles in a triangle are in the ratio 1:3:5. Find them.

17. The three angles in a triangle are in the ratio 3:4:5. Find them.

18. The product of two consecutive odd numbers is 10 more than the square of the smaller number. Find the smaller number.

19. The product of two consecutive even numbers is 12 more than the square of the smaller number. Find the numbers.

20. The sum of three numbers is 66. The second number is twice the first and six less than the third. Find the numbers.

21. The sum of three numbers is 28. The second number is three times the first and the third is 7 less than the second. What are the numbers?

22. David weighs 5 kg less than John, who in turn is 8 kg lighter than Paul. If their total weight is 197 kg, how heavy is each person?

23. Brian is 2 years older than Bob who is 7 years older than Mark. If their combined age is 61 years, find the age of each person.

24. Richard has four times as many marbles as John. If Richard gave 18 to John they would have the same number. How many marbles has each?

25. Stella has five times as many books as Tina. If Stella gave 16 books to Tina, they would each have the same number. How many books did each girl have?

26. The result of trebling a number is the same as adding 12 to it. What is the number?

27. Find the area of the rectangle if the perimeter is 52 cm.

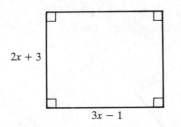

28. The result of trebling a number and subtracting 5 is the same as doubling the number and adding 9. What is the number?

29. Two girls have 76p between them. If the first gave the second 7p they would each have the same amount of money. How much did each girl have?

30. A tennis racket costs £12 more than a hockey stick. If the price of the two is £31, find the cost of the tennis racket.

Example 3

A man goes out at 16·42 h and arrives at a pillar box, 6 km away, at 17·30 h. He walked part of the way at 5 km/h and then, realising the time, he ran the rest of the way at 10 km/h. How far did he have to run?

(a) Let the distance he ran be x km. Then the distance he walked $= (6 - x)$ km.

(b) Time taken to walk $(6 - x)$ km at 5 km/h
$$= \frac{(6 - x)}{5} \text{ hours.}$$

Time taken to run x km at 10 km/h
$$= \frac{x}{10} \text{ hours.}$$

Total time taken = 48 minutes
$$= \frac{4}{5} \text{ hour}$$

$$\therefore \quad \frac{(6 - x)}{5} + \frac{x}{10} = \frac{4}{5}$$

(c) Multiply by 10:
$$2(6 - x) + x = 8$$
$$12 - 2x + x = 8$$
$$4 = x$$
He ran a distance of 4 km.

(d) Check:

Time to run 4 km $= \dfrac{4}{10} = \dfrac{2}{5}$ hour.

Time to walk 2 km $= \dfrac{2}{5}$ hour.

Total time taken $= \left(\dfrac{2}{5} + \dfrac{2}{5}\right) \text{ h} = \dfrac{4}{5} \text{ h}$

Exercise 16

1. Every year a man is paid £500 more than the previous year. If he receives £17 800 over four years, what was he paid in the first year?

2. A man buys x cans of beer at 30p each and $(x + 4)$ cans of lager at 35p each. The total cost was £3·35. Find x.

3. The sides of a rectangle measure 6 cm and x cm. If the diagonal is of length $(x + 2)$ cm, find x.

4. The length of a straight line ABC is 5 m. If AB : BC = 2 : 5, find the length of AB.

5. The opposite angles of a cyclic quadrilateral are $(3x + 10)°$ and $(2x + 20)°$. Find the angles.

6. The interior angles of a hexagon are in the ratio 1:2:3:4:5:9. Find the angles. This is an example of a concave hexagon. Try to sketch the hexagon.

7. A man is 32 years older than his son. Ten years ago he was three times as old as his son was then. Find the present age of each.

8. A man runs to a telephone and back in 15 minutes. His speed on the way to the telephone is 5 m/s and his speed on the way back is 4 m/s. Find the distance to the telephone.

9. A car completes a journey in 10 minutes. For the first half of the distance the speed was 60 km/h and for the second half the speed was 40 km/h. How far is the journey?

10. A lemming runs from a point A to a cliff at 4 m/s, jumps over the edge at B and falls to C at an average speed of 25 m/s. If the total distance from A to C is 500 m and the time taken for the journey is 41 seconds, find the height BC of the cliff.

11. A bus is travelling with 48 passengers. When it arrives at a stop, x passengers get off and 3 get on. At the next stop half the passengers get off and 7 get on. There are now 22 passengers. Find x.

12. A bus is travelling with 52 passengers. When it arrives at a stop, y passengers get off and 4 get on. At the next stop one third of the passengers get off and 3 get on. There are now 25 passengers. Find y.

13. Mr Lee left his fortune to his 3 sons, 4 daughters and his wife. Each son received twice as much as each daughter and his wife received £6000, which was a quarter of the money. How much did each son receive?

14. In a regular polygon with n sides each interior angle is $180 - \dfrac{360}{n}$ degrees. How many sides does a polygon have if each angle is 156°?

15. A sparrow flies to see a friend at a speed of 4 km/h. His friend is out, so the sparrow immediately returns home at a speed of 5 km/h. The complete journey took 54 minutes. How far away does his friend live?

16. Consider the equation $an^2 = 182$ where a is any number between 2 and 5 and n is a positive integer. What are the possible values of n?

17. Consider the equation $\dfrac{k}{x} = 12$ where k is any number between 20 and 65 and x is a positive integer. What are the possible values of x?

2.6 SIMULTANEOUS EQUATIONS

To find the value of two unknowns in a problem, *two* different equations must be given that relate the unknowns to each other. These two equations are called *simultaneous* equations.

(a) Substitution method

This method is used when one equation contains a unit quantity of one of the unknowns, as in equation [2] of Example 1 below.

Example 1

$$3x - 2y = 0 \qquad \ldots[1]$$
$$2x + y = 7 \qquad \ldots[2]$$

(a) Label the equations so that the working is made clear.

(b) In *this* case, write y in terms of x from equation [2].

(c) Substitute this expression for y in equation [1] and solve to find x.

(d) Find y from equation [2] using this value of x.

$$2x + y = 7 \qquad \ldots[2]$$
$$y = 7 - 2x$$

Substituting in [1]
$$3x - 2(7 - 2x) = 0$$
$$3x - 14 + 4x = 0$$
$$7x = 14$$
$$x = 2$$

Substituting in [2]
$$2 \times 2 + y = 7$$
$$y = 3$$

The solutions are $x = 2$, $y = 3$.

These values of x and y are the only pair which simultaneously satisfy *both* equations.

Exercise 17

Use the substitution method to solve the following:

1. $2x + y = 5$
   $x + 3y = 5$

2. $x + 2y = 8$
   $2x + 3y = 14$

3. $3x + y = 10$
   $x - y = 2$

4. $2x + y = -3$
   $x - y = -3$

5. $4x + y = 14$
   $x + 5y = 13$

6. $x + 2y = 1$
   $2x + 3y = 4$

7. $2x + y = 5$
   $3x - 2y = 4$

8. $2x + y = 13$
   $5x - 4y = 13$

9. $7x + 2y = 19$
   $x - y = 4$

10. $b - a = -5$
    $a + b = -1$

11. $a + 4b = 6$
    $8b - a = -3$

12. $a + b = 4$
    $2a + b = 5$

13. $3m = 2n - 6\frac{1}{2}$
    $4m + n = 6$

14. $2w + 3x - 13 = 0$
    $x + 5w - 13 = 0$

15. $x + 2(y - 6) = 0$
    $3x + 4y = 30$

16. $2x = 4 + z$
    $6x - 5z = 18$

17. $3m - n = 5$
    $2m + 5n = 7$

18. $5c - d - 11 = 0$
    $4d + 3c = -5$

It is useful, at this point to revise the operations of addition and subtraction with negative numbers.

Example 2

Simplify:
(a) $-7 + -4 = -7 - 4 = -11$
(b) $-3x + (-4x) = -3x - 4x = -7x$
(c) $4y - (-3y) = 4y + 3y = 7y$
(d) $3a + (-3a) = 3a - 3a = 0$

Exercise 18

Evaluate

1. $7 + (-6)$
2. $8 + (-11)$
3. $5 - (+7)$
4. $6 - (-9)$
5. $-8 + (-4)$
6. $-7 - (-4)$
7. $10 + (-12)$
8. $-7 - (+4)$
9. $-10 - (+11)$
10. $-3 - (-4)$
11. $4 - (+4)$
12. $8 - (-7)$
13. $-5 - (+5)$
14. $-7 - (-10)$
15. $16 - (+10)$
16. $-7 - (+4)$
17. $-6 - (-8)$
18. $10 - (+5)$
19. $-12 + (-7)$
20. $7 + (-11)$

Simplify

21. $3x + (-2x)$
22. $4x + (-7x)$
23. $6x - (+2x)$
24. $10y - (+6y)$
25. $6y - (-3y)$
26. $7x + (-4x)$
27. $-5x + (-3x)$
28. $-3x - (-7x)$
29. $5x - (+3x)$
30. $-7y - (-10y)$

(b) **Elimination method**

Use this method when the first method is unsuitable (some prefer to use it for every question).

Example 3

$$x + 2y = 8 \qquad \ldots [1]$$
$$2x + 3y = 14 \qquad \ldots [2]$$

(a) Label the equations so that the working is made clear.
(b) Choose an unknown in one of the equations and multiply the equations by a factor or factors so that this unknown has the same coefficient in both equations.
(c) Eliminate this unknown from the two equations by subtracting them, then solve for the remaining unknown.
(d) Substitute in the first equation and solve for the eliminated unknown.

$$x + 2y = 8 \qquad \ldots [1]$$
$$[1] \times 2 \quad 2x + 4y = 16 \qquad \ldots [3]$$
$$2x + 3y = 14 \qquad \ldots [2]$$

Subtract [2] from [3]
$$y = 2$$

Substituting in [1]
$$x + 2 \times 2 = 8$$
$$x = 8 - 4$$
$$x = 4$$

The solutions are $x = 4$, $y = 2$.

Example 4

$$2x + 3y = 5 \qquad \ldots [1]$$
$$5x - 2y = -16 \qquad \ldots [2]$$

$$[1] \times 5 \quad 10x + 15y = 25 \qquad \ldots [3]$$
$$[2] \times 2 \quad 10x - 4y = -32 \qquad \ldots [4]$$

$$[3] - [4] \quad 15y - (-4y) = 25 - (-32)$$
$$19y = 57$$
$$y = 3$$

Substitute in [1]
$$2x + 3 \times 3 = 5$$
$$2x = 5 - 9 = -4$$
$$x = -2$$

The solutions are $x = -2$, $y = 3$.

Exercise 19

Use the elimination method to solve the following:

1. $2x + 5y = 24$
$4x + 3y = 20$

2. $5x + 2y = 13$
$2x + 6y = 26$

3. $3x + y = 11$
$9x + 2y = 28$

4. $x + 2y = 17$
$8x + 3y = 45$

5. $3x + 2y = 19$
$x + 8y = 21$

6. $2a + 3b = 9$
$4a + b = 13$

7. $2x + 3y = 11$
$3x + 4y = 15$

8. $3x + 8y = 27$
$4x + 3y = 13$

9. $2x + 7y = 17$
$5x + 3y = -1$

10. $5x + 3y = 23$
$2x + 4y = 12$

11. $7x + 5y = 32$
$3x + 4y = 23$

12. $3x + 2y = 4$
$4x + 5y = 10$

13. $3x + 2y = 11$
$2x - y = -3$

14. $3x + 2y = 7$
$2x - 3y = -4$

15. $x - 2y = -4$
$3x + y = 9$

16. $5x - 7y = 27$
$3x - 4y = 16$

17. $3x - 2y = 7$
$4x + y = 13$

18. $x - y = -1$
$2x - y = 0$

19. $y - x = -1$
$3x - y = 5$

20. $x - 3y = -5$
$2y + 3x + 4 = 0$

21. $x + 3y - 7 = 0$
$2y - x - 3 = 0$

22. $3a - b = 9$
$2a + 2b = 14$

23. $3x - y = 9$
$4x - y = -14$

24. $x + 2y = 4$
$3x + y = 9\frac{1}{2}$

25. $2x - y = 5$
$\dfrac{x}{4} + \dfrac{y}{3} = 2$

26. $3x - y = 17$
$\dfrac{x}{5} + \dfrac{y}{2} = 0$

27. $3x - 2y = 5$
$\dfrac{2x}{3} + \dfrac{y}{2} = -\dfrac{7}{9}$

28. $2x = 11 - y$
$\dfrac{x}{5} - \dfrac{y}{4} = 1$

29. $4x - 0{\cdot}5y = 12{\cdot}5$
$3x + 0{\cdot}8y = 8{\cdot}2$

30. $0{\cdot}4x + 3y = 2{\cdot}6$
$x - 2y = 4{\cdot}6$

2.7   PROBLEMS SOLVED BY SIMULTANEOUS EQUATIONS

Example 1

A motorist buys 24 litres of petrol and 5 litres of oil for £10·70, while another motorist buys 18 litres of petrol and 10 litres of oil for £12·40. Find the cost of 1 litre of petrol and 1 litre of oil at this garage.

Let cost of 1 litre of petrol be x pence.
Let cost of 1 litre of oil be y pence.

We have, $24x + 5y = 1070$ ...[1]
              $18x + 10y = 1240$ ...[2]

(a) Multiply [1] by 2,
        $48x + 10y = 2140$ ...[3]

(b) Subtract [2] from [3],
              $30x = 900$
                $x = 30$

(c) Substitute $x = 30$ into equation [2]
        $18(30) + 10y = 1240$
              $10y = 1240 - 540$
              $10y = 700$
                $y = 70$

1 litre of petrol costs 30 pence.
1 litre of oil costs 70 pence.

Exercise 20

Solve each problem by forming a pair of simultaneous equations.

1. Find two numbers with a sum of 15 and a difference of 4.
2. Twice one number added to three times another gives 21. Find the numbers, if the difference between them is 3.
3. The average of two numbers is 7, and three times the difference between them is 18. Find the numbers.
4. The line, with equation $y + ax = c$, passes through the points (1, 5) and (3, 1). Find a and c.
   Hint: For the point (1, 5) put $x = 1$ and $y = 5$ into $y + ax = c$, etc.
5. The line $y = mx + c$ passes through (2, 5) and (4, 13). Find m and c.
6. The curve $y = ax^2 + bx$ passes through (2, 0) and (4, 8). Find a and b.
7. A fishing enthusiast buys fifty maggots and twenty worms for £1·10 and her mother buys thirty maggots and forty worms for £1·50. Find the cost of one maggot and one worm.

8. A television addict can buy either two televisions and three video-recorders for £1750 or four televisions and one video-recorder for £1250. Find the cost of one of each.

9. Half the difference between two numbers is 2. The sum of the greater number and twice the smaller number is 13. Find the numbers.

10. A pigeon can lay either white or brown eggs. Three white eggs and two brown eggs weigh 13 ounces, while five white eggs and four brown eggs weigh 24 ounces. Find the weight of a brown egg and of a white egg.

11. A tortoise makes a journey in two parts; it can either walk at 4 m/s or crawl at 3 m/s. If the tortoise walks the first part and crawls the second, it takes 110 seconds. If it crawls the first part and walks the second, it takes 100 seconds. Find the lengths of the two parts of the journey.

12. A cyclist completes a journey of 500 m in 22 seconds, part of the way at 10 m/s and the remainder at 50 m/s. How far does she travel at each speed?

13. A bag contains forty coins, all of them either 2p or 5p coins. If the value of the money in the bag is £1·55, find the number of each kind.

14. A slot machine takes only 10p and 50p coins and contains a total of twenty-one coins altogether. If the value of the coins is £4·90, find the number of coins of each value.

15. Thirty tickets were sold for a concert, some at 60p and the rest at £1. If the total raised was £22, how many had the cheaper tickets?

16. The wage bill for five men and six women workers is £670, while the bill for eight men and three women is £610. Find the wage for a man and for a woman.

17. A kipper can swim at 14 m/s with the current and at 6 m/s against it. Find the speed of the current and the speed of the kipper in still water.

18. If the numerator and denominator of a fraction are both decreased by one the fraction becomes $\frac{2}{3}$. If the numerator and denominator are both increased by one the fraction becomes $\frac{3}{4}$. Find the original fraction.

19. The denominator of a fraction is 2 more than the numerator. If both denominator and numerator are increased by 1 the fraction becomes $\frac{2}{3}$. Find the original fraction.

20. In three years time a pet mouse will be as old as his owner was four years ago. Their present ages total 13 years. Find the age of each now.

21. Find two numbers where three times the smaller number exceeds the larger by 5 and the sum of the numbers is 11.

22. A straight line passes through the points (2, 4) and (−1, −5). Find its equation.

23. A spider can walk at a certain speed and run at another speed. If she walks for 10 seconds and runs for 9 seconds she travels 85 m. If she walks for 30 seconds and runs for 2 seconds she travels 130 m. Find her speeds of walking and running.

24. A wallet containing £40 has three times as many £1 coins as £5 notes. Find the number of each kind.

25. At the present time a man is four times as old as his son. Six years ago he was 10 times as old. Find their present ages.

26. A submarine can travel at 25 knots with the wind and at 16 knots against it. Find the speed of the wind and the speed of the submarine in still air.

27. The curve $y = ax^2 + bx + c$ passes through the points (1, 8), (0, 5) and (3, 20). Find the values of a, b and c and hence the equation of the curve.

28. The curve $y = ax^2 + bx + c$ passes through the points (1, 4), (−2, 19) and (0, 5). Find the equation of the curve.

29. The curve $y = ax^2 + bx + c$ passes through (1, 8), (−1, 2) and (2, 14). Find the equation of the curve.

30. The curve $y = ax^2 + bx + c$ passes through (2, 5), (3, 12) and (−1, −4). Find the equation of the curve.

2.8 FACTORISING

Earlier in this section we expanded expressions such as $x(3x-1)$ to give $3x^2 - x$.
The reverse of this process is called *factorising*.

Example 1

Factorise: (a) $x^2 + 7x$ (b) $3y^2 - 12y$
  (c) $6a^2b - 10ab^2$

(a) x is common to x^2 and $7x$.
  $\therefore$ $x^2 + 7x = x(x+7)$.

  The factors are x and $(x+7)$.

(b) $3y$ is common
  $\therefore$ $3y^2 - 12y = 3y(y-4)$

(c) $2ab$ is common
  $\therefore$ $6a^2b - 10ab^2 = 2ab(3a - 5b)$

Exercise 21

Factorise the following expressions completely.

1. $x^2 + 5x$
2. $x^2 - 6x$
3. $7x - x^2$
4. $y^2 + 8y$
5. $2y^2 + 3y$
6. $6y^2 - 4y$
7. $3x^2 - 21x$
8. $16a - 2a^2$
9. $6c^2 - 21c$
10. $15x - 9x^2$

11. $56y - 21y^2$
12. $ax + bx + 2cx$
13. $x^2 + xy + 3xz$
14. $x^2y + y^3 + z^2y$
15. $3a^2b + 2ab^2$
16. $x^2y + xy^2$
17. $6a^2 + 4ab + 2ac$
18. $ma + 2bm + m^2$
19. $2kx + 6ky + 4kz$
20. $ax^2 + ay + 2ab$

21. $x^2k + xk^2$
22. $a^3b + 2ab^2$
23. $abc - 3b^2c$
24. $2a^2e - 5ae^2$
25. $a^3b + ab^3$
26. $x^3y + x^2y^2$
27. $6xy^2 - 4x^2y$
28. $3ab^3 - 3a^3b$
29. $2a^3b + 5a^2b^2$
30. $ax^2y - 2ax^2z$
31. $2abx + 2ab^2 + 2a^2b$
32. $ayx + yx^3 - 2y^2x^2$

Example 2

Factorise $ah + ak + bh + bk$.

(a) Divide into pairs, $ah + ak\,|\,+ bh + bk$.

(b) a is common to the first pair
  b is common to the second pair
  $$a(h+k) + b(h+k)$$

(c) $(h+k)$ is common to both terms.
  Thus we have $(h+k)(a+b)$

Example 3

Factorise $6mx - 3nx + 2my - ny$.

(a) $6mx - 3nx\,|\,+ 2my - ny$

(b) $= 3x(2m - n) + y(2m - n)$

(c) $= (2m - n)(3x + y)$

Exercise 22

Factorise the following expressions:

1. $ax + ay + bx + by$
2. $ay + az + by + bz$
3. $xb + xc + yb + yc$
4. $xh + xk + yh + yk$
5. $xm + xn + my + ny$
6. $ah - ak + bh - bk$
7. $ax - ay + bx - by$
8. $am - bm + an - bn$
9. $hs + ht + ks + kt$
10. $xs - xt + ys - yt$
11. $ax - ay - bx + by$
12. $xs - xt - ys + yt$
13. $as - ay - xs + xy$
14. $hx - hy - bx + by$
15. $am - bm - an + bn$
16. $xk - xm - kz + mz$
17. $2ax + 6ay + bx + 3by$
18. $2ax + 2ay + bx + by$
19. $2mh - 2mk + nh - nk$
20. $2mh + 3mk - 2nh - 3nk$
21. $6ax + 2bx + 3ay + by$
22. $2ax - 2ay - bx + by$
23. $x^2a + x^2b + ya + yb$
24. $ms + 2mt^2 - ns - 2nt^2$

Example 4

Factorise $x^2 + 6x + 8$

(a) Find two numbers which multiply to give 8 and add up to 6.
  In this case the numbers are 4 and 2.

(b) Put these numbers into brackets.
  So $x^2 + 6x + 8 = (x+4)(x+2)$.

Example 5

Factorise (a) $x^2 + 2x - 15$
  (b) $x^2 - 6x + 8$

(a) Two numbers which multiply to give -15 and add up to $+2$ are -3 and 5.
  $\therefore$ $x^2 + 2x - 15 = (x-3)(x+5)$.

(b) Two numbers which multiply to give $+8$ and add up to -6 are -2 and -4.
  $\therefore$ $x^2 - 6x + 8 = (x-2)(x-4)$.

Exercise 23

Factorise the following:

1. $x^2 + 7x + 10$ 2. $x^2 + 7x + 12$
3. $x^2 + 8x + 15$ 4. $x^2 + 10x + 21$
5. $x^2 + 8x + 12$ 6. $y^2 + 12y + 35$
7. $y^2 + 11y + 24$ 8. $y^2 + 10y + 25$
9. $y^2 + 15y + 36$ 10. $a^2 - 3a - 10$
11. $a^2 - a - 12$ 12. $z^2 + z - 6$
13. $x^2 - 2x - 35$ 14. $x^2 - 5x - 24$
15. $x^2 - 6x + 8$ 16. $y^2 - 5y + 6$
17. $x^2 - 8x + 15$ 18. $a^2 - a - 6$
19. $a^2 + 14a + 45$ 20. $b^2 - 4b - 21$
21. $x^2 - 8x + 16$ 22. $y^2 + 2y + 1$
23. $y^2 - 3y - 28$ 24. $x^2 - x - 20$
25. $x^2 - 8x - 240$ 26. $x^2 - 26x + 165$
27. $y^2 + 3y - 108$ 28. $x^2 - 49$
29. $x^2 - 9$ 30. $x^2 - 16$

Example 6

Factorise $3x^2 + 13x + 4$

(a) Find two numbers which multiply to give 12
    and add up to 13.
    In this case the numbers are 1 and 12.

(b) Split the '$13x$' term,
    $3x^2 + x + 12x + 4$

(c) Factorise in pairs,
    $x(3x + 1) + 4(3x + 1)$

(d) $(3x + 1)$ is common,
    $(3x + 1)(x + 4)$

Exercise 24

Factorise the following:

1. $2x^2 + 5x + 3$ 2. $2x^2 + 7x + 3$
3. $3x^2 + 7x + 2$ 4. $2x^2 + 11x + 12$
5. $3x^2 + 8x + 4$ 6. $2x^2 + 7x + 5$
7. $3x^2 - 5x - 2$ 8. $2x^2 - x - 15$
9. $2x^2 + x - 21$ 10. $3x^2 - 17x - 28$
11. $6x^2 + 7x + 2$ 12. $12x^2 + 23x + 10$
13. $3x^2 - 11x + 6$ 14. $3y^2 - 11y + 10$
15. $4y^2 - 23y + 15$ 16. $6y^2 + 7y - 3$
17. $6x^2 - 27x + 30$ 18. $10x^2 + 9x + 2$
19. $6x^2 - 19x + 3$ 20. $8x^2 - 10x - 3$
21. $12x^2 + 4x - 5$ 22. $16x^2 + 19x + 3$
23. $4a^2 - 4a + 1$ 24. $12x^2 + 17x - 14$
25. $15x^2 + 44x - 3$ 26. $48x^2 + 46x + 5$
27. $64y^2 + 4y - 3$ 28. $120x^2 + 67x - 5$
29. $9x^2 - 1$ 30. $4a^2 - 9$

The difference of two squares

$$x^2 - y^2 = (x - y)(x + y)$$

Remember this result.

Example 7

Factorise (a) $4a^2 - b^2$
           (b) $3x^2 - 27y^2$

(a) $4a^2 - b^2 = (2a)^2 - b^2$
    $\qquad\quad = (2a - b)(2a + b)$

(b) $3x^2 - 27y^2 = 3(x^2 - 9y^2)$
    $\qquad\qquad\quad = 3[x^2 - (3y)^2]$
    $\qquad\qquad\quad = 3(x - 3y)(x + 3y).$

Exercise 25

Factorise the following:

1. $y^2 - a^2$ 2. $m^2 - n^2$
3. $x^2 - t^2$ 4. $y^2 - 1$
5. $x^2 - 9$ 6. $a^2 - 25$
7. $x^2 - \dfrac{1}{4}$ 8. $x^2 - \dfrac{1}{9}$
9. $4x^2 - y^2$ 10. $a^2 - 4b^2$
11. $25x^2 - 4y^2$ 12. $9x^2 - 16y^2$
13. $x^2 - \dfrac{y^2}{4}$ 14. $9m^2 - \dfrac{4}{9}n^2$
15. $16t^2 - \dfrac{4}{25}s^2$ 16. $4x^2 - \dfrac{z^2}{100}$
17. $x^3 - x$ 18. $a^3 - ab^2$
19. $4x^3 - x$ 20. $8x^3 - 2xy^2$
21. $12x^3 - 3xy^2$ 22. $18m^3 - 8mn^2$
23. $5x^2 - 1\frac{1}{4}$ 24. $50a^3 - 18ab^2$
25. $12x^2y - 3yz^2$ 26. $36a^3b - 4ab^3$
27. $50a^5 - 8a^3b^2$ 28. $36x^3y - 225xy^3$

Evaluate the following

29. $81^2 - 80^2$ 30. $102^2 - 100^2$
31. $225^2 - 215^2$ 32. $1211^2 - 1210^2$
33. $723^2 - 720^2$ 34. $3 \cdot 8^2 - 3 \cdot 7^2$
35. $5 \cdot 24^2 - 4 \cdot 76^2$ 36. $1234^2 - 1235^2$
37. $3 \cdot 81^2 - 3 \cdot 8^2$ 38. $540^2 - 550^2$
39. $7 \cdot 68^2 - 2 \cdot 32^2$ 40. $0 \cdot 003^2 - 0 \cdot 002^2$

2.9 QUADRATIC EQUATIONS

So far, we have met linear equations which have one solution only.

Quadratic equations always have an x^2 term, and often an x term and a number term, and generally have two different solutions.

(a) Solution by factors

Consider the equation $a \times b = 0$, where a and b are numbers. The product $a \times b$ can only be zero if either a or b (or both) is equal to zero.

Can you think of other possible pairs of numbers which multiply together to give zero?

Example 1

Solve the equation $x^2 + x - 12 = 0$

Factorising, $(x - 3)(x + 4) = 0$

either $\quad x - 3 = 0 \quad$ or $\quad x + 4 = 0$

$\qquad\qquad x = 3 \qquad\qquad\quad x = -4$

Example 2

Solve the equation $6x^2 + x - 2 = 0$

Factorising, $(2x - 1)(3x + 2) = 0$

either $\quad 2x - 1 = 0 \quad$ or $\quad 3x + 2 = 0$

$\qquad\qquad 2x - 1 \qquad\qquad\quad 3x = -2$

$\qquad\qquad x = \frac{1}{2} \qquad\qquad\quad x = -\frac{2}{3}$

Exercise 26

Solve the following equations

1. $x^2 + 7x + 12 = 0$
2. $x^2 + 7x + 10 = 0$
3. $x^2 + 2x - 15 = 0$
4. $x^2 + x - 6 = 0$
5. $x^2 - 8x + 12 = 0$
6. $x^2 + 10x + 21 = 0$
7. $x^2 - 5x + 6 = 0$
8. $x^2 - 4x - 5 = 0$
9. $x^2 + 5x - 14 = 0$
10. $2x^2 - 3x - 2 = 0$

11. $3x^2 + 10x - 8 = 0$
12. $2x^2 + 7x - 15 = 0$
13. $6x^2 - 13x + 6 = 0$
14. $4x^2 - 29x + 7 = 0$
15. $10x^2 - x - 3 = 0$
16. $y^2 - 15y + 56 = 0$
17. $12y^2 - 16y + 5 = 0$
18. $y^2 + 2y - 63 = 0$
19. $x^2 + 2x + 1 = 0$
20. $x^2 - 6x + 9 = 0$

21. $x^2 + 10x + 25 = 0$
22. $x^2 - 14x + 49 = 0$
23. $6a^2 - a - 1 = 0$
24. $4a^2 - 3a - 10 = 0$
25. $z^2 - 8z - 65 = 0$
26. $6x^2 + 17x - 3 = 0$
27. $10k^2 + 19k - 2 = 0$
28. $y^2 - 2y + 1 = 0$
29. $36x^2 + x - 2 = 0$
30. $20x^2 - 7x - 3 = 0$

Example 3

Solve the equation $x^2 - 7x = 0$

Factorising, $x(x - 7) = 0$

either $\quad x = 0 \quad$ or $\quad x - 7 = 0$

$\qquad\qquad\qquad\qquad\qquad x = 7$

The solutions are $x = 0$ and $x = 7$.

Example 4

Solve the equation $4x^2 - 9 = 0$

(a) Factorising, $(2x - 3)(2x + 3) = 0$

either $\quad 2x - 3 = 0 \quad$ or $\quad 2x + 3 = 0$

$\qquad\qquad 2x = 3 \qquad\qquad\qquad 2x = -3$

$\qquad\qquad x = \frac{3}{2} \qquad\qquad\qquad x = -\frac{3}{2}$

(b) Alternative method

$\qquad 4x^2 - 9 = 0$

$\qquad\quad 4x^2 = 9$

$\qquad\quad\ x^2 = \frac{9}{4}$

$\qquad x = +\frac{3}{2} \quad$ or $\quad -\frac{3}{2}$.

Exercise 27

1. $x^2 - 3x = 0$
2. $x^2 + 7x = 0$
3. $2x^2 - 2x = 0$
4. $3x^2 - x = 0$
5. $x^2 - 16 = 0$
6. $x^2 - 49 = 0$
7. $4x^2 - 1 = 0$
8. $9x^2 - 4 = 0$
9. $6y^2 + 9y = 0$
10. $6a^2 - 9a = 0$

11. $10x^2 - 55x = 0$
12. $16x^2 - 1 = 0$
13. $y^2 - \frac{1}{4} = 0$
14. $56x^2 - 35x = 0$
15. $36x^2 - 3x = 0$
16. $x^2 = 6x$
17. $x^2 = 11x$
18. $2x^2 = 3x$
19. $x^2 = x$
20. $4x = x^2$

21. $3x - x^2 = 0$
22. $4x^2 = 1$
23. $9x^2 = 16$
24. $x^2 = 9$
25. $12x = 5x^2$
26. $1 - 9x^2 = 0$

27. $x^2 = \dfrac{x}{4}$
28. $2x^2 = \dfrac{x}{3}$

29. $4x^2 = \dfrac{1}{4}$
30. $\dfrac{x}{5} - x^2 = 0$

(b) Solution by formula

The solutions of the quadratic equation
$$ax^2 + bx + c = 0$$
are given by the formula
$$x = \frac{-b \pm \sqrt{(b^2 - 4ac)}}{2a}$$

Use this formula only after trying (and failing) to factorise.

Example 5

Solve the equation $2x^2 - 3x - 4 = 0$.
In this case $a = 2$, $b = -3$, $c = -4$.

$$x = \frac{-(-3) \pm \sqrt{[(-3)^2 - (4 \times 2 \times -4)]}}{2 \times 2}$$

$$x = \frac{3 \pm \sqrt{[9 + 32]}}{4} = \frac{3 \pm \sqrt{41}}{4}$$

$$x = \frac{3 \pm 6 \cdot 403}{4}$$

either $x = \dfrac{3 + 6 \cdot 403}{4} = 2 \cdot 35$ (2 decimal places)

or $x = \dfrac{3 - 6 \cdot 403}{4} = \dfrac{-3 \cdot 403}{4}$

$$= -0 \cdot 85 \text{ (2 decimal places)}.$$

Exercise 28

Solve the following, giving answers to two decimal places where necessary.

1. $2x^2 + 11x + 5 = 0$
2. $3x^2 + 11x + 6 = 0$
3. $6x^2 + 7x + 2 = 0$
4. $3x^2 - 10x + 3 = 0$
5. $5x^2 - 7x + 2 = 0$
6. $6x^2 - 11x + 3 = 0$
7. $2x^2 + 6x + 3 = 0$
8. $x^2 + 4x + 1 = 0$
9. $5x^2 - 5x + 1 = 0$
10. $x^2 - 7x + 2 = 0$
11. $2x^2 + 5x - 1 = 0$
12. $3x^2 + x - 3 = 0$

13. $3x^2 + 8x - 6 = 0$
14. $3x^2 - 7x - 20 = 0$
15. $2x^2 - 7x - 15 = 0$
16. $x^2 - 3x - 2 = 0$
17. $2x^2 + 6x - 1 = 0$
18. $6x^2 - 11x - 7 = 0$
19. $3x^2 + 25x + 8 = 0$
20. $3y^2 - 2y - 5 = 0$
21. $2y^2 - 5y + 1 = 0$
22. $\frac{1}{2}y^2 + 3y + 1 = 0$
23. $2 - x - 6x^2 = 0$
24. $3 + 4x - 2x^2 = 0$

25. $1 - 5x - 2x^2 = 0$
26. $3x^2 - 1 + 4x = 0$
27. $5x - x^2 + 2 = 0$
28. $24x^2 - 22x - 35 = 0$
29. $36x^2 - 17x - 35 = 0$
30. $20x^2 + 17x - 63 = 0$
31. $x^2 + 2 \cdot 5x - 6 = 0$
32. $0 \cdot 3y^2 + 0 \cdot 4y - 1 \cdot 5 = 0$
33. $10 - x - 3x^2 = 0$
34. $x^2 + 3 \cdot 3x - 0 \cdot 7 = 0$
35. $12 - 5x^2 - 11x = 0$
36. $5x - 2x^2 + 187 = 0$

The solution to a problem can involve an equation which does not at first appear to be quadratic. The terms in the equation may need to be rearranged as shown below.

Example 6

Solve:
$$2x(x - 1) = (x + 1)^2 - 5$$
$$2x^2 - 2x = x^2 + 2x + 1 - 5$$
$$2x^2 - 2x - x^2 - 2x - 1 + 5 = 0$$
$$x^2 - 4x + 4 = 0$$
$$(x - 2)(x - 2) = 0$$
$$x = 2$$

In this example the quadratic has a repeated root of $x = 2$.

Exercise 29

Solve the following equations, giving answers to two decimal places where necessary.

1. $x^2 = 6 - x$
2. $x(x + 10) = -21$
3. $3x + 2 = 2x^2$
4. $x^2 + 4 = 5x$
5. $6x(x + 1) = 5 - x$
6. $(2x)^2 = x(x - 14) - 5$
7. $(x - 3)^2 = 10$
8. $(x + 1)^2 - 10 = 2x(x - 2)$
9. $(2x - 1)^2 = (x - 1)^2 + 8$
10. $3x(x + 2) - x(x - 2) + 6 = 0$

11. $x = \dfrac{15}{x} - 22$ 12. $x + 5 = \dfrac{14}{x}$

13. $4x + \dfrac{7}{x} = 29$ 14. $10x = 1 + \dfrac{3}{x}$

15. $2x^2 = 7x$ 16. $16 = \dfrac{1}{x^2}$

17. $2x + 2 = \dfrac{7}{x} - 1$ 18. $\dfrac{2}{x} + \dfrac{2}{x + 1} = 3$

19. $\dfrac{3}{x - 1} + \dfrac{3}{x + 1} = 4$ 20. $\dfrac{2}{x - 2} + \dfrac{4}{x + 1} = 3$

21. One of the solutions published by Cardan in 1545 for the solution of cubic equations is given below. For an equation in the form $x^3 + px = q$

$$x = \sqrt[3]{\left[\sqrt{\left(\frac{p}{3}\right)^3 + \left(\frac{q}{2}\right)^2} + \frac{q}{2} \right]} - \sqrt[3]{\left[\sqrt{\left(\frac{p}{3}\right)^3 + \left(\frac{q}{2}\right)^2} - \frac{q}{2} \right]}$$

Use the formula to solve the following equations, giving answers to 4 sig. fig. where necessary.
(a) $x^3 + 7x = -8$
(b) $x^3 + 6x = 4$
(c) $x^3 + 3x = 2$
(d) $x^3 + 9x - 2 = 0$

2.10 PROBLEMS SOLVED BY QUADRATIC EQUATIONS

Example 1

The perimeter of a rectangle is 42 cm. If the diagonal is 15 cm, find the width of the rectangle.

Let the width of the rectangle be x cm.

Since the perimeter is 42 cm, the sum of the length and the width is 21 cm.

$\therefore$ length of rectangle $= (21 - x)$ cm

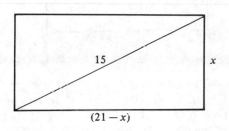

By Pythagoras' theorem

$$x^2 + (21 - x)^2 = 15^2$$
$$x^2 + (21 - x)(21 - x) = 15^2$$
$$x^2 + 441 - 42x + x^2 = 225$$
$$2x^2 - 42x + 216 = 0$$
$$x^2 - 21x + 108 = 0$$
$$(x - 12)(x - 9) = 0$$
$$x = 12$$
$$\text{or} \quad x = 9$$

Note that the dimensions of the rectangle are 9 cm and 12 cm, whichever value of x is taken.

$\therefore$ The width of the rectangle is 9 cm.

Example 2

A man bought a certain number of golf balls for £20. If each ball had cost 20p less, he could have bought five more for the same money. How many golf balls did he buy?

Let the number of balls bought be x.

Cost of each ball $= \dfrac{2000}{x}$ pence

If five more balls had been bought

Cost of each ball now $= \dfrac{2000}{(x + 5)}$ pence

The new price is 20p less than the original price.

$\therefore \qquad \dfrac{2000}{x} - \dfrac{2000}{(x + 5)} = 20$

(multiply by x)

$$x \cdot \dfrac{2000}{x} - x \cdot \dfrac{2000}{(x + 5)} = 20x$$

(multiply by $(x + 5)$)

$$2000(x + 5) - x\dfrac{2000}{(x + 5)}(x + 5) = 20x(x + 5)$$
$$2000x + 10\,000 - 2000x = 20x^2 + 100x$$
$$20x^2 + 100x - 10\,000 = 0$$
$$x^2 + 5x - 500 = 0$$
$$(x - 20)(x + 25) = 0$$
$$\therefore \qquad x = 20$$
$$\text{or} \quad x = -25$$

We discard $x = -25$ as meaningless.
The number of balls bought $= 20$.

Exercise 30

Solve by forming a quadratic equation.
1. Two numbers, which differ by 3, have a product of 88. Find them.
2. The product of two consecutive odd numbers is 143. Find the numbers. (Hint: If the first odd number is x, what is the next odd number?)
3. The length of a rectangle exceeds the width by 7 cm. If the area is 60 cm^2, find the length of the rectangle.
4. The length of a rectangle exceeds the width by 2 cm. If the diagonal is 10 cm long, find the width of the rectangle.
5. The area of the rectangle exceeds the area of the square by 24 m^2. Find x.

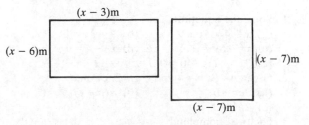

6. The perimeter of a rectangle is 68 cm. If the diagonal is 26 cm, find the dimensions of the rectangle.

7. A man walks a certain distance due North and then the same distance plus a further 7 km due East. If the final distance from the starting point is 17 km, find the distances he walks North and East.

8. A farmer makes a profit of x pence on each of the $(x + 5)$ eggs her hen lays. If her total profit was 84 pence, find the number of eggs the hen lays.

9. A boy buys x eggs at $(x - 8)$ pence each and $(x - 2)$ rashers of bacon at $(x - 3)$ pence each. If the total bill is £1·75, how many eggs does he buy?

10. A number exceeds four times its reciprocal by 3. Find the number.

11. Two numbers differ by 3. The sum of their reciprocals is $\frac{7}{10}$; find the numbers.

12. A cyclist travels 40 km at a speed x km/h. Find the time taken in terms of x. Find the time taken when his speed is reduced by 2 km/h. If the difference between the times is 1 hour, find the original speed x.

13. An increase of speed of 4 km/h on a journey of 32 km reduces the time taken by 4 hours. Find the original speed.

14. A train normally travels 60 miles at a certain speed. One day, due to bad weather, the train's speed is reduced by 10 mph so that the journey takes 3 hours longer. Find the normal speed.

15. The speed of a sparrow is x mph in still air. When the wind is blowing at 1 mph, the sparrow takes 5 hours to fly 12 miles to its nest and 12 miles back again. She goes out directly into the wind and returns with the wind behind her. Find her speed in still air.

16. An aircraft flies a certain distance on a bearing of 135° and then twice the distance on a bearing of 225°. Its distance from the starting point is then 350 km. Find the length of the first part of the journey.

17. In Fig. 1, ABCD is a rectangle with AB = 12 cm and BC = 7 cm. AK = BL = CM = DN = x cm. If the area of KLMN is 54 cm² find x.

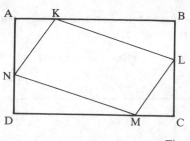

Fig. 1

18. In Fig. 1, AB = 14 cm, BC = 11 cm and AK = BL = CM = DN = x cm. If the area of KLMN is now 97 cm², find x.

19. The numerator of a fraction is 1 less than the denominator. When both numerator and denominator are increased by 2, the fraction is increased by $\frac{1}{12}$. Find the original fraction.

20. The perimeters of a square and a rectangle are equal. The length of the rectangle is 11 cm and the area of the square is 4 cm² more than the area of the rectangle. Find the side of the square.

REVISION EXERCISE 2A

1. Solve the equations
   (a) $x + 4 = 3x + 9$ (b) $9 - 3a = 1$
   (c) $y^2 + 5y = 0$ (d) $x^2 - 4 = 0$
   (e) $3x^2 + 7x - 40 = 0$

2. Given $a = 3$, $b = 4$ and $c = -2$, evaluate
   (a) $2a^2 - b$ (b) $a(b - c)$
   (c) $2b^2 - c^2$

3. Factorise completely
   (a) $4x^2 - y^2$ (b) $2x^2 + 8x + 6$
   (c) $6m + 4n - 9km - 6kn$
   (d) $2x^2 - 5x - 3$

4. Solve the simultaneous equations
   (a) $3x + 2y = 5$ (b) $2m - n = 6$
   $\quad 2x - y = 8$ $\quad 2m + 3n = -6$
   (c) $3x - 4y = 19$ (d) $3x - 7y = 11$
   $\quad x + 6y = 10$ $\quad 2x - 3y = 4$

5. Given that $x = 4$, $y = 3$, $z = -2$, evaluate
   (a) $2x(y + z)$ (b) $(xy)^2 - z^2$
   (c) $x^2 + y^2 + z^2$ (d) $(x + y)(x - z)$
   (e) $\sqrt{[x(1 - 4z)]}$ (f) $\dfrac{xy}{z}$

6. (a) Simplify $3(2x - 5) - 2(2x + 3)$
   (b) Factorise $2a - 3b - 4xa + 6xb$
   (c) Solve the equation $\dfrac{x - 11}{2} - \dfrac{x - 3}{5} = 2$.

7. Solve the equations
   (a) $5 - 7x = 4 - 6x$ (b) $\dfrac{7}{x} = \dfrac{2}{3}$
   (c) $2x^2 - 7x = 0$ (d) $x^2 + 5x + 6 = 0$
   (e) $\dfrac{1}{x} + \dfrac{1}{4} = \dfrac{1}{3}$

8. Factorise completely
   (a) $z^3 - 16z$ (b) $x^2y^2 + x^2 + y^2 + 1$
   (c) $2x^2 + 11x + 12$

9. Find the value of $\dfrac{2x - 3y}{5x + 2y}$ when $x = 2a$ and
   $y = -a$.

10. Solve the simultaneous equations
   (a) $7c + 3d = 29$ (b) $2x - 3y = 7$
       $5c - 4d = 33$ $2y - 3x = -8$
   (c) $5x = 3(1 - y)$ (d) $5s + 3t = 16$
       $3x + 2y + 1 = 0$ $11s + 7t = 34$

11. Solve the equations,
   (a) $4(y + 1) = \dfrac{3}{1 - y}$
   (b) $4(2x - 1) - 3(1 - x) = 0$
   (c) $\dfrac{x + 3}{x} = 2$
   (d) $x^2 = 5x$

12. Solve the following, giving your answers correct to 2 decimal places.
   (a) $2x^2 - 3x - 1 = 0$ (b) $x^2 - x - 1 = 0$
   (c) $3x^2 + 2x - 4 = 0$ (d) $x + 3 = \dfrac{7}{x}$

13. Find x by forming a suitable equation.

(a)   (b)
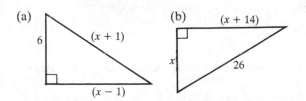

14. Given that $m = -2$, $n = 4$, evaluate
   (a) $5m + 3n$ (b) $5 + 2m - m^2$
   (c) $m^2 + 2n^2$ (d) $(2m + n)(2m - n)$
   (e) $(n - m)^2$ (f) $n - mn - 2m^2$

15. A car travels for x hours at a speed of $(x + 2)$ miles per hour. If the distance travelled is 15 miles, write down an equation for x and solve it to find the speed of the car.

16. ABCD is a rectangle, where $AB = x$ cm and BC is 1·5 cm less than AB.

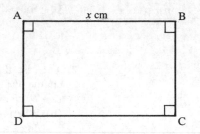

If the area of the rectangle is 52 cm², form an equation in x and solve it to find the dimensions of the rectangle.

17. Solve the equations
   (a) $(2x + 1)^2 = (x + 5)^2$
   (b) $\dfrac{x + 2}{2} - \dfrac{x - 1}{3} = \dfrac{x}{4}$
   (c) $x^2 - 7x + 5 = 0$, giving the answers correct to two decimal places.

18. Solve the equation
   $$\dfrac{x}{x + 1} - \dfrac{x + 1}{3x - 1} = \dfrac{1}{4}$$

19. Given that $a + b = 2$ and that $a^2 + b^2 = 6$, prove that $2ab = -2$. Find also the value of $(a - b)^2$.

20. The sides of a right-angled triangle have lengths $(x - 3)$ cm, $(x + 11)$ cm and $2x$ cm, where $2x$ is the hypotenuse. Find x.

21. A piggy-bank contains 50 coins, all either 2p or 5p. The total value of the coins in £1·87. How many 2p coins are there?

22. Pat bought 45 stamps, some for 10p and some for 18p. If he spent £6·66 altogether, how many 10p stamps did he buy?

23. When each edge of a cube is decreased by 1 cm, its volume is decreased by 91 cm³. Find the length of a side of the original cube.

24. One solution of the equation $2x^2 - 7x + k = 0$ is $x = -\frac{1}{2}$. Find the value of k.

EXAMINATION EXERCISE 2B

```
┌─────────────────┐
│   ADMISSION     │
│   ━━━━━▶         │
│   CHILDREN      │
│   HALF          │
│   PRICE         │
└─────────────────┘
      ╻╻
```

1. Three children and their parents visit a castle. The children are charged half the adult price of admission. The total paid by their father is £5·25.
   (a) Using x as the price of admission for an adult, form an equation in x.
   (b) Solve this equation and write down the adult admission charge.   [M]

2. The cost of hiring a cleaning machine is a fixed charge of £C with an additional charge of £A for each day of the hire period.
   (a) Write down an expression for the cost of hiring the machine for 3 days.
   (b) When the number of days in the hire period is n, the total cost is £T. Write down a formula for T in terms of C, n and A.   [N]

3. (a) Simplify
   (i) $\dfrac{2x^2}{x}$,   (ii) $\dfrac{3y-6}{3}$
   (b) Multiply out $(3p - q)(p + 2q)$ and simplify your answer
   (c) Factorise completely
   (i) $x^2 - x - 20$ (ii) $3k^2 - 192$
   (d) Solve each of the following equations:
   (i) $3(2x - 5) = 14$
   (ii) $\frac{1}{4}(x - 7) + \frac{1}{2}(x + 1) = 4$
   (e) Solve the simultaneous equations:
   $4x - 5y = 3$
   $2x - 3y = 7$ [MEI]

4. The diagram shows a row of 3 white tiles surrounded by 12 shaded tiles

   (a) How many shaded tiles are needed to surround a row of
   (i) 2 white tiles,
   (ii) 4 white tiles,
   (iii) 20 white tiles?
   (b) Find a rule which gives the numbers of shaded tiles needed to surround a row of w white tiles.
   (c) Use your rule to find how many shaded tiles are needed to surround a row of 100 white tiles.
   (d) How many white tiles in a row are surrounded by 626 shaded tiles?   [W]

5. In her will, granny leaves all her money to be shared by her three grandchildren. Anne is to have £400 more than Beatrice. Clarissa is to have three times as much as Beatrice.
   (a) If Beatrice receives £x, then Anne receives £$(x + 400)$. Write an expression for the amount Clarissa receives in terms of x.
   (b) Suppose Clarissa receives £300. How much does Anne receive?
   (c) When granny died, Anne and Clarissa received the same amount of money. How much did granny leave?   [S]

6. A gardener always plants his white and red rose bushes in a special pattern. Below are three of his arrangements.

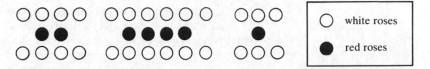

(a) Draw another way he could plant bushes which still fits his pattern.
(b) How many white rose bushes would he need for an arrangement which used 27 red ones?
(c) How many red bushes would he need if he planted 100 white ones?
(d) If R stands for the number of red bushes used and W stands for the number of white ones used, write a formula connecting R and W for this pattern. ($W =$)
(e) Another gardener always plants her rose bushes using the rule $W = 3R + 4$. Find the TOTAL number of bushes used if there are 10 red ones.
(f) Find the value of R given by the formula if W is 20.
(g) What does this tell you about a pattern using 20 white roses?   [S]

7. (a) At the moment, Paul is six times as old as his daughter Jane.
   (i) If Jane's present age is x years, write down, in terms of x, the age Paul will be in 12 years' time.
   (ii) In 12 years' time, Paul will be three times as old as Jane. Write down an equation for x and hence find Jane's present age.
(b) Solve the equation $(x - 1)^2 = 3$, giving your answers correct to two decimal places.   [M]

8. The diagram shows a number-rectangle with a square marked on it. This is called the '15 square' since 15 is the number in the top left hand corner of the square.

1	2	3	4
5	6	7	8
9	10	11	12
13	14	15	16
17	18	19	20
21	22	23	24
25	26	27	28
29	30	31	32
33	34	35	36
37	38	39	40
41	42	43	44

(a) Find the total of the numbers in the '26 square'.
(b) Write down, in terms of x, the other three numbers in the 'x square' and find the total of the four numbers.
   Show that the total number is $4x + 10$.
(c) Which square has a total of 126?
(d) Explain why the total of the four numbers in any square could not be
   (i) 93
   (ii) 106   [W]

9. A function $f(x)$ is defined as
$$f(x) = x^2 + 5x - 24$$
(a) Evaluate
   (i) $f(-2)$,
   (ii) $f(-3)$.
(b) Find the values of x for which
   (i) $f(x) = 0$,
   (ii) $f(x) = -24$.   [S]

10. The average speed of a train travelling the 90 miles from Porthmadog to Shrewsbury is 15 miles per hour faster than the average speed of a car doing the same journey. The journey by car takes $\frac{1}{2}$ hour longer than the journey by train.
    (a) Copy the table below, and complete it to show the average speed and the time taken by the car and the train in terms of u.
    (b) Form an equation in u and solve it to obtain the average speed of the car.

	Distance in miles	Average speed in miles per hour	Time in hours
Car	90	u	
Train	90		

[W]

11. A road tanker carries 30 tonnes of oil. When cold the oil can be pumped out at a rate of x tonnes per minute.
    (a) Write down an expression for the time, in minutes, taken to empty the tanker.

    If the oil is heated then an extra 0·5 tonnes can be pumped out per minute.
    (b) Write down an expression for the time taken to empty the tanker when the oil is heated.

    If the oil has been heated then the time taken to empty the tanker is reduced by two minutes.
    (c) Show that the equation for x can be expressed in the form
    $$2x^2 + x - 15 = 0$$
    Solve this equation for x, and hence find the time taken to empty the tanker when the oil is cold. [NI]

12. A plastics firm is asked to make a small open rectangular container with a square base. The side of the base is x mm and the height of the container is h mm.

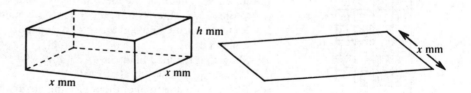

The pieces making up the faces of the container are cut from a single rectangular strip of plastic of width x mm.
    (a) Write down, in terms of x and h,
       (i) the length of the piece of plastic,
       (ii) the area of the piece of plastic.
    The area of plastic used for one container is 2500 mm².
    (b) Use your answer in (a) (ii) to find an expression for h in terms of x.
    (c) Find the value of x when the height is 20 mm.
    (d) Write down, to the nearest millimetre, the dimensions of the strip of plastic.

[L]

3 Mensuration

Archimedes of Samos (287–212 B.C.) studied at Alexandria as a young man. One of the first to apply scientific thinking to everyday problems, he was a practical man of common sense. He gave proofs for finding the area, the volume and the centre of gravity of circles, spheres, conics and spirals. By drawing polygons with many sides, he arrived at a value of π between $3\frac{10}{71}$ and $3\frac{10}{70}$. He was killed in the siege of Syracuse at the age of 75.

3.1 AREA

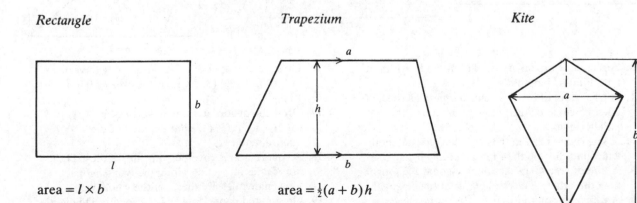

Rectangle

area $= l \times b$

Trapezium

area $= \frac{1}{2}(a + b)h$

Kite

area $= \frac{1}{2}a \times b$
$\qquad = \frac{1}{2} \times$ (**product of diagonals**)

Exercise 1

For questions **1** to **7**, find the area of each shape. Decide which information to use:
you may not need all of it.

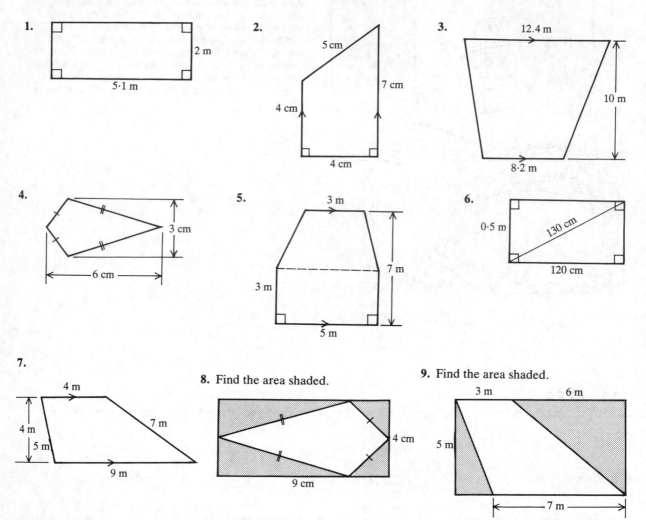

1.

2 m

5·1 m

2.

5 cm

7 cm

4 cm

4 cm

3.

12.4 m

10 m

8·2 m

4.

3 cm

6 cm

5.

3 m

7 m

3 m

5 m

6.

0·5 m

130 cm

120 cm

7.

4 m

4 m

7 m

5 m

9 m

8. Find the area shaded.

4 cm

9 cm

9. Find the area shaded.

3 m   6 m

5 m

7 m

10. A rectangle has an area of 117 m² and a width of 9 m. Find its length.

11. A trapezium of area 105 cm² has parallel sides of length 5 cm and 9 cm. How far apart are the parallel sides?

12. A kite of area 252 m² has one diagonal of length 9 m. Find the length of the other diagonal.

13. A kite of area 40 m² has one diagonal 2 m longer than the other. Find the lengths of the diagonals.

14. A trapezium of area 140 cm² has parallel sides 10 cm apart and one of these sides is 16 cm long. Find the length of the other parallel side.

15. A floor 5 m by 20 m is covered by square tiles of side 20 cm. How many tiles are needed?

16. On squared paper draw the triangle with vertices at (1, 1), (5, 3), (3, 5). Find the area of the triangle.

17. Draw the quadrilateral with vertices at (1, 1), (6, 2), (5, 5), (3, 6). Find the area of the quadrilateral.

18. A square wall is covered with square tiles. There are 85 tiles altogether along the two diagonals. How many tiles are there on the whole wall?

19. On squared paper draw a 7 × 7 square. Divide it up into nine smaller squares.

20. A rectangular field, 400 m long, has an area of 6 hectares. Calculate the perimeter of the field. [1 hectare = 10 000 m²].

Triangle

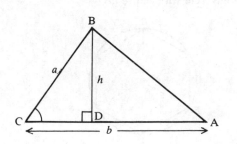

area $= \frac{1}{2} \times b \times h$.

In triangle BCD, $\sin C = \dfrac{h}{a}$

$\therefore$ $\qquad\qquad h = a \sin C$

$\therefore$ area of triangle $= \frac{1}{2} \times b \times a \sin C$.

This formula is useful when *two sides* and the *included* angle are known.

Example 1

Find the area of the triangle shown.

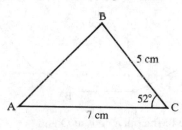

Area $= \frac{1}{2} ab \sin C$

$\quad = \frac{1}{2} \times 5 \times 7 \times \sin 52°$

$\quad = 13 \cdot 8 \text{ cm}^2$ (3 sig. fig.)

Parallelogram

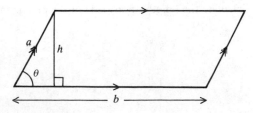

area $= b \times h$

area $= ba \sin \theta$

Exercise 2

In questions **1** to **12** find the area of $\triangle ABC$ where $AB = c$, $AC = b$ and $BC = a$. (Sketch the triangle in each case.)

1. $a = 7$ cm, $b = 14$ cm, $\hat{C} = 80°$.
2. $b = 11$ cm, $a = 9$ cm, $\hat{C} = 35°$.
3. $c = 12$ m, $b = 12$ m, $\hat{A} = 67 \cdot 2°$.
4. $a = 5$ cm, $c = 6$ cm, $\hat{B} = 11 \cdot 8°$.
5. $b = 4 \cdot 2$ cm, $a = 10$ cm, $\hat{C} = 120°$.
6. $a = 5$ cm, $c = 8$ cm, $\hat{B} = 142°$.
7. $b = 3 \cdot 2$ cm, $c = 1 \cdot 8$ cm, $\hat{B} = 10°$, $\hat{C} = 65°$.
8. $a = 7$ m, $b = 14$ m, $\hat{A} = 32°$, $\hat{B} = 100°$.
9. $a = b = c = 12$ m.
10. $a = c = 8$ m, $\hat{B} = 72°$.
11. $b = c = 10$ cm, $\hat{B} = 32°$.
12. $a = b = c = 0 \cdot 8$ m.

In questions **13** to **20**, find the area of each shape.

13.

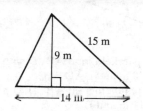

14.

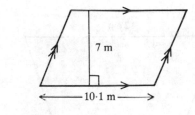

15.

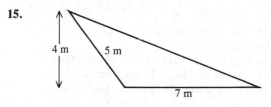

16.

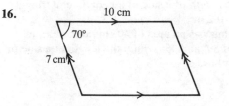

17.

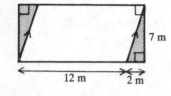

5 cm

65°

8 cm

18.

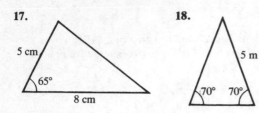

5 m

70° 70°

19. Find the area shaded.

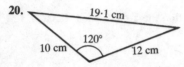

7 m

12 m   2 m

20.

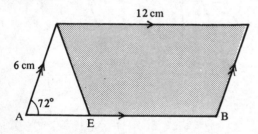

19·1 cm

120°

10 cm   12 cm

21. Find the area of a parallelogram ABCD with AB = 7 m, AD = 20 m and BÂD = 62°.

22. Find the area of a parallelogram ABCD with AD = 7 m, CD = 11 m and BÂD = 65°.

23. In the diagram if $AE = \frac{1}{3} AB$, find the area shaded.

12 cm

6 cm

72°

A   E   B

24. The area of an equilateral triangle ABC is 50 cm^2. Find AB.

25. The area of a triangle ABC is 64 cm^2. Given AB = 11 cm and BC = 15 cm, find AB̂C.

26. The area of a triangle XYZ is 11 m^2. Given YZ = 7 m and XŶZ = 130°, find XY.

27. Find the length of a side of an equilateral triangle of area 10·2 m^2.

28. A rhombus has an area of 40 cm^2 and adjacent angles of 50° and 130°. Find the length of a side of the rhombus.

29. A regular hexagon is circumscribed by a circle of radius 3 cm with centre O.

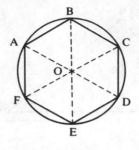

B

A   C

O *

F   D

E

(a) What is angle EOD?

(b) Find the area of triangle EOD and hence find the area of the hexagon ABCDEF.

30. Hexagonal tiles of side 20 cm are used to tile a room which measures 6·25 m by 4·85 m. Assuming we complete the edges by cutting up tiles, how many tiles are needed?

31. Find the area of a regular pentagon of side 8 cm.

32. The diagram shows a part of the perimeter of a regular polygon with n sides

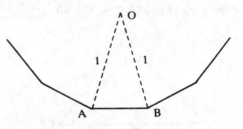

O

1   1

A   B

The centre of the polygon is at O and OA = OB = 1 unit.

(a) What is the angle AOB in terms of n?

(b) Work out an expression in terms of n for the area of the polygon.

(c) Find the area of polygons where $n = 6, 10, 300, 1000, 10\,000$. What do you notice?

33. The area of a regular pentagon is 600 cm^2. Calculate the length of one side of the pentagon.

3.2 THE CIRCLE

For any circle, the ratio $\left(\dfrac{\text{circumference}}{\text{diameter}}\right)$ is equal to π.

The value of π is usually taken to be 3·14, but this is not an exact value. Through the centuries, mathematicians have been trying to obtain a better value for π.

For example, in the third century A.D., the Chinese mathematician Liu Hui obtained the value 3·14159 by considering a regular polygon having 3072 sides! Ludolph van Ceulen (1540–1610) worked even harder to produce a value correct to 35 significant figures. He was so proud of his work that he had this value of π engraved on his tombstone.

Electronic computers are now able to calculate the value of π to many thousands of figures, but its value is still not exact. It was shown in 1761 that π is an *irrational number* which, like $\sqrt{2}$ or $\sqrt{3}$ cannot be expressed exactly as a fraction.

The first fifteen significant figures of π can be remembered from the number of letters in each word of the following sentence.

How I need a drink, cherryade of course, after the silly lectures involving Italian kangaroos.
There remain a lot of unanswered questions concerning π, and many mathematicians today are still working on them.

The following formulae should be memorised.

circumference $= \pi d$
$\qquad\qquad = 2\pi r$
area $= \pi r^2$

Example 1

Find the circumference and area of a circle of diameter 8 cm. (Take $\pi = 3\cdot14$.)

Circumference $= \pi d$
$\qquad\qquad = 3\cdot14 \times 8$
$\qquad\qquad = 25\cdot1$ cm (3 s.f.)

$\quad$ Area $= \pi r^2$
$\qquad\quad = 3\cdot14 \times 4^2$
$\qquad\quad = 50\cdot2$ cm^2 (3 s.f.)

Exercise 3

For each shape find (a) the perimeter, (b) the area. All lengths are in cm. Use the π button on a calculator or take $\pi = 3\cdot142$. All the arcs are either semi-circles or quarter circles.

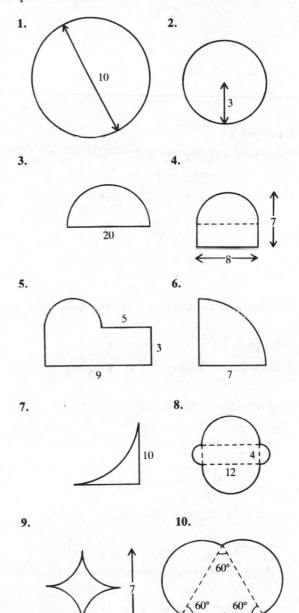

11.

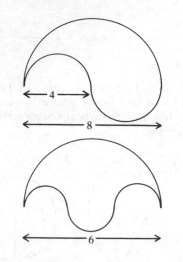

12.

Example 2

A circle has a circumference of 20 m. Find the radius of the circle. (Take $\pi = 3 \cdot 14$.)

Let the radius of the circle be r m.

$$\text{Circumference} = 2\pi r$$
$$\therefore \qquad 2\pi r = 20$$
$$\therefore \qquad r = \frac{20}{2\pi}$$
$$r = 3 \cdot 18$$

The radius of the circle is $3 \cdot 18$ m (3 S.F.).

Example 3

A circle has an area of 45 cm². Find the radius of the circle. (Take $\pi = 3 \cdot 14$.)

Let the radius of the circle be r cm.

$$\pi r^2 = 45$$
$$r^2 = \frac{45}{\pi}$$
$$r = \sqrt{\left(\frac{45}{\pi}\right)} = 3 \cdot 78 \text{ (3 S.F.)}$$

The radius of the circle is $3 \cdot 78$ cm.

Exercise 4

Use the 'π' button on a calculator and give answers to 3 S.F.

1. A circle has an area of 15 cm². Find its radius.
2. A circle has a circumference of 190 m. Find its radius.
3. Find the radius of a circle of area 22 km².

4. Find the radius of a circle of circumference 58·6 cm.
5. A circle has an area of 16 mm². Find its circumference.
6. A circle has a circumference of 2500 km. Find its area.
7. A circle of radius 5 cm is inscribed inside a square as shown. Find the area shaded.

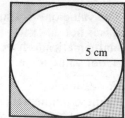

5 cm

8. A circular pond of radius 6 m is surrounded by a path of width 1 m.
   (a) Find the area of the path.
   (b) The path is resurfaced with astroturf which is bought in packs each containing enough to cover an area of 7 m². How many containers are required?

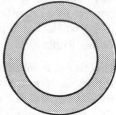

9. Discs of radius 4 cm are cut from a rectangular plastic sheet of length 84 cm and width 24 cm. How many complete discs can be cut out? Find
   (a) the total area of the discs cut
   (b) the area of the sheet wasted.
10. The tyre of a car wheel has an outer diameter of 30 cm. How many times will the wheel rotate on a journey of 5 km?
11. A golf ball of diameter 1·68 inches rolls a distance of 4 m in a straight line. How many times does the ball rotate completely?
    (1 inch = 2·54 cm)
12. 100 yards of cotton is wound without stretching onto a reel of diameter 3 cm. How many times does the reel rotate?
    (1 yard = 0·914 m. Ignore the thickness of the cotton)
13. A rectangular metal plate has a length of 65 cm and a width of 35 cm. It is melted down and recast into circular discs of the same thickness. How many complete discs can be formed if
    (a) the radius of each disc is 3 cm?
    (b) the radius of each disc is 10 cm?
14. Calculate the radius of a circle whose area is equal to the sum of the areas of three circles of radii 2 cm, 3 cm and 4 cm respectively.

15. The diameter of a circle is given as 10 cm, correct to the nearest cm. Calculate
    (a) the maximum possible circumference
    (b) the minimum possible area of the circle consistent with this data.

16. A square is inscribed in a circle of radius 7 cm. Find
    (a) the area of the square
    (b) the area shaded.

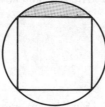

17. An archery target has three concentric regions. The diameters of the regions are in the ratio 1:2:3. Find the ratio of their areas.

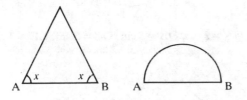

18. The governor of a prison has 100 m of wire fencing. What area can he enclose if he makes a circular compound?

19. The semi-circle and the isosceles triangle have the same base AB and the same area. Find the angle x.

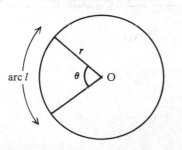

20. Mr Gibson decided to measure the circumference of the Earth using a very long tape measure. For reasons best known to himself he held the tape measure 1 m from the surface of the (perfectly spherical) Earth all the way round. When he had finished Mrs Gibson told him that his measurement gave too large an answer. She suggested taking off 6 m. Was she correct? [Take the radius of the Earth to be 6400 km (if you need it)]

21. The large circle has a radius of 10 cm. Find the radius of the largest circle which will fit in the middle.

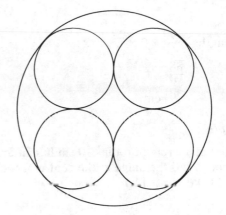

3.3 ARC LENGTH AND SECTOR AREA

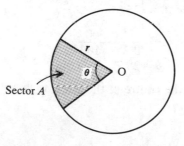

Arc length, $l = \dfrac{\theta}{360} \times 2\pi r$

We take a fraction of the whole circumference depending on the angle at the centre of the circle.

Sector area, $A = \dfrac{\theta}{360} \times \pi r^2$

We take a fraction of the whole area depending on the angle at the centre of the circle.

Example 1

Find the length of an arc which subtends an angle of 140° at the centre of a circle of radius 12 cm. (Take $\pi = \frac{22}{7}$.)

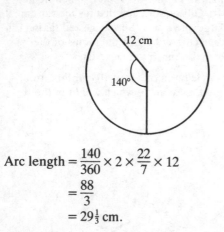

Arc length $= \dfrac{140}{360} \times 2 \times \dfrac{22}{7} \times 12$

$\qquad = \dfrac{88}{3}$

$\qquad = 29\frac{1}{3}$ cm.

Example 2

A sector of a circle of radius 10 cm has an area of 25 cm². Find the angle at the centre of the circle. (Take $\pi = 3 \cdot 14$.)

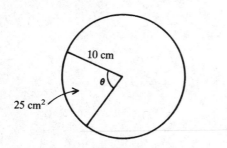

Let the angle at the centre of the circle be θ.

$$\dfrac{\theta}{360} \times \pi \times 10^2 = 25$$

$\therefore \qquad\qquad \theta = \dfrac{25 \times 360}{3 \cdot 14 \times 100}$

$\qquad\qquad\qquad \theta = 28 \cdot 7°$ (3 s.f.)

The angle at the centre of the circle is $28 \cdot 7°$.

Exercise 5

[Use the π button on a calculator unless told otherwise]

1. Arc AB subtends an angle θ at the centre of circle radius r.

Find the arc length and sector area when
(a) $r = 4$ cm, $\theta = 30°$
(b) $r = 10$ cm, $\theta = 45°$
(c) $r = 2$ cm, $\theta = 235°$.

In questions **2** and **3** find the total area of the shape.

2.

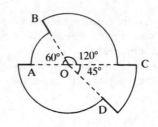

OA = 2 cm, OB = 3 cm, OC = 5 cm, OD = 3 cm.

3.

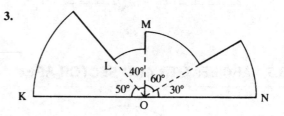

ON = 6 cm, OM = 3 cm, OL = 2 cm, OK = 6 cm.

4. Find the shaded area.

(a)   (b)

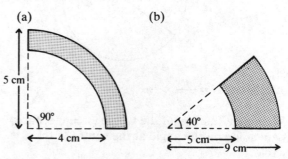

5. In the diagram the arc length is l and the sector area is A.

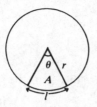

   (a) Find θ, when $r = 5$ cm and $l = 7 \cdot 5$ cm
   (b) Find θ, when $r = 2$ m and $A = 2$ m^2
   (c) Find r, when $\theta = 55°$ and $l = 6$ cm.

6. The length of the minor arc AB of a circle, centre O, is 2π cm and the length of the major arc is 22π cm. Find
   (a) the radius of the circle,
   (b) the acute angle AOB.

7. The lengths of the minor and major arcs of a circle are $5 \cdot 2$ cm and $19 \cdot 8$ respectively. Find
   (a) the radius of the circle
   (b) the angle subtended at the centre by the minor arc.

8. A wheel of radius 10 cm is turning at a rate of 5 revolutions per minute. Calculate
   (a) the angle through which the wheel turns in 1 second
   (b) the distance moved by a point on the rim in 2 seconds.

9. The length of an arc of a circle is 12 cm. The corresponding sector area is 108 cm^2. Find
   (a) the radius of the circle
   (b) the angle subtended at the centre of the circle by the arc.

10. The length of an arc of a circle is $7 \cdot 5$ cm. The corresponding sector area is $37 \cdot 5$ cm^2. Find
   (a) the radius of the circle
   (b) the angle subtended at the centre of the circle by the arc.

11. In the diagram the arc length is l and the sector area is A.

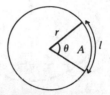

   (a) Find l, when $\theta = 72°$ and $A = 15$ cm^2
   (b) Find l, when $\theta = 135°$ and $A = 162$ m^2
   (c) Find A, when $l = 11$ cm and $r = 5 \cdot 2$ cm

12. A long time ago Mr Gibson found an island shaped as a triangle with three straight shores of length 3 km, 4 km and 5 km. He declared an 'exclusion zone' around his island and forbade anyone to came within 1 km of his shore. What was the area of his exclusion zone?

3.4   CHORD OF A CIRCLE

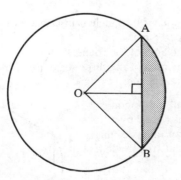

The line AB is a chord. The area of a circle cut off by a chord is called a *segment*. In the diagram the *minor* segment is shaded and the *major* segment is unshaded.

   (a) The line from the centre of a circle to the mid-point of a chord *bisects* the chord at *right angles*.
   (b) The line from the centre of a circle to the mid-point of a chord bisects the angle subtended by the chord at the centre of the circle.

Example 1

XY is a chord of length 12 cm of a circle of radius 10 cm, centre O. Calculate
(a) the angle XOY
(b) the area of the minor segment cut off by the chord XY.

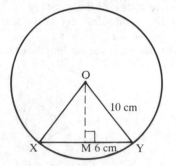

Let the mid-point of XY be M

$\therefore$ $\qquad$ MY = 6 cm

$$\sin M\hat{O}Y = \frac{6}{10}$$

$\therefore$ $\qquad$ $M\hat{O}Y = 36\cdot87°$

$\therefore$ $\qquad$ $X\hat{O}Y = 2 \times 36\cdot87$
$\qquad\qquad\quad = 73\cdot74°$

area of minor segment =
$\qquad$ area of sector XOY − area of $\triangle$XOY

$$\text{area of sector XOY} = \frac{73\cdot74}{360} \times \pi \times 10^2$$
$$= 64\cdot32 \text{ cm}^2.$$

$$\text{area of } \triangle \text{XOY} = \tfrac{1}{2} \times 10 \times 10 \times \sin 73\cdot74°$$
$$= 48\cdot00 \text{ cm}^2$$

$\therefore$ $\quad$ Area of minor segment = 64·32 − 48·00
$\qquad\qquad\qquad\qquad\quad = 16\cdot3 \text{ cm}^2 \text{ (3 s.f.)}$

Exercise 6

Use the 'π' button on a calculator

1. The chord AB subtends an angle of 130° at the centre O. The radius of the circle is 8 cm. Find
   (a) the length of AB,
   (b) the area of sector OAB,
   (c) the area of triangle OAB,
   (d) the area of the minor segment (shown shaded).

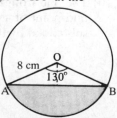

2. Find the shaded area when,
   (a) $r = 6$ cm, $\theta = 70°$
   (b) $r = 14$ cm, $\theta = 104°$
   (c) $r = 5$ cm, $\theta = 80°$

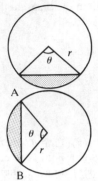

3. Find θ and hence the shaded area when,
   (a) AB = 10 cm, $r = 10$ cm
   (b) AB = 8 cm, $r = 5$ cm

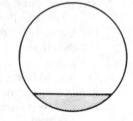

4. How far is a chord of length 8 cm from the centre of a circle of radius 5 cm?

5. How far is a chord of length 9 cm from the centre of a circle of radius 6 cm?

6. The diagram shows the cross section of a cylindrical pipe with water lying in the bottom.
   (a) If the maximum depth of the water is 2 cm and the radius of the pipe is 7 cm, find the area shaded.
   (b) What is the *volume* of water in a length of 30 cm?

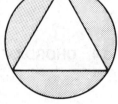

7. An equilateral triangle is inscribed in a circle of radius 10 cm. Find
   (a) the area of the triangle.
   (b) the area shaded.

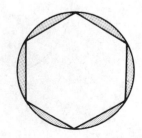

8. An equilateral triangle is inscribed in a circle of radius 18·8 cm. Find
   (a) the area of the triangle.
   (b) the area of the three segments surrounding the triangle.

9. A regular hexagon is circumscribed by a circle of radius 6 cm. Find the area shaded.

10. A regular octagon is circumscribed by a circle of radius r cm. Find the area enclosed between the circle and the octagon. (Give the answer in terms of r.)

11. Find the radius of the circle
    (a) when $\theta = 90°$, $A = 20$ cm^2
    (b) when $\theta = 30°$, $A = 35$ cm^2
    (c) when $\theta = 150°$, $A = 114$ cm^2

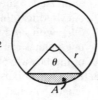

12. (Harder) The diagram shows a regular pentagon of side 10 cm with a star inside. Calculate the area of the star.

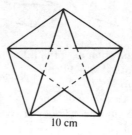

10 cm

3.5   VOLUME

Prism

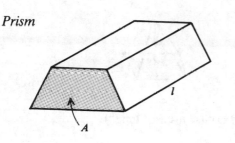

A prism is an object with the same cross section throughout its length.

Volume of prism = (area of cross section) × length
$$= A \times l.$$

A *cuboid* is a prism whose six faces are all rectangles. A cube is a special case of a cuboid in which all six faces are squares.

Cylinder

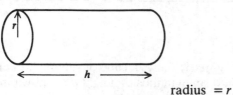

radius $= r$
height $= h$

A cylinder is a prism whose cross section is a circle.

Volume of cylinder
        = (area of cross section) × length
    Volume $= \pi r^2 h$

Example 1

Calculate the height of a cylinder of volume 500 cm^3 and base radius 8 cm. Let the height of the cylinder be h cm.

$$\pi r^2 h = 500$$
$$3 \cdot 14 \times 8^2 \times h = 500$$
$$h = \frac{500}{3 \cdot 14 \times 64}$$
$$h = 2 \cdot 49 \ (3 \ \text{s.f.})$$

The height of the cylinder is $2 \cdot 49$ cm.

Exercise 7

1. Calculate the volume of the prisms. All lengths are in cm.

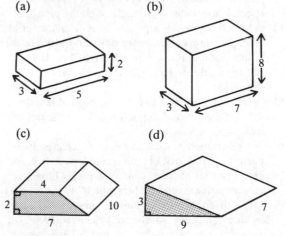

(a)
(b)
(c)
(d)

(e)   (f)

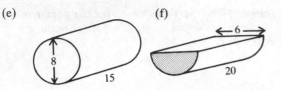

2. Calculate the volume of the following cylinders
   (a) $r = 4$ cm, $h = 10$ cm
   (b) $r = 11$ m, $h = 2$ m
   (c) $r = 2 \cdot 1$ cm, $h = 0 \cdot 9$ cm
3. Find the height of a cylinder of volume 200 cm^3 and radius 4 cm.
4. Find the length of a cylinder of volume 2 litres and radius 10 cm.
5. Find the radius of a cylinder of volume 45 cm^3 and length 4 cm.
6. A prism has volume 100 cm^3 and length 8 cm. If the cross section is an equilateral triangle, find the length of a side of the triangle.
7. When 3 litres of oil is removed from an upright cylindrical can the level falls by 10 cm.
   Find the radius of the can.
8. A solid cylinder of radius 4 cm and length 8 cm is melted down and recast into a solid cube.
   Find the side of the cube.
9. A solid rectangular block of copper 5 cm by 4 cm by 2 cm is drawn out to make a cylindrical wire of diameter 2 mm. Calculate the length of the wire.
10. Water flows through a circular pipe of internal diameter 3 cm at a speed of 10 cm/s. If the pipe is full, how much water issues from the pipe in one minute? (answer in litres)
11. Water issues from a hose-pipe of internal diameter 1 cm at a rate of 5 litres per minute. At what speed is the water flowing through the pipe?
12. A cylindrical metal pipe has external diameter of 6 cm and internal diameter of 4 cm. Calculate the volume of metal in a pipe of length 1 m. If 1 cm^3 of the metal weighs 8 g, find the weight of the pipe.
13. For two cylinders A and B, the ratio of lengths is 3:1 and the ratio of diameters is 1:2. Calculate the ratio of their volumes.
14. A well-trained hen can lay eggs which are either perfect cylinders of diameter and length 4 cm, or perfect cubes of side 5 cm. Which eggs have the greater volume, and by how much? (Take $\pi = 3$.)
15. Mrs Gibson decided to build a garage and began by calculating the number of bricks required. The garage was to be 6 m by 4 m and 2·5 m in height. Each brick measures 22 cm by 10 cm by 7 cm. Mrs Gibson estimated that she would need about 40 000 bricks. Is this a reasonable estimate?

16. A cylindrical can of internal radius 20 cm stands upright on a flat surface. It contains water to a depth of 20 cm. Calculate the rise in the level of the water when a brick of volume 1500 cm^3 is immersed in the water.
17. A cylindrical tin of height 15 cm and radius 4 cm is filled with sand from a rectangular box. How many times can the tin be filled if the dimensions of the box are 50 cm by 40 cm by 20 cm?
18. Rain which falls onto a flat rectangular surface of length 6 m and width 4 m is collected in a cylinder of internal radius 20 cm. What is the depth of water in the cylinder after a storm in which 1 cm of rain fell?

Pyramid

Volume $= \frac{1}{3}$ (base area) × height.

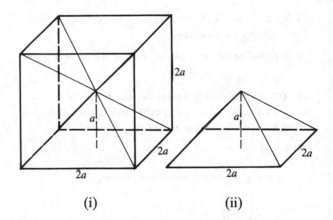

(i)   (ii)

Figure (i) shows a cube of side $2a$ broken down into six pyramids of height a as shown in figure (ii).
   If the volume of each pyramid is V.

then $6V = 2a \times 2a \times 2a$
        $V = \frac{1}{6} \times (2a)^2 \times 2a$
so $V = \frac{1}{3} \times (2a)^2 \times a$
        $V = \frac{1}{3}$ (base area) × height.

Cone

Volume $= \frac{1}{3}\pi r^2 h$
(note the similarity with
the pyramid)

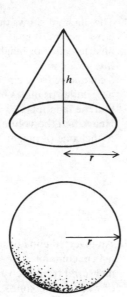

Sphere

Volume $= \frac{4}{3}\pi r^3$

Example 2

A pyramid has a square base of side 5 m and
vertical height 4 m. Find its volume.

Volume of pyramid $= \frac{1}{3}(5 \times 5) \times 4$
$\qquad\qquad\qquad = 33\frac{1}{3}$ m^3.

Example 3

Calculate the radius of a sphere of volume
500 cm^3.

Let the radius of the sphere be r cm

$$\frac{4}{3}\pi r^3 = 500$$

$$r^3 = \frac{3 \times 500}{4\pi}$$

$$r = \sqrt[3]{\left(\frac{3 \times 500}{4\pi}\right)} = 4\cdot 92 \ (3 \text{ S.F.})$$

The radius of the sphere is 4·92 cm.

Exercise 8

Find the volumes of the following objects:

1. cone: height $= 5$ cm, radius $= 2$ cm
2. sphere: radius $= 5$ cm
3. sphere: radius $= 10$ cm
4. cone: height $= 6$ cm, radius $= 4$ cm
5. sphere: diameter $= 8$ cm
6. cone: height $= x$ cm, radius $= 2x$ cm
7. sphere: radius $= 0\cdot 1$ m
8. cone: height $= \frac{1}{\pi}$ cm, radius $= 3$ cm

9. pyramid: rectangular base 7 cm by 8 cm;
   height $= 5$ cm
10. pyramid: square base of side 4 m, height $= 9$ m
11. pyramid: equilateral triangular base of side $=$
    8 cm, height $= 10$ cm
12. Find the volume of a hemisphere of radius 5 cm.
13. A cone is attached to a
    hemisphere of radius 4 cm.
    If the total height of the
    object is 10 cm, find its
    volume.

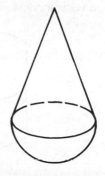

14. A toy consists of a cylinder
    of diameter 6 cm 'sandwiched'
    between a hemisphere and a
    cone of the same diameter.
    If the cone is of height 8 cm
    and the cylinder is of height
    10 cm, find the total volume
    of the toy.

15. Find the height of a pyramid of volume 20 m^3 and
    base area 12 m^2.
16. Find the radius of a sphere of volume 60 cm^3.
17. Find the height of a cone of volume 2·5 litre and
    radius 10 cm.
18. Six square-based pyramids fit exactly onto the six
    faces of a cube of side 4 cm. If the volume of the
    object formed is 256 cm^3, find the height of each
    of the pyramids.
19. A solid metal cube of side 6 cm is recast into a
    solid sphere. Find the radius of the sphere.
20. A hollow spherical vessel has internal and external
    radii of 6 cm and 6·4 cm respectively. Calculate
    the weight of the vessel if it is made of metal of
    density 10 g/cm^3.
21. Water is flowing into an inverted cone, of
    diameter and height 30 cm, at a rate of 4 litres per
    minute. How long, in seconds, will it take to fill
    the cone?

22. A solid metal sphere is recast into many smaller spheres. Calculate the number of the smaller spheres if the initial and final radii are as follows:
   (a) initial radius = 10 cm, final radius = 2 cm
   (b) initial radius = 7 cm,   final radius = $\frac{1}{2}$ cm
   (c) initial radius = 1 m,   final radius = $\frac{1}{3}$ cm.

23. A spherical ball is immersed in water contained in a vertical cylinder.

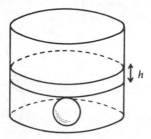

Assuming the water covers the ball, calculate the rise in the water level if
   (a) sphere radius = 3 cm, cylinder radius = 10 cm
   (b) sphere radius = 2 cm, cylinder radius = 5 cm.

24. A spherical ball is immersed in water contained in a vertical cylinder. The rise in water level is measured in order to calculate the radius of the spherical ball. Calculate the radius of the ball in the following cases:
   (a) cylinder of radius 10 cm, water level rises 4 cm.
   (b) cylinder of radius 100 cm, water level rises 8 cm.

25. One corner of a solid cube of side 8 cm is removed by cutting through the mid-points of three adjacent sides. Calculate the volume of the piece removed.

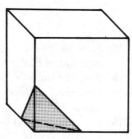

26. The cylindrical end of a pencil is sharpened to produce a perfect cone at the end with no overall loss of length. If the diameter of the pencil is 1 cm, and the cone is of length 2 cm, calculate the volume of the shavings.

27. Metal spheres of radius 2 cm are packed into a rectangular box of internal dimensions 16 cm × 8 cm × 8 cm. When 16 spheres are packed the box is filled with a preservative liquid. Find the volume of this liquid.

28. The diagram shows the cross section of an inverted cone of height MC = 12 cm. If AB = 6 cm and XY = 2 cm, use similar triangles to find the length NC. Hence find the volume of the cone of height NC.

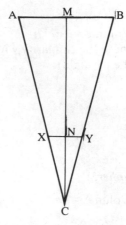

29. An inverted cone of height 10 cm and base radius 6·4 cm contains water to a depth of 5 cm, measured from the vertex. Calculate the volume of water in the cone.

30. An inverted cone of height 15 cm and base radius 4 cm contains water to a depth of 10 cm. Calculate the volume of water in the cone.

31. An inverted cone of height 12 cm and base radius 6 cm contains 20 cm³ of water. Calculate the depth of water in the cone, measured from the vertex.

32. A frustrum is a cone with 'the end chopped off'. A bucket in the shape of a frustrum as shown has diameters of 10 cm and 4 cm at its ends and a depth of 3 cm. Calculate the volume of the bucket.

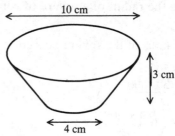

33. Find the volume of a frustrum with end diameters of 60 cm and 20 cm and a depth of 40 cm.

34. The diagram shows a sector of a circle of radius 10 cm.
   (a) Find, as a multiple of π, the arc length of the sector.
The straight edges are brought together to make a cone.
Calculate (b) the radius of the base of the cone,
          (c) the vertical height of the cone.

35. Calculate the volume of a regular octahedron whose edges are all 10 cm.

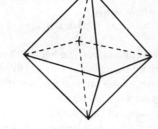

36. A sphere passes through the eight corners of a cube of side 10 cm. Find the volume of the sphere.

37. (Harder) Find the volume of a regular tetrahedron of side 20 cm.

38. Find the volume of a regular tetrahedron of side 35 cm.

3.6   SURFACE AREA

We are concerned here with the surface areas of the *curved* parts of cylinders, spheres and cones. The areas of the plane faces are easier to find.

(a) Cylinder
Curved surface area
= $2\pi rh$.

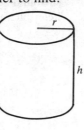

(b) Sphere
Surface area = $4\pi r^2$.

(c) Cone
Curved surface area
= πrl
where l is the slant height.

Example 1

Find the *total* surface area of a solid cone of radius 4 cm and vertical height 3 cm.

Let the slant height of the cone be l cm.

$l^2 = 3^2 + 4^2$
$l = 5$.

Curved surface area = $\pi \times 4 \times 5$
$= 20\pi$ cm^2
Area of end face = $\pi \times 4^2 = 16\pi$ cm^2

∴   Total surface area = $20\pi + 16\pi$
$= 36\pi$ cm^2
$= 113$ cm^2 to 3 s.f.

Exercise 9

Use π on a calculator unless otherwise instructed.

1. Copy the table and find the quantities marked *.
(Leave π in your answers.)

solid object	radius	vertical height	curved surface area	total surface area
(a) sphere	3 cm		*	
(b) cylinder	4 cm	5 cm		*
(c) cone	6 cm	8 cm	*	
(d) cylinder	0·7 m	1 m		*
(e) sphere	10 m		*	
(f) cone	5 cm	12 cm	*	
(g) cylinder	6 mm	10 mm		*
(h) cone	2·1 cm	4·4 cm	*	
(i) sphere	0·01 m		*	
(j) hemisphere	7 cm		*	*

2. Find the radius of a sphere of surface area 34 cm^2.

3. Find the slant height of a cone of curved surface area 20 cm^2 and radius 3 cm.

4. Find the height of a solid cylinder of radius 1 cm and *total* surface area 28 cm^2.

5. Copy the table and find the quantities marked *.
(Take $\pi = 3$.)

object	radius	vertical height	curved surface area	total surface area
(a) cylinder	4 cm	*	72 cm²	
(b) sphere	*		192 cm²	
(c) cone	4 cm	*	60 cm²	
(d) sphere	*			0·48 m²
(e) cylinder	5 cm	*		330 cm²
(f) cone	6 cm	*		225 cm²
(g) cylinder	2 m	*		108 m²

6. A solid wooden cylinder of height 8 cm and radius 3 cm is cut in two along a vertical axis of symmetry. Calculate the total surface area of the two pieces.

7. A tin of paint covers a surface area of 60 m² and costs £4·50. Find the cost of painting the outside surface of a hemispherical dome of radius 50 m. (Just the curved part.)

8. A solid cylinder of height 10 cm and radius 4 cm is to be plated with material costing £11 per cm². Find the cost of the plating.

9. Find the volume of a sphere of surface area 100 cm².

10. Find the surface area of a sphere of volume 28 cm³.

11. Calculate the total surface area of the combined cone/cylinder/hemisphere.

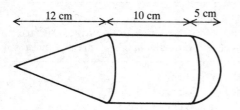

12. A man is determined to spray the entire surface of the Earth (including the oceans) with a revolutionary new weed killer. If it takes him 10 seconds to spray 1 m², how long will it take to spray the whole world?
(radius of the Earth = 6370 km; ignore leap years.)

13. An inverted cone of vertical height 12 cm and base radius 9 cm contains water to a depth of 4 cm. Find the area of the interior surface of the cone not in contact with the water.

14. A circular paper of radius 20 cm is cut in half and each half is made into a hollow cone by joining the straight edges. Find the slant height and base radius of each cone.

15. A golf ball has a diameter of 4·1 cm and the surface has 150 dimples of radius 2 mm. Calculate the total surface area which is exposed to the surroundings. (Assume the 'dimples' are hemispherical.)

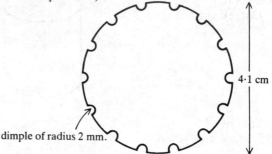

dimple of radius 2 mm.

16. A cone of radius 3 cm and slant height 6 cm is cut into four identical pieces. Calculate the total surface area of the four pieces.

REVISION EXERCISE 3A

1. Find the area of the following shapes:

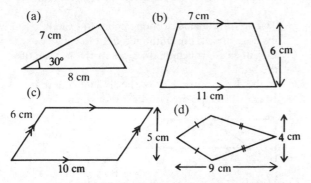

(a)

7 cm
30°
8 cm

(b)

7 cm
6 cm
11 cm

(c)

6 cm
5 cm
10 cm

(d)

4 cm
9 cm

2. (a) A circle has radius 9 m. Find its circumference and area.
  (b) A circle has circumference 34 cm. Find its diameter.
  (c) A circle has area 50 cm². Find its radius.

3. A target consists of concentric circles of radii 3 cm and 9 cm.

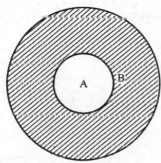

  (a) Find the area of A, in terms of π
  (b) Find the ratio $\dfrac{\text{area of B}}{\text{area of A}}$.

4. In Figure 1 a circle of radius 4 cm is inscribed in a square. In Figure 2 a square is inscribed in a circle of radius 4 cm.
Calculate the shaded area in each diagram.

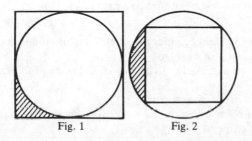

Fig. 1   Fig. 2

5. Given that OA = 10 cm and AÔB = 70° (where O is the centre of the circle), calculate
  (a) the arc length AB
  (b) the area of minor sector AOB.

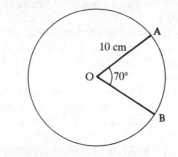

A
10 cm
O 70°
B

6. The points X and Y lie on the circumference of a circle, of centre O and radius 8 cm, where XÔY = 80°. Calculate
  (a) the length of the minor arc XY
  (b) the length of the chord XY
  (c) the area of sector XOY
  (d) the area of triangle XOY
  (e) the area of the minor segment of the circle cut off by XY.

7. Given that ON = 10 cm and minor arc MN = 18 cm, calculate the angle MÔN (shown as $x°$).

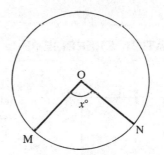

O
$x°$
M   N

8. A cylinder of radius 8 cm has a volume of 2 litres. Calculate the height of the cylinder.

9. Calculate
  (a) the volume of a sphere of radius 6 cm
  (b) the radius of a sphere whose volume is 800 cm³.

10. A sphere of radius 5 cm is melted down and made into a solid cube. Find the length of a side of the cube.

11. The curved surface area of a solid circular cylinder of height 8 cm is 100 cm². Calculate the volume of the cylinder.

12. A cone has base radius 5 cm and vertical height 10 cm, correct to the nearest cm. Calculate the maximum and minimum possible volumes of the cone, consistent with this data.

13. Calculate the radius of a hemispherical solid whose total surface area is 48π cm^2.

14. Calculate:
    (a) the area of an equilateral triangle of side 6 cm.
    (b) the area of a regular hexagon of side 6 cm.
    (c) the volume of a regular hexagonal prism of length 10 cm, where the side of the hexagon is 12 cm.

15. Ten spheres of radius 1 cm are immersed in liquid contained in a vertical cylinder of radius 6 cm. Calculate the rise in the level of the liquid in the cylinder.

16. A cube of side 10 cm is melted down and made into ten identical spheres. Calculate the surface area of one of the spheres.

17. The square has sides of length 3 cm and the arcs have centres at the corners. Find the shaded area.

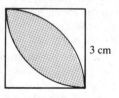

3 cm

18. A copper pipe has external diameter 18 mm and thickness 2 mm. The density of copper is 9 g/cm^3 and the price of copper is £150 per tonne. What is the cost of the copper in a length of 5 m of this pipe?

19. Twenty-seven small wooden cubes fit exactly inside a cubical box without a lid. How many of the cubes are touching the sides or the bottom of the box?

20. In the diagram the area of the smaller square is 10 cm^2. Find the area of the larger square.

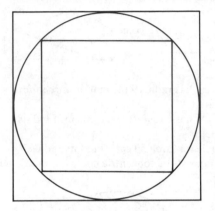

EXAMINATION EXERCISE 3B

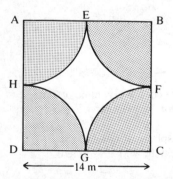

1. The diagram shows a square garden of side 14 m. E, F, G and H are the mid-points of AB, BC, CD and DA respectively and EF, FG, GH and HE are arcs of a circle of radius 7 m.
   The middle section of the garden is to be grassed as a lawn and the four side sections (shown shaded) are to be flower beds.
   Calculate
   (a) the total area of the flower beds [take π as $\frac{22}{7}$ or 3·14],
   (b) the area of the lawn,
   (c) the amount of grass seed needed given that 80 grams of seed are needed per square metre. (W)

2. A metal ingot is in the form of a solid cylinder of length 7 cm and radius 3 cm.
   (a) Calculate the volume, in cm^3, of the ingot.
   The ingot is to be melted down and used to make cylindrical coins of thickness 3 mm and radius 12 mm.
   (b) Calculate the volume, in mm^3, of each coin.
   (c) Calculate the number of coins which can be made from the ingot, assuming that there is no wastage of metal. [M]

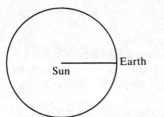

3. The orbit of the Earth around the Sun is approximately a circle of radius 1.5×10^8 km.
  (a) Light travels at 3×10^5 km/sec.
    How long does it take for a flash of light from the Sun to reach the Earth?
  (b) Calculate the circumference of the orbit (assumed circular).
  (c) Calculate the speed in km/sec with which the earth travels around the Sun given that the time taken is $365\frac{1}{4}$ days. [W]

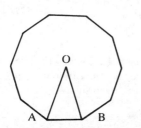

4. The diagram represents a regular 10-sided figure, centre O, where AO = 6 cm. Calculate, giving all your answers to 3 sig figs, and giving the units:
  (a) the angle AOB
  (b) the length of the perpendicular from O to AB
  (c) the length of AB
  (d) the area of the triangle AOB
  (e) the area of the whole figure. [MEI]

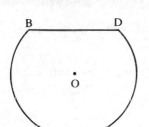

5. The shape of the top of a table in a cafe is shown by the part of the circle, centre O, in the diagram. The angle BOD = 90° and BO = OD = 50 cm.

The top of the table is made of formica and a thin metal strip is fixed to its perimeter.
  (a) Calculate
    (i) the length of the metal strip,
    (ii) the area of the formica surface.
  (b) If the table top is to be cut from a square of formica of area 1 m², what percentage of formica is wasted? [S]

6. [In this question take π to be 3.142 and give each answer correct to three significant figures.]
A solid silver sphere has a radius of 0.7 cm.
  (a) Calculate
    (i) the surface area of the sphere,
    (ii) the volume of the sphere.
  (b) A silversmith is asked to make a solid pyramid with a vertical height of 25 cm and a square base. To make the pyramid, the silversmith has to melt down 1000 of the silver spheres.
    Assuming that none of the silver is wasted, calculate the total surface area of the pyramid. [M]

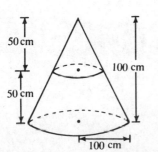

7. [In this question you should treat all measurements as exact and give volume answers in cm³ as a multiple or a fraction of π.]
A heap of sand is in the shape of a cone. The radius of the base is 100 cm. The top portion is removed, leaving an upper circular face parallel to the base and 50 cm above it.
  (a) State the radius of the upper circular face.
  (b) Find the volume of the original heap of sand.
  (c) Find the ratio of the volume of the portion removed to the volume of the original heap, in the form 1:n. [N]

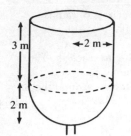

8. The shape of an oil-tank is that of a hemisphere of radius 2 m surmounted by a circular cylinder, also of radius 2 m whose height is 3 m.
   (a) Calculate the capacity of the tank in cubic metres.
   A tap at the base of the tank controls a cylindrical outflow pipe of radius 4 cm. When the tap is open the oil flows down the pipe at an average speed of 1·5 m/s.
   (b) Calculate the average volume of oil discharged per second when the tap is open.
   (c) In these circumstances, how many hours will it take to empty a full tank of oil?
   Give all answers correct to 3 significant figures.   [MEI]

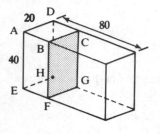

9. A rectangular block of ice cream measures 80 cm by 40 cm by 20 cm. It is to be cut into smaller blocks of the same shape as (i.e., similar to) the original block. One possible cut is shown shaded in the diagram, and the resulting block is ABCDEFGH.
   (a) State the length of AB.
   (b) State the number of blocks of this new size which can be produced from the original block.
   These blocks are further divided into similar blocks each with a longest edge of 5 cm.
   (c) State the other two dimensions of such a block.
   (d) Calculate how many blocks of this size can be cut from the block 80 cm by 40 cm by 20 cm.   [N]

10. A mathematical D-I-Y enthusiast plans to re-cover a lampshade of circular cross-section. The dimensions of the lampshade are shown in Fig. 4. The fabric is to be cut from a rectangular piece of material, EFGH, to the pattern shown in Fig. 5. The arcs AB and CD are from separate circles with the same centre O and each arc subtends the same angle $\theta°$ at O.
   (a) Using the measurements given in Fig. 4 write down, as a multiple of π, the lengths of
   (i) arc AB,   (ii) arc CD.
   (b) Using the measurements given in Fig. 5 write down, in terms of r, θ and π, the lengths of
   (i) arc AB,   (ii) arc CD.
   (c) Use the results of (a) and (b) to find
   (i) r,   (ii) θ.   [L]

Fig. 4

Fig. 5

4 Geometry

Pythagoras (569–500 B.C.) was one of the first of the great mathematical names in Greek antiquity. He settled in southern Italy and formed a mysterious brotherhood with his students who were bound by an oath not to reveal the secrets of numbers and who exercised great influence. They laid the foundations of arithmetic through geometry but failed to resolve the concept of irrational numbers. The work of these and others was brought together by Euclid at Alexandria in a book called 'The Elements' which was still studied in English schools as recently as 1900.

4.1 FUNDAMENTAL RESULTS

The student should already be familiar with the following results. They are used later in this section and are quoted here for reference.

The angles on a straight line add up to 180°:

$$\hat{x} + \hat{y} + \hat{z} = 180°$$

The angles at a point add up to 360°:

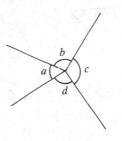

$$\hat{a} + \hat{b} + \hat{c} + \hat{d} = 360°$$

(i) The angle sum of a triangle is 180°.
(ii) The angle sum of a quadrilateral is 360°.

Exercise 1

Find the angles marked with letters. (AB is always a straight line.)

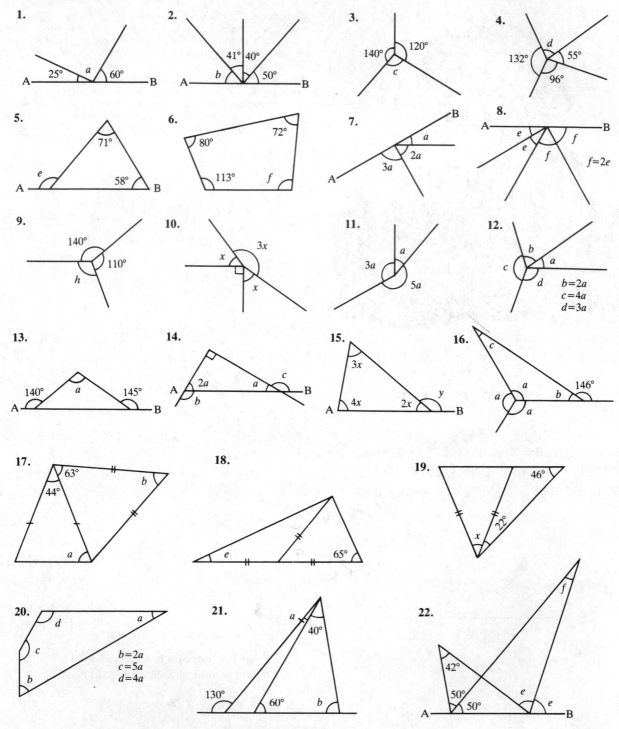

1. 25°   a   60°   A ——— B

2. 41° 40°   b   50°   A ——— B

3. 140° 120°   c

4. 132°   d   55°   96°

5. 71°   e   58°   A ——— B

6. 80°   72°   113°   f

7. B   a   2a   3a   A

8. A ——— B   e   e   f   f   f=2e

9. 140°   110°   h

10. 3x   x   x

11. a   3a   5a

12. b   a   c   d   b=2a   c=4a   d=3a

13. 140°   a   145°   A ——— B

14. 2a   a   c   b   A ——— B

15. 3x   4x   2x   y   A ——— B

16. c   a   a   b   146°

17. 63°   b   44°   a

18. e   65°

19. 46°   22°   x

20. d   a   c   b   b=2a   c=5a   d=4a

21. a   40°   130°   60°   b

22. f   42°   50°   50°   e   e   A ——— B

23. Calculate the largest angle of a triangle in which one angle is eight times each of the others.
24. In $\triangle ABC$, $\hat{A}$ is a right angle and D is a point on AC such that BD bisects $\hat{B}$. If $B\hat{D}C = 100°$, calculate $\hat{C}$.
25. WXYZ is a quadrilateral in which $\hat{W} = 108°$, $\hat{X} = 88°$, $\hat{Y} = 57°$ and $W\hat{X}Z = 31°$. Calculate $W\hat{Z}X$ and $X\hat{Z}Y$.
26. In quadrilateral ABCD, AB produced is perpendicular to DC produced. If $\hat{A} = 44°$ and $\hat{C} = 148°$, calculate $\hat{D}$ and $\hat{B}$.
27. Triangles ABD, CBD and ADC are all isosceles. Find the angle x.

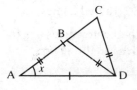

Polygons

(i) The exterior angles of a polygon add up to $360°$ ($\hat{a} + \hat{b} + \hat{c} + \hat{d} + \hat{e} = 360°$)

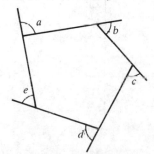

(ii) The sum of the interior angles of a polygon is $(2n - 4) \times 90°$ where n is the number of sides of the polygon.
This result is investigated in question **3** in the next exercise.

Example 1

Find the angles marked

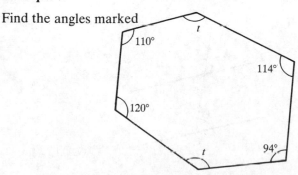

The sum of the interior angles $= (2n - 4) \times 90°$

where n is the number of sides of the polygon.

In this case $n = 6$.

$$\therefore \quad 110 + 120 + 94 + 114 + 2t = (2 \times 6 - 4) \times 90$$
$$438 + 2t = 720$$
$$2t = 282$$
$$t = 141°$$

Exercise 2

1. Find angles a and b for the regular pentagon below.

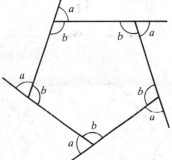

2. Find x and y.

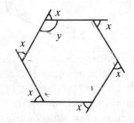

3. Consider the pentagon below which has been divided into three triangles.

$\hat{A} = a + f + g$, $\hat{B} = b$
$\hat{C} = c + d$, $\hat{D} = e + i$
$\hat{E} = h$

Now $a + b + c = d + e + f = g + h + i = 180°$.
$\therefore$ $\hat{A} + \hat{B} + \hat{C} + \hat{D} + \hat{E} = a + b + c + d + e + f + g + h + i$
$$= 3 \times 180°$$
$$= 6 \times 90°$$

Draw further polygons and make a table of results.

Number of sides n	5	6	7	8...
Sum of interior angles	$6 \times 90°$			

What is the sum of the interior angles for a polygon with n sides?

4. Find *a*.

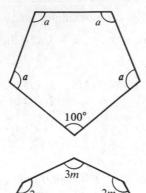

5. Find *m*.

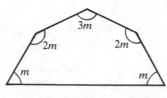

6. Find *a*.

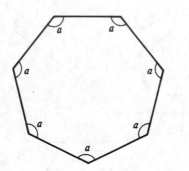

7. Calculate the number of sides of a regular polygon whose interior angles are each 156°.
8. Calculate the number of sides of a regular polygon whose interior angles are each 150°.
9. Calculate the number of sides of a regular polygon whose exterior angles are each 40°.
10. In a regular polygon each interior angle is 140° greater than each exterior angle.
11. In a regular polygon each interior angle is 120° greater than each exterior angle. Calculate the number of sides of the polygon.
12. Two sides of a regular pentagon are produced to form angle *x*. What is *x*?

Parallel lines

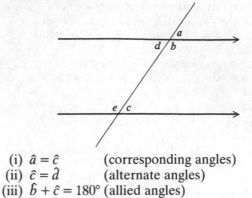

(i) $\hat{a} = \hat{c}$ (corresponding angles)
(ii) $\hat{c} = \hat{d}$ (alternate angles)
(iii) $\hat{b} + \hat{c} = 180°$ (allied angles)

Remember: 'The acute angles are the same and the obtuse angles are the same.'

Exercise 3

In questions **1** to **9** find the angles marked.

1.

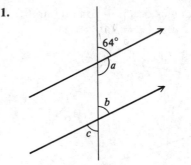

2.

3.

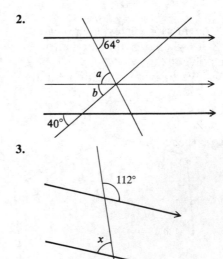

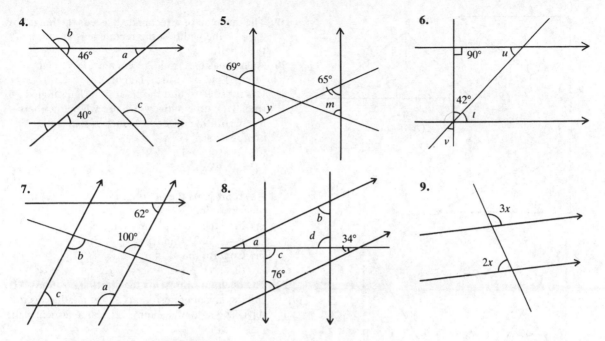

4.

b

46°   a

40°   c

5.

69°

65°

y   m

6.

90°   u

42°   t

v

7.

62°

100°

b

b

c   a

8.

b

a   d   34°

c

76°

9.

3x

2x

4.2   PYTHAGORAS' THEOREM

In a right-angled triangle
the square on the
hypotenuse is equal to
the sum of the squares
on the other two sides.

$$a^2 + b^2 = c^2$$

a   c

b

Example 1

Find the side marked d.

4 cm

d

7 cm

$$d^2 + 4^2 = 7^2$$
$$d^2 = 49 - 16$$
$$d = \sqrt{33} = 5 \cdot 74 \text{ cm (3 s.f.)}$$

The *converse* is also true:
'If the square on one side of a triangle is equal to
the sum of the squares on the other two sides,
then the triangle is right-angled.'

Exercise 4

In questions **1** to **10**, find x. All the lengths are in cm.

1.

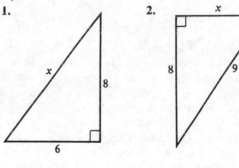

x

8

6

2.

x

8   9

3.

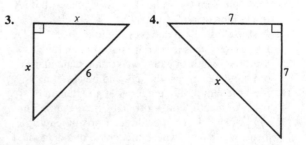

x

x

6

4.

7

x

7

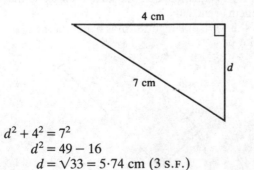

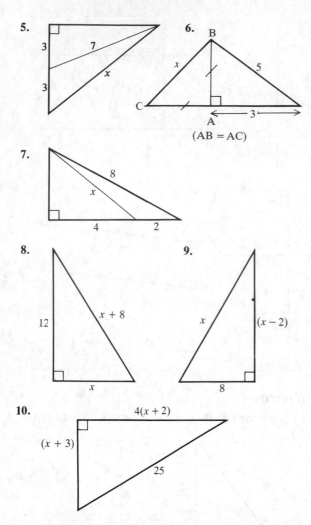

5.

6. (AB = AC)

7.

8.

9.

10.

11. Find the length of a diagonal of a rectangle of length 9 cm and width 4 cm.

12. A square has diagonals of length 10 cm. Find the sides of the square.

13. A 4 m ladder rests against a vertical wall with its foot 2 m from the wall. How far up the wall does the ladder reach?

14. A ship sails 20 km due North and then 35 km due East. How far is it from its starting point?

15. Find the length of a diagonal of a rectangular box of length 12 cm, width 5 cm and height 4 cm.

16. Find the length of a diagonal of a rectangular room of length 5 m, width 3 m and height 2·5 m.

17. Find the height of a rectangular box of length 8 cm, width 6 cm where the length of a diagonal is 11 cm.

18. An aircraft flies equal distances South-East and then South-West to finish 120 km due South of its starting-point. How long is each part of its journey?

19. The diagonal of a rectangle exceeds the length by 2 cm. If the width of the rectangle is 10 cm, find the length.

20. A cone has base radius 5 cm and *slant* height 11 cm. Find its vertical height.

21. It is possible to find the sides of a right-angled triangle, with lengths which are whole numbers, by substituting different values of x into the expressions:
    (a) $2x^2 + 2x + 1$
    (b) $2x^2 + 2x$
    (c) $2x + 1$
    ((a) represents the hypotenuse, (b) and (c) the other two sides.)
    (i) Find the sides of the triangles when $x = 1, 2, 3, 4$ and 5.
    (ii) Confirm that
    $$(2x + 1)^2 + (2x^2 + 2x)^2 = (2x^2 + 2x + 1)^2$$

22. The diagram represents the starting position (AB) and the finishing position (CD) of a ladder as it slips. The ladder is leaning against a vertical wall.

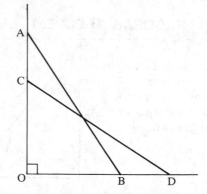

Given: AC $= x$, OC $= 4$AC, BD $= 2$AC and OB $= 5$ m.
Form an equation in x, find x and hence find the length of the ladder.

23. A thin wire of length 18 cm is bent into the shape shown.

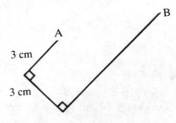

Calculate the length from A to B.

24. An aircraft is vertically above a point which is 10 km West and 15 km North of a control tower. If the aircraft is 4000 m above the ground, how far is it from the control tower?

4.3 SYMMETRY

(a) Line symmetry

The letter A has one line of symmetry, shown dotted.

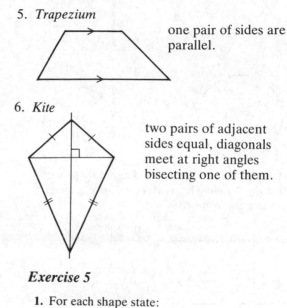

(b) Rotational symmetry

The shape may be turned about O into three identical positions. It has rotational symmetry of order 3.

(c) Quadrilaterals

1. *Square*

all sides are equal, all angles 90°, opposite sides parallel; diagonals bisect at right angles.

2. *Rectangle*

opposite sides parallel and equal, all angles 90°, diagonals bisect each other.

3. *Parallelogram*

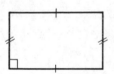

opposite sides parallel and equal, opposite angles equal, diagonals bisect each other (but not equal)

4. *Rhombus*

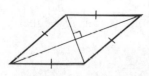

a parallelogram with all sides equal, diagonals bisect each other at right angles and bisect angles.

5. *Trapezium*

one pair of sides are parallel.

6. *Kite*

two pairs of adjacent sides equal, diagonals meet at right angles bisecting one of them.

Exercise 5

1. For each shape state:
   (a) the number of lines of symmetry
   (b) the order of rotational symmetry

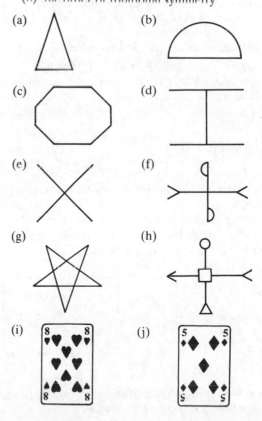

(k)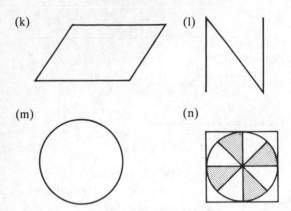

(l)

(m)

(n)

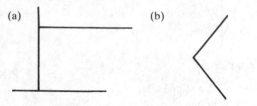

2. Add one line to each of the diagrams below so that the resulting figure has rotational symmetry but not line symmetry.

(a)   (b)

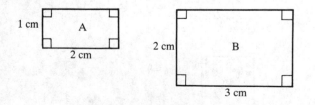

3. Draw a hexagon with just two lines of symmetry.
4. For each of the following shapes, find:
   (a) the number of lines of symmetry
   (b) the order of rotational symmetry.

   square; rectangle; parallelogram; rhombus; trapezium; kite; equilateral triangle; regular hexagon.

In questions **5** to **15**, begin by drawing a diagram.

5. In a rectangle KLMN, $L\hat{N}M = 34°$. Calculate:
   (a) $K\hat{L}N$ (b) $K\hat{M}L$
6. In a trapezium ABCD; $A\hat{B}D = 35°$, $B\hat{A}D = 110°$ and AB is parallel to DC. Calculate:
   (a) $A\hat{D}B$ (b) $B\hat{D}C$.
7. In a parallelogram WXYZ, $W\hat{X}Y = 72°$, $Z\hat{W}Y = 80°$. Calculate:
   (a) $W\hat{Z}Y$ (b) $X\hat{W}Z$ (c) $W\hat{Y}X$
8. In a kite ABCD, AB = AD; BC = CD; $C\hat{A}D = 40°$ and $C\hat{B}D = 60°$. Calculate:
   (a) $B\hat{A}C$ (b) $B\hat{C}A$ (c) $A\hat{D}C$
9. In a rhombus ABCD, $A\hat{B}C = 64°$. Calculate:
   (a) $B\hat{C}D$ (b) $A\hat{D}B$ (c) $B\hat{A}C$
10. In a rectangle WXYZ, M is the mid-point of WX and $Z\hat{M}Y = 70°$. Calculate:
    (a) $M\hat{Z}Y$ (b) $Y\hat{M}X$
11. In a trapezium ABCD, AB is parallel to DC, AB = AD, BD = DC and $B\hat{A}D = 128°$. Find:
    (a) $A\hat{B}D$ (b) $B\hat{D}C$ (c) $B\hat{C}D$
12. In a parallelogram KLMN, KL = KM and $K\hat{M}L = 64°$. Find:
    (a) $M\hat{K}L$ (b) $K\hat{N}M$ (c) $L\hat{M}N$
13. In a kite PQRS with PQ = PS and RQ = RS, $Q\hat{R}S = 40°$ and $Q\hat{P}S = 100°$. Find:
    (a) $Q\hat{S}R$ (b) $P\hat{S}Q$ (c) $P\hat{Q}R$
14. In a rhombus PQRS, $R\hat{P}Q = 54°$. Find:
    (a) $P\hat{R}Q$ (b) $P\hat{S}R$ (c) $R\hat{Q}S$
15. In a kite PQRS, $R\hat{P}S = 2\,P\hat{R}S$, PQ = QS = PS and QR = RS. Find:
    (a) $Q\hat{P}S$ (b) $P\hat{R}S$ (c) $Q\hat{S}R$ (d) $P\hat{Q}R$

4.4 SIMILARITY

Two triangles are similar if they have the same angles.

For other shapes, not only must corresponding angles be equal, but also corresponding sides must be in the same proportion.

The two rectangles A and B are *not* similar even though they have the same angles.

Example 1

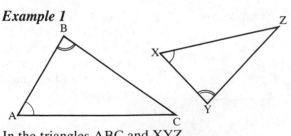

In the triangles ABC and XYZ

   $\hat{A} = \hat{X}$ and $\hat{B} = \hat{Y}$

so the triangles are similar. ($\hat{C}$ must be equal to $\hat{Z}$.)

We have $\dfrac{BC}{YZ} = \dfrac{AC}{XZ} = \dfrac{AB}{XY}$

Exercise 6

Find the sides marked with letters in questions **1** to **11**; all lengths are given in centimetres

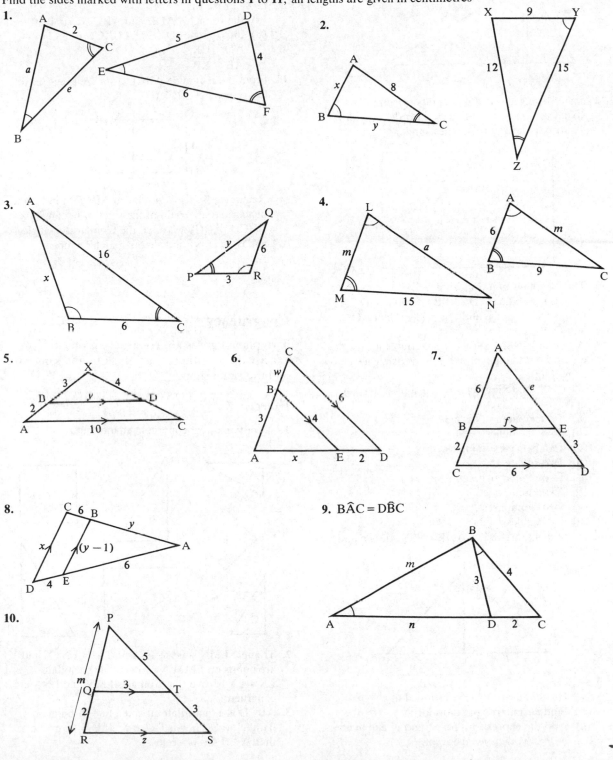

11.

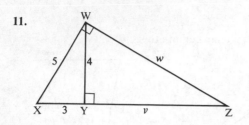

12. The drawing shows a rectangular picture 16 cm × 8 cm surrounded by a border of width 4 cm. Are the two rectangles similar?

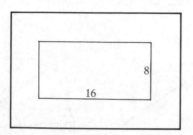

13. The diagonals of a trapezium ABCD intersect at O. AB is parallel to DC, AB = 3 cm and DC = 6 cm. If CO = 4 cm and OB = 3 cm, find AO and DO.

14. A tree of height 4 m casts a shadow of length 6·5 m. Find the height of a house casting a shadow 26 m long.

15. Which of the following *must* be similar to each other.
(a) Two equilateral triangles.
(b) Two rectangles.
(c) Two isosceles triangles.
(d) Two squares.
(e) Two regular pentagons.
(f) Two kites.
(g) Two rhombuses.
(h) Two circles.

16. In the diagram $A\hat{B}C = A\hat{D}B = 90°$, AD = p and DC = q.

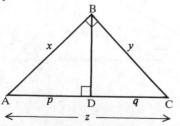

(a) Use similar triangles to show that $x^2 = pz$
(b) Find a similar expression for y^2
(c) Add the expressions for x^2 and y^2 and hence prove Pythagoras' theorem.

17. In a triangle ABC, a line is drawn parallel to BC to meet AB at D and AC at E. DC and BE meet at X. Prove that
(a) the triangles ADE and ABC are similar
(b) the triangles DXE and BXC are similar
(c) $\dfrac{AD}{AB} = \dfrac{EX}{XB}$

18. From the rectangle ABCD a square is cut off to leave rectangle BCEF.

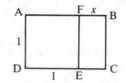

Rectangle BCEF is similar to ABCD. Find x and hence state the ratio of the sides of rectangle ABCD. ABCD is called the Golden Rectangle and is an important shape in architecture.

Congruency

Two plane figures are congruent if one fits exactly on the other. They must be the same size and the same shape.

Exercise 7

1. Identify pairs of congruent shapes below.

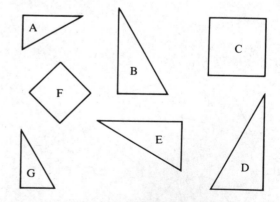

2. Triangle LMN is isosceles with LM = LN; X and Y are points on LM, LN respectively such that LX = LY. Prove that triangles LMY and LNX are congruent.

3. ABCD is a quadrilateral and a line through A parallel to BC meets DC at X. If $\hat{D} = \hat{C}$, prove that △ADX is isosceles.

4. In the diagram, N lies on a side of the square ABCD, AM and LC are perpendicular to DN. Prove that

(a) $A\hat{D}N = L\hat{C}D$ (b) $AM = LD$

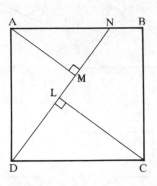

5. Points L and M on the side YZ of a triangle XYZ are drawn so that L is between Y and M. Given that $XY = XZ$ and $Y\hat{X}L = M\hat{X}Z$, prove that $YL = MZ$.

6. Squares AMNB and AOPC are drawn on the sides of triangle ABC, so that they lie outside the triangle. Prove that $MC = OB$.

7. In the diagram, $L\hat{M}N = O\hat{N}M = 90°$. P is the midpoint of MN, $MN = 2ML$ and $MN = NO$. Prove that

(a) the triangles MNL and NOP are congruent
(b) $O\hat{P}N = L\hat{N}O$
(c) $L\hat{Q}O = 90°$.

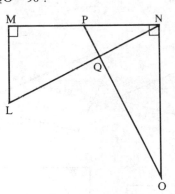

8. PQRS is a parallelogram in which the bisectors of the angles P and Q meet at X. Prove that the angle PXQ is a right angle.

Areas of similar shapes

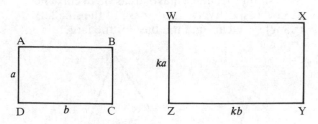

The two rectangles are similar, the ratio of corresponding sides being k.

area of $ABCD = ab$
area of $WXYZ = ka \cdot kb = k^2ab$.

$$\therefore \quad \frac{\text{area } WXYZ}{\text{area } ABCD} = \frac{k^2ab}{ab} = k^2$$

This illustrates an important general rule for all similar shapes:

If two figures are similar and the ratio of corresponding sides is k, then the ratio of their areas is k^2.

Note. k is sometimes called the *linear scale factor*.
This result also applies for the surface areas of similar three dimensional objects.

Example 2

XY is parallel to BC

$$\frac{AB}{AX} = \frac{3}{2}$$

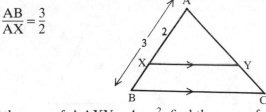

If the area of $\triangle AXY = 4 \text{ cm}^2$, find the area of $\triangle ABC$.

The triangles ABC and AXY are similar.

Ratio of corresponding sides $(k) = \frac{3}{2}$

$\therefore$ Ratio of areas $(k^2) = \frac{9}{4}$

$\therefore$ Area of $\triangle ABC = \frac{9}{4} \times (\text{area of } \triangle AXY)$
                $= \frac{9}{4} \times (4) = 9 \text{ cm}^2$.

Example 3

Two similar triangles have areas of 18 cm² and 32 cm² respectively. If the base of the smaller triangle is 6 cm, find the base of the larger triangle.

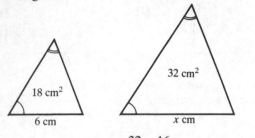

Ratio of areas $(k^2) = \dfrac{32}{18} = \dfrac{16}{9}$

∴   Ratio of corresponding sides $(k) = \sqrt{\left(\dfrac{16}{9}\right)}$

$$= \dfrac{4}{3}$$

∴   Base of larger triangle $= 6 \times \dfrac{4}{3} = 8$ cm.

Exercise 8

In this exercise a number written inside a figure represents the area of the shape in cm². Numbers on the outside give linear dimensions in cm.
In questions **1** to **6** find the unknown area A. In each case the shapes are similar.

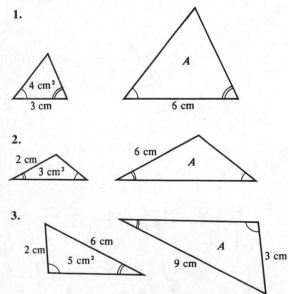

1.

2.

3.

4.

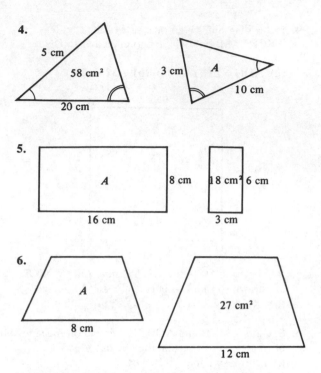

5.

6.

In questions **7** to **12**, find the lengths marked for each pair of similar shapes.

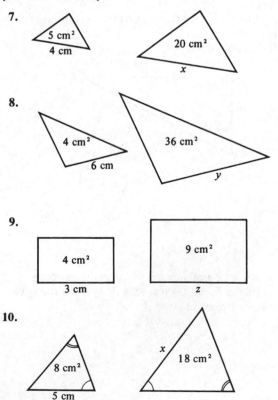

7.

8.

9.

10.

11.

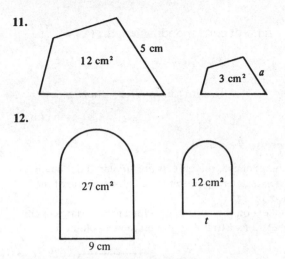

12.

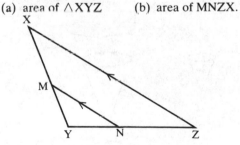

13. Given AD = 3 cm, AB = 5 cm and area of
△ADE = 6 cm².
Find:
(a) area of △ABC
(b) area of DECB.

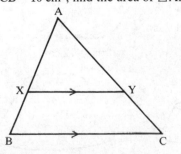

14. Given XY = 5 cm, MY = 2 cm and area of
△MYN = 4 cm². Find:
(a) area of △XYZ   (b) area of MNZX.

15. Given XY = 2 cm, BC = 3 cm and area of
XYCB = 10 cm², find the area of △AXY.

16. Given KP = 3 cm, area of △KOP = 2 cm² and
area of OPML = 16 cm², find the length of PM.

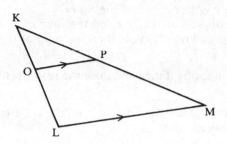

17. The triangles ABC and EBD are similar (AC and
DE are *not* parallel).
If AB = 8 cm, BE = 4 cm
and the area of
△DBE = 6 cm²,
find the area of △ABC.

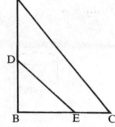

18. Given AZ = 3 cm, ZC = 2 cm, MC = 5 cm,
BM = 3 cm. Find:
(a) XY
(b) YZ
(c) the ratio of
areas AXY : AYZ
(d) the ratio of
areas AXY : ABM

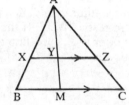

19. A floor is covered by 600 tiles which are 10 cm by
10 cm. How many 20 cm by 20 cm tiles are needed
to cover the same floor?

20. A wall is covered by 160 tiles which are 15 cm by
15 cm. How many 10 cm by 10 cm tiles are needed
to cover the same wall?

21. When potatoes are peeled do you lose more peel
or less when big potatoes are used as opposed to
small ones?

Volumes of similar objects

When solid objects are similar, one is an accurate enlargement of the other.
If two objects are similar and the ratio of corresponding sides is k, then the ratio of their volumes is k^3.

A line has one dimension, and the scale factor is used once.

An area has two dimensions, and the scale factor is used twice.

A volume has three dimensions, and the scale factor is used three times.

Example 4

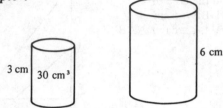

Two similar cylinders have heights of 3 cm and 6 cm respectively. If the volume of the smaller cylinder is 30 cm³, find the volume of the larger cylinder.

$$\text{ratio of heights } (k) = \frac{6}{3}$$
$$\text{(linear scale factor)}$$
$$= 2$$

$$\therefore \quad \text{ratio of volumes } (k^3) = 2^3$$
$$= 8$$
$$\text{and volume of larger cylinder} = 8 \times 30$$
$$= 240 \text{ cm}^3.$$

Example 5

Two similar spheres made of the same material have weights of 32 kg and 108 kg respectively. If the radius of the larger sphere is 9 cm, find the radius of the smaller sphere.

We may take the ratio of weights to be the same as the ratio of volumes.

$$\text{ratio of volumes } (k^3) = \frac{32}{108}$$
$$= \frac{8}{27}$$

$$\text{ratio of corresponding lengths } (k) = \sqrt[3]{\left(\frac{8}{27}\right)}$$
$$= \frac{2}{3}.$$

$$\therefore \quad \text{Radius of smaller sphere} = \frac{2}{3} \times 9$$
$$= 6 \text{ cm}.$$

Exercise 9

In this exercise, the objects are similar and a number written inside a figure represents the volume of the object in cm³.
Numbers on the outside give linear dimensions in cm.
In questions **1** to **8**, find the unknown volume V.

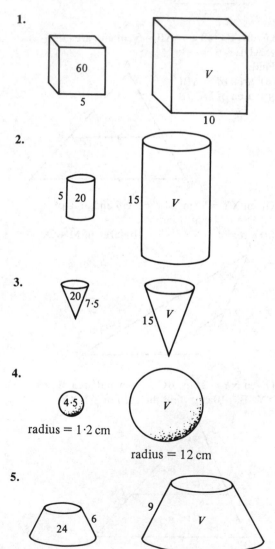

1.

2.

3.

4.
radius = 1·2 cm
radius = 12 cm

5.

6.

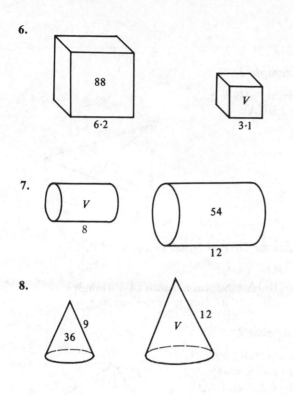

7.

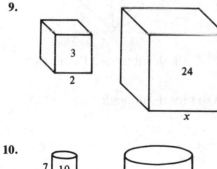

8.

In questions **9** to **14**, find the lengths marked by a letter.

9.

10.

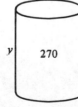

11.

12.

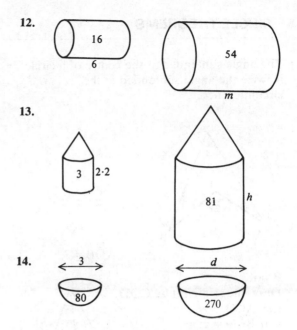

13.

14.

15. Two similar jugs have heights of 4 cm and 6 cm respectively. If the capacity of the smaller jug is 50 cm³, find the capacity of the larger jug.

16. Two similar cylindrical tins have base radii of 6 cm and 8 cm respectively. If the capacity of the larger tin is 252 cm³, find the capacity of the small tin.

17. Two solid metal spheres have masses of 5 kg and 135 kg respectively. If the radius of the smaller one is 4 cm, find the radius of the larger one.

18. Two similar cones have surface areas in the ratio 4:9. Find the ratio of:
(a) their lengths,   (b) their volumes.

19. The area of the bases of two similar glasses are in the ratio 4:25. Find the ratio of their volumes.

20. Two similar solids have volumes V_1 and V_2 and corresponding sides of length x_1 and x_2. State the ratio $V_1:V_2$ in terms of x_1 and x_2.

21. Two solid spheres have surface areas of 5 cm² and 45 cm² respectively and the mass of the smaller sphere is 2 kg. Find the mass of the larger sphere.

22. The masses of two similar objects are 24 kg and 81 kg respectively. If the surface area of the larger object is 540 cm², find the surface area of the smaller object.

23. A cylindrical can has a circumference of 40 cm and a capacity of 4·8 litres. Find the capacity of a similar cylinder of circumference 50 cm.

24. A container has a surface area of 5000 cm² and a capacity of 12·8 litres. Find the surface area of a similar container which has a capacity of 5·4 litres.

4.5   CIRCLE THEOREMS

(a) The angle subtended at the centre of a circle is twice the angle subtended at the circumference.

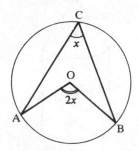

$$A\hat{O}B = 2 \times A\hat{C}B$$

Proof:
Draw the straight line COD.
Let $A\hat{C}O = y$
and $B\hat{C}O = z$.

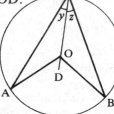

In triangle AOC,

$$AO = OC \quad \text{(radii)}$$
$$\therefore \quad O\hat{C}A = O\hat{A}C \quad \text{(isosceles triangle)}$$
$$\therefore \quad C\hat{O}A = 180 - 2y \quad \text{(angle sum of triangle)}$$
$$\therefore \quad A\hat{O}D = 2y \quad \text{(angles on a straight line)}$$

Similarly from triangle COB, we find

$$D\hat{O}B = 2z$$
Now $\quad A\hat{C}B = y + z$
and $\quad A\hat{O}B = 2y + 2z$
$\therefore \quad A\hat{O}B = 2 \times A\hat{C}B$ as required.

(b) Angles subtended by an arc in the same segment of a circle are equal.

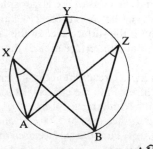

$$A\hat{X}B = A\hat{Y}B = A\hat{Z}B$$

Example 1

Given $A\hat{B}O = 50°$,
find $B\hat{C}A$.

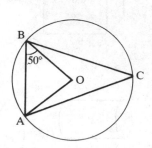

Triangle OBA is isosceles (OA = OB).

$$\therefore \quad O\hat{A}B = 50°$$
$$\therefore \quad B\hat{O}A = 80° \quad \text{(angle sum of a triangle)}$$
$$\therefore \quad B\hat{C}A = 40° \quad \text{(angle at the centre)}$$

Example 2

Given $B\hat{D}C = 62°$
and $D\hat{C}A = 44°$,
find $B\hat{A}C$ and $A\hat{B}D$.

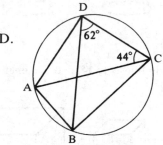

$$B\hat{D}C = B\hat{A}C \quad \text{(both subtended by arc BC)}$$
$$\therefore \quad B\hat{A}C = 62°$$

$$D\hat{C}A = A\hat{B}D \quad \text{(both subtended by arc DA)}$$
$$\therefore \quad A\hat{B}D = 44°.$$

Exercise 10

Find the angles marked with letters.   A line passes through the centre only when point O is shown.

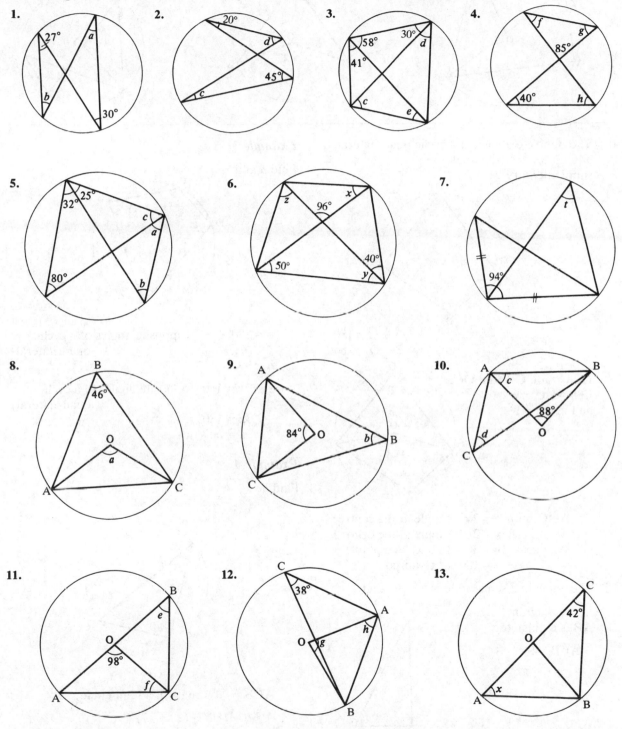

14.

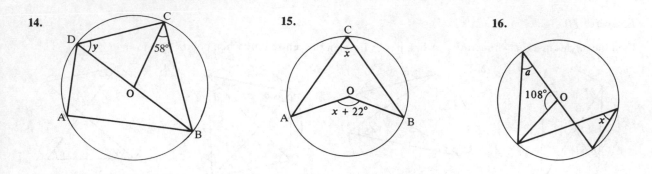

15.

16.

(c) The opposite angles in a cyclic quadrilateral add up to 180° (the angles are supplementary).

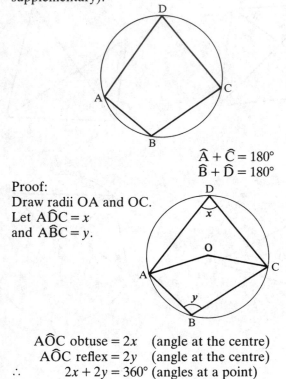

$$\hat{A} + \hat{C} = 180°$$
$$\hat{B} + \hat{D} = 180°$$

Proof:
Draw radii OA and OC.
Let $A\hat{D}C = x$
and $A\hat{B}C = y$.

$A\hat{O}C$ obtuse $= 2x$ (angle at the centre)
$A\hat{O}C$ reflex $= 2y$ (angle at the centre)
∴ $2x + 2y = 360°$ (angles at a point)
∴ $x + y = 180°$ as required.

(d) The angle in a semi-circle is a right angle.

In the diagram,
AB is a diameter.

$A\hat{C}B = 90°$.

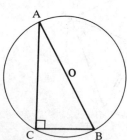

Example 3

Find a and x.

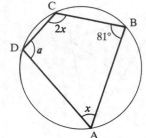

$a = 180° - 81°$ (opposite angles of a cyclic
                                    quadrilateral)
∴ $a = 99°$

$x + 2x = 180°$ (opposite angles of a cyclic
                                    quadrilateral)
$$3x = 180°$$
$$x = 60°$$

Example 4

Find b.

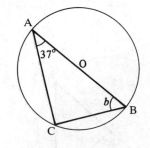

$A\hat{C}B = 90°$ (angle in a semi-circle)
∴ $b = 180° - (90 + 37)°$
          $= 53°$.

Exercise 11

Find the angles marked with a letter.

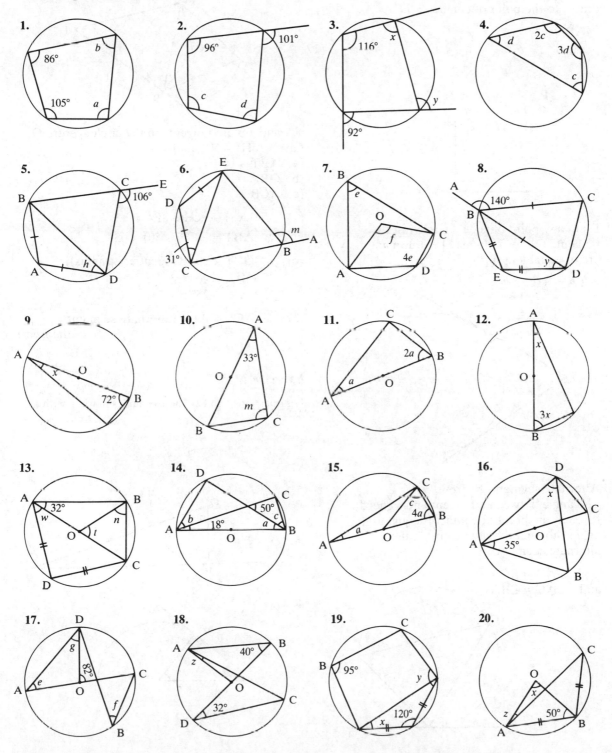

Tangents to circles

(a) The angle between a tangent and the radius drawn to the point of contact is 90°.

$$A\hat{B}O = 90°$$

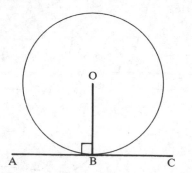

(b) From any point outside a circle just two tangents to the circle may be drawn and they are of equal length.

$$TA = TB$$

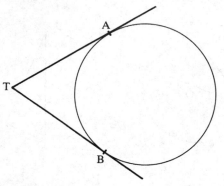

(c) Alternate segment theorem.
The angle between a tangent and a chord through the point of contact is equal to the angle subtended by the chord in the alternate segment.

$$T\hat{A}B = B\hat{C}A$$
$$\text{and} \quad S\hat{A}C = C\hat{B}A$$

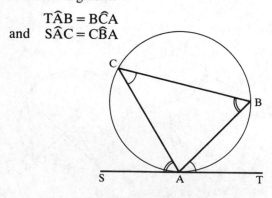

Example 1

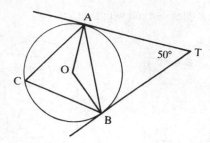

TA and TB are tangents to the circle, centre O.
Given $A\hat{T}B = 50°$, find
(a) $A\hat{B}T$
(b) $O\hat{B}A$
(c) $A\hat{C}B$

(a) $\triangle TBA$ is isosceles $(TA = TB)$
∴ $A\hat{B}T = \frac{1}{2}(180 - 50) = 65°$

(b) $O\hat{B}T = 90°$ (tangent and radius)
∴ $O\hat{B}A = 90 - 65$
$= 25°$

(c) $A\hat{C}B = A\hat{B}T$ (alternate segment
theorem)
$A\hat{C}B = 65°$.

Exercise 12

For questions **1** to **12**, find the angles marked with a letter.

1.

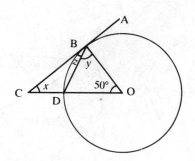

2.

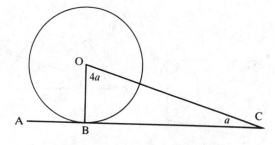

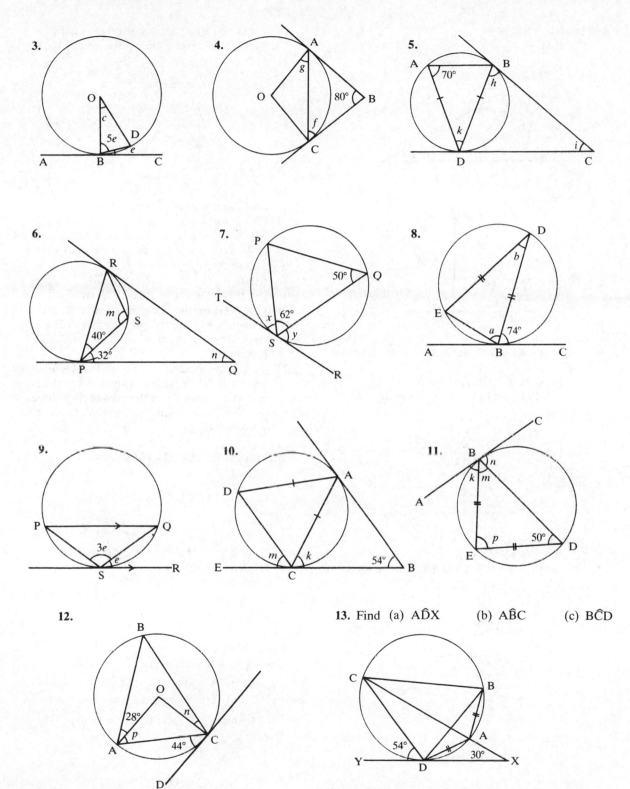

3.

4.

5.

6.

7.

8.

9.

10.

11.

12.

13. Find (a) AD̂X (b) AB̂C (c) BĈD

14. Find, in terms of p,
    (a) $B\hat{A}C$ (b) $X\hat{C}A$ (c) $A\hat{C}O$

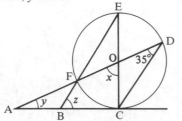

15. Find x, y and z.

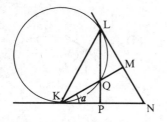

16. Given that $KL = LN$, LP bisects $K\hat{L}N$ and $M\hat{K}N = a$
    (a) prove that $\triangle KLQ$ is isosceles
    (b) find $L\hat{Q}M$ and $L\hat{M}Q$ in terms of a.

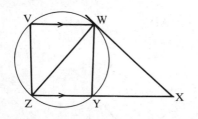

17. Show that:
    (a) $Y\hat{W}X = V\hat{W}Z$
    (b) the triangles VWZ and YWX are similar
    (c) $VW . WX = YW . WZ$.

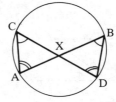

18. Given that BOC is a diameter and that $A\hat{D}C = 90°$, prove that AC bisects $B\hat{C}D$.

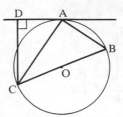

19. The angles of a triangle are 50°, 60° and 70°, and a circle touches the sides at A, B, C. Calculate the angles of triangle ABC.

20. The tangents at A and B on a circle intersect at T, and C is any point on the major arc AB.
    (a) If $A\hat{T}B = 52°$, calculate $A\hat{C}B$.
    (b) If $A\hat{C}B = x$, find $A\hat{T}B$ in terms of x.

21. Line ATB touches a circle at T and TC is a diameter. AC and BC cut the circle at D and E respectively. Prove that the quadrilateral ADEB is cyclic.

22. Two circles touch externally at T. A chord of the first circle XY is produced and touches the other at Z. The chord ZT of the second circle, when produced, cuts the first circle at W. Prove that $X\hat{T}W = Y\hat{T}Z$.

Intersecting chords theorem

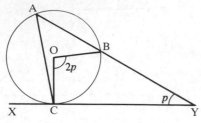

$$AX . BX = CX . DX$$

Proof:
In triangles AXC and BXD:

    $A\hat{C}X = D\hat{B}X$ (same segment)
    $C\hat{A}X = B\hat{D}X$ (same segment)

∴ the triangles AXC and BXD are similar.

$$\frac{AX}{DX} = \frac{CX}{BX}$$

or $AX . BX = CX . DX$

Exercise 13

1. Draw several circles with intersecting chords and confirm that $AX \cdot XB = DX \cdot XC$

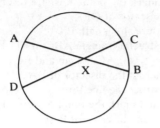

2. Find x.

3. Find x.

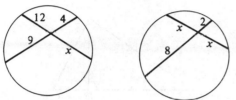

4. In the diagram AB is a diameter of the circle which bisects chord CD. $XB = 2$ cm and $CD = 12$ cm. Find the radius of the circle.

5. Show that the triangles ATC and BTC are similar. Hence show that $AT \cdot BT = CT^2$.

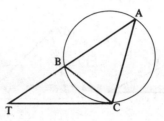

This is the *secant/tangent theorem*.

4.6 CONSTRUCTIONS AND LOCI

When the word 'construct' is used, the diagram in question should be drawn using *ruler* and *compasses* only.
Three basic constructions are shown below.

(a) Perpendicular bisector of a line joining two points.

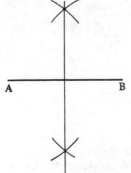

(b) Bisector of an angle.

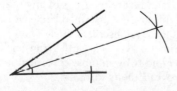

(c) 60° angle construction.

Exercise 14

1. Construct a triangle ABC in which $AB = 8$ cm, $AC = 6$ cm and $BC = 5$ cm. Measure the angle $A\hat{C}B$.

2. Construct a triangle PQR in which $PQ = 10$ cm, $PR = 7$ cm and $RQ = 6$ cm. Measure the angle $R\hat{P}Q$.

3. Construct an equilateral triangle of side 7 cm.

4. Draw a line AB of length 10 cm. Construct the perpendicular bisector of AB.

5. Draw two lines AB and AC of length 8 cm, where $B\hat{A}C$ is approximately 40°. Construct the line which bisects $B\hat{A}C$.

6. Draw a line AB of length 12 cm and draw a point X approximately 6 cm above the middle of the line. Construct the line through X which is perpendicular to AB.

7. Construct an equilateral triangle ABC of side 9 cm. Construct a line through A to meet BC at 90° at the point D. Measure the length AD.

8. Construct the triangles shown and measure the length x.

(a)

(b)

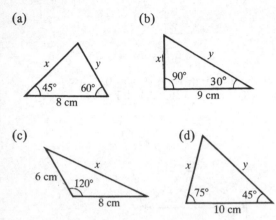

(c)

(d)

9. (a) Construct a triangle ABC in which AB = 8 cm AC = 6 cm and BC = 5 cm.
   (b) Construct the perpendicular bisectors of AB and AC and hence construct the circumcircle of triangle ABC.

10. (a) Construct a triangle XYZ in which XZ = 10 cm, XY = 8 cm and YZ = 7 cm.
    (b) Construct the circumcircle of triangle XYZ and measure the radius of the circle.

11. (a) Construct a triangle PQR in which PQ = 11 cm, PR = 9 cm and RQ = 7 cm.
    (b) Construct the bisectors of angles QPR and RQP and hence draw the inscribed circle of triangle PQR.

12. (a) Construct a triangle XYZ in which XY = 10 cm, XZ = 9 cm and YZ = 8 cm.
    (b) Construct the inscribed circle of triangle XYZ.
    (c) Construct the circumscribed circle of triangle XYZ.

13. Construct a parallelogram WXYZ in which WX = 10 cm, WZ = 6 cm and XŴZ = 60°. By construction, find the point A that lies on ZY and is equidistant from lines WZ and WX. Measure the length WA.

14. (a) Draw a line OX = 10 cm and construct an angle XOY = 60°.
    (b) Bisect the angle XOY and mark a point A on the bisector so that OA = 7 cm.
    (c) Construct a circle with centre A to touch OX and OY and measure the radius of the circle.

15. (a) Construct a triangle PQR with PQ = 8 cm, PR = 12 cm and PQ̂R = 90°.
    (b) Construct the bisector of QP̂R.
    (c) Construct the perpendicular bisector of PR and mark the point X where this line meets the bisector of QP̂R.
    (d) Measure the length PX.

16. (a) Construct a triangle ABC in which AB = 8 cm, AC = 6 cm and BC = 9 cm.
    (b) Construct the bisector of BÂC.
    (c) Construct the line through C perpendicular to CA and mark the point X where this line meets the bisector of BÂC.
    (d) Measure the lengths CX and AX.

17. AD is an altitude of triangle ABC.

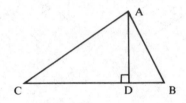

Draw any triangle and construct the three altitudes. They should all pass through a single point.

18. Draw four lines so that they form four triangles as in the diagram.

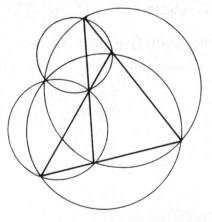

Construct the circumcircle of each of the four triangles to demonstrate Wallace's theorem that the four circles always pass through one point.

19. Leonard Euler found the formula $4rRs = abc$ where a, b and c are the sides of any triangle, r is the radius of the inscribed circle, R is the radius of the circumscribed circle and s is the semi-perimeter $[\frac{1}{2}(a + b + c)]$. Draw any triangle and find r and R by suitable construction. Confirm (or disprove) Euler's formula.

The locus of a point

The locus of a point is the path which it describes as it moves.

Example 1

Draw a line AB of length 8 cm.
Construct the locus of a point P which moves so that $\widehat{BAP} = 90°$.

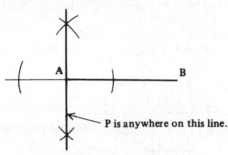

P is anywhere on this line.

Construct the perpendicular at A.
This line is the locus of P.

Exercise 15

1. Draw a line XY of length 10 cm. Construct the locus of a point which is equidistant from X and Y.

2. Draw two lines AB and AC of length 8 cm, where $\widehat{BAC}$ is approximately 70°. Construct the locus of a point which is equidistant from the lines AB and AC.

3. Draw a circle, centre O, of radius 5 cm and draw a radius OA. Construct the locus of a point P which moves so that $\widehat{OAP} = 90°$.

4. Draw a line AB of length 10 cm and construct the circle with diameter AB. Indicate the locus of a point P which moves so that $\widehat{APB} = 90°$.

5. Construct triangle ABP where AB = 8 cm, $\widehat{ABP} = 45°$ and $\widehat{BAP} = 60°$. Measure $\widehat{APB}$. Construct the locus of a point X which moves so that $\widehat{AXB} = 75°$. (Hint: construct the circumcircle of △ABP.)

6. (a) Construct a triangle PQR where PQ = 10 cm, $\widehat{QPR} = 30°$ and $\widehat{PQR} = 90°$.
   (b) Construct the locus of a point X which moves so that $\widehat{PXQ} = 60°$.

7. Construct a triangle PQR with PQ = 10 cm, $\widehat{QPR} = 30°$ and $\widehat{PQR} = 60°$. Construct the locus of a point X such that $\widehat{PXQ} = 90°$.

8. (a) Draw a line AB 8 cm in length and construct the segment of a circle to contain an angle of 45°.
   (b) Construct a line through A to cut the segment of the circle such that it makes an angle of 60° with AB.
   (c) Hence draw triangle ABC with AB = 8 cm, $\widehat{BAC} = 60°$ and $\widehat{ACB} = 45°$.
   (d) Measure and state the length BC.

4.7 NETS

If the cube below was made of cardboard, and you cut along some of the edges and laid it out flat, you would have the *net* of the cube.

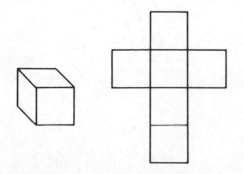

Exercise 16

1. Which of the nets below can be used to make a cube?

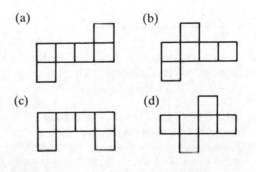

(a)   (b)

(c)   (d)

2. The diagram shows the net of a closed rectangular
box. All lengths are in cm.

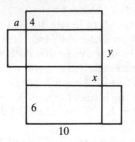

(a) Find the lengths a, x, y
(b) Calculate the volume of the box.

3. The diagram shows the net of a pyramid. The base
in shaded. The lengths are in cm.

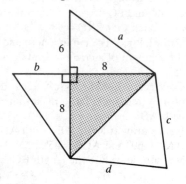

(a) Find the lengths a, b, c, d
(b) Find the volume of the pyramid.

4. The diagram shows the net of a prism.

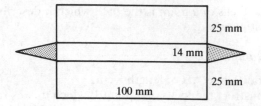

(a) Find the area of one of the triangular faces
(shown shaded).
(b) Find the volume of the prism.

5. This is the net of a square-based pyramid.

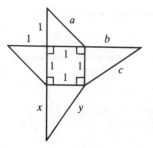

What are the lengths a, b, c, x, y?

6. Sketch nets for the following
(a) Closed rectangular box 7 cm × 9 cm × 5 cm.
(b) Closed cylinder: length 10 cm, radius 6 cm.
(c) Prism of length 12 cm, cross-section an
equilateral triangle of side 4 cm.

REVISION EXERCISE 4A

1. ABCD is a parallelogram and AE bisects angle A.
Prove that DE = BC.

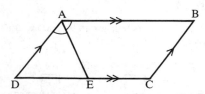

2. In a triangle PQR, $P\hat{Q}R = 50°$ and point X lies on
PQ such QX = XR. Calculate $Q\hat{X}R$.

3. (a) ABCDEF is a regular hexagon. Calculate
$F\hat{D}E$.
(b) ABCDEFGH is a regular eight-sided
polygon. Calculate $A\hat{G}H$.
(c) Each interior angle of a regular polygon
measures 150°. How many sides has the polygon?

4. In the quadrilateral PQRS, PQ = QS = QR, PS is
parallel to QR and $Q\hat{R}S = 70°$. Calculate
(a) $R\hat{Q}S$
(b) $P\hat{Q}S$.

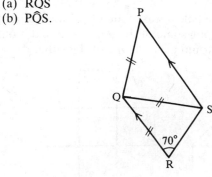

5. Find x.

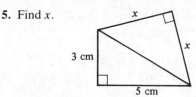

6. In the triangle ABC, AB = 7 cm, BC = 8 cm and
A$\hat{B}$C = 90°. Point P lies inside the triangle such
that BP = PC = 5 cm. Calculate
(i) the perpendicular distance from P to BC
(ii) the length AP.

7. In triangle PQR the bisector of P$\hat{Q}$R meets PR at
S and the point T lies on PQ such that ST is
parallel to RQ.
(a) Prove that QT = TS
(b) Prove that the triangles PTS and PQR are
similar
(c) Given that PT = 5 cm and TQ = 2 cm,
calculate the length of QR.

8. In the quadrilateral ABCD, AB is parallel to DC
and D$\hat{A}$B = D$\hat{B}$C.
(a) Prove that the triangles ABD and DBC are
similar.
(b) If AB = 4 cm and DC = 9 cm, calculate the
length of BD.

9. A rectangle 11 cm by 6 cm is similar to a rectangle
2 cm by x cm. Find the two possible values of x.

10. In the diagram, triangles ABC and EBD are
similar but DE is *not* parallel to AC. Given that
AD = 5 cm, DB = 3 cm and BE = 4 cm, calculate
the length of BC.

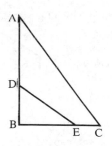

11. The radii of two spheres are in the ratio 2:5. The
volume of the smaller sphere is 16 cm³. Calculate
the volume of the larger sphere.

12. The surface areas of two similar jugs are 50 cm²
and 450 cm² respectively.
(a) If the height of the larger jug is 10 cm, find
the height of the smaller jug.
(b) If the volume of the smaller jug is 60 cm³, find
the volume of the larger jug.

13. A car is an enlargement of a model, the scale
factor being 10.
(a) If the windscreen of the model has an area of
100 cm², find the area of the windscreen on
the actual car (answer in m²).
(b) If the capacity of the boot of the car is 1 m³,
find the capacity of the boot on the model
(answer in cm³).

14. Find the angles marked with letters. (O is the
centre of the circle.)

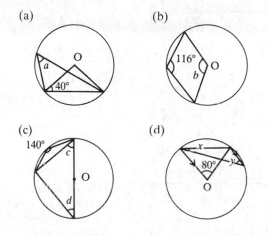

15. ABCD is a cyclic quadrilateral in which AB = BC
and A$\hat{B}$C = 70°. AD produced meets BC
produced at the point P, where A$\hat{P}$B = 30°.
Calculate
(a) A$\hat{D}$B   (b) A$\hat{B}$D.

16. Using ruler and compasses only,
(a) Construct the triangle ABC in which
AB = 7 cm, BC = 5 cm and AC = 6 cm.
(b) Construct the circle which passes through A,
B and C and measure the radius of this circle.

17. Construct:
(a) the triangle XYZ in which XY = 10 cm,
YZ = 11 cm and XZ = 9 cm.
(b) the locus of points, inside the triangle, which
are equidistant from the lines XZ and YZ.
(c) the locus of points which are equidistant from
Y and Z.
(d) the circle which touches YZ at its mid-point
and also touches XZ.

EXAMINATION EXERCISE 4B

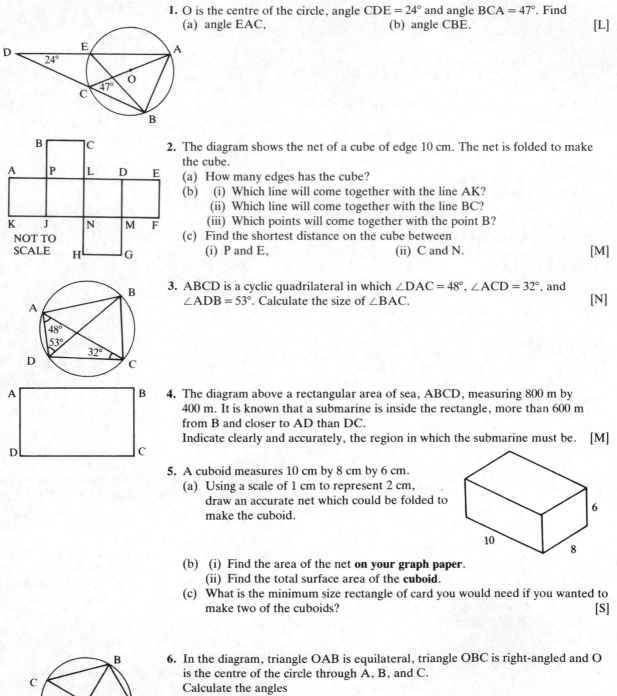

1. O is the centre of the circle, angle CDE = 24° and angle BCA = 47°. Find
(a) angle EAC, (b) angle CBE. [L]

2. The diagram shows the net of a cube of edge 10 cm. The net is folded to make the cube.
(a) How many edges has the cube?
(b) (i) Which line will come together with the line AK?
    (ii) Which line will come together with the line BC?
    (iii) Which points will come together with the point B?
(c) Find the shortest distance on the cube between
    (i) P and E, (ii) C and N. [M]

3. ABCD is a cyclic quadrilateral in which ∠DAC = 48°, ∠ACD = 32°, and ∠ADB = 53°. Calculate the size of ∠BAC. [N]

4. The diagram above a rectangular area of sea, ABCD, measuring 800 m by 400 m. It is known that a submarine is inside the rectangle, more than 600 m from B and closer to AD than DC.
Indicate clearly and accurately, the region in which the submarine must be. [M]

5. A cuboid measures 10 cm by 8 cm by 6 cm.
(a) Using a scale of 1 cm to represent 2 cm, draw an accurate net which could be folded to make the cuboid.
(b) (i) Find the area of the net **on your graph paper**.
    (ii) Find the total surface area of the **cuboid**.
(c) What is the minimum size rectangle of card you would need if you wanted to make two of the cuboids? [S]

6. In the diagram, triangle OAB is equilateral, triangle OBC is right-angled and O is the centre of the circle through A, B, and C.
Calculate the angles
(a) AOB
(b) ACB
(c) ACO [NI]

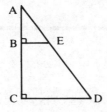

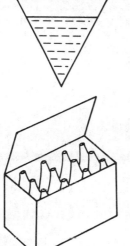

7. $\angle ABE = \angle ACD = 90°$.
   AC = 4·00 m, BC = 2·50 m, CD = 3·00 m.
   Calculate BE.   [L]

8. A sealed hollow cone with vertex downwards is partially filled with water. The
   volume of water is 200 cm³ and the depth of the water is 50 mm. Find the
   volume of the water which must be added to increase the depth to 70 mm.   [NI]

9. Two bottles of sauce of circular cross section are completely similar in every
   respect. One is 24 cm high and the other is 36 cm.
   (a) Calculate the diameter of the outside base of the smaller bottle given that
       the corresponding diameter of the larger is 6 cm.
   (b) The smaller bottle can hold 256 cm³ of sauce. Calculate the volume of sauce
       that the larger bottle can hold.
   (c) Twelve of the larger bottles are placed as four rows of three bottles in a
       rectangular box. If there is no room for the bottles to move in the box,
       calculate the area of cardboard needed to make a closed box.
   (d) What area is needed to make a smaller box for twelve small bottles packed
       in the same way?   [W]

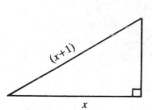

10. The following sets of whole numbers represent the lengths of the sides of right-
    angled triangles:

    $$5, 4, 3: \quad 13, 12, 5: \quad 25, 24, 7.$$

    (a) Taking $x + 1$ for the longest side and x for the next side write down an
        expression for the length of the third side and simplify it.
    (b) Use your result to construct the next two sets of numbers in the sequence
        started above.   [M]

11. A, B, C and D are four points on the circumference of a circle and AD is
    parallel to BC. Prove, showing each step clearly and giving reasons, that
    (a) $D\hat{A}C = A\hat{D}B$
    (b) $B\hat{A}D = A\hat{D}C$ [W]

12. A field is in the shape of a quadrilateral ABCD with AB = 80 m, BC = 70 m and
    CD = 110 m. Angle ABC = 80° and the angle BCD = 120°.
    Using a scale of 1 cm to represent 10 m make an accurate scale drawing of the
    field.
    (a) Use your scale drawing to find the length of the side DA.
    (b) A tree is at the point of intersection of the bisector of the angle CDA and
        the perpendicular bisector of the side BC. Using only ruler and compass
        construct and indicate the position of the tree.
    (c) Find the distance of the tree from the corner B of the field.   [N]

5 Algebra 2

Girolamo Cardan (1501–1576) was a colourful character who became Professor of Mathematics at Milan. As well as being a distinguished academic, he was an astrologer, a physician, a gambler and a heretic, yet he received a pension from the Pope. His mathematical genius enabled him to open up the general theory of cubic and quartic equations, although a method for solving cubic equations which he claimed as his own was pirated from Niccolo Tartaglia.

5.1 ALGEBRAIC FRACTIONS

Simplifying fractions

Example 1

Simplify: (a) $\dfrac{32}{56}$; (b) $\dfrac{3a}{5a^2}$ (c) $\dfrac{3y+y^2}{6y}$

(a)
$$\frac{32}{56} = \frac{8 \times 4}{8 \times 7}$$
$$= \frac{4}{7}$$

(b)
$$\frac{3a}{5a^2} = \frac{3 \times a}{5 \times a \times a}$$
$$= \frac{3}{5a}$$

(c) $\dfrac{y(3+y)}{6y} = \dfrac{3+y}{6}$

Exercise 1

Simplify as far as possible, where you can.

1. $\dfrac{25}{35}$

2. $\dfrac{84}{96}$

3. $\dfrac{5y^2}{y}$

4. $\dfrac{y}{2y}$

5. $\dfrac{8x^2}{2x^2}$

6. $\dfrac{2x}{4y}$

7. $\dfrac{6y}{3y}$

8. $\dfrac{5ab}{10b}$

9. $\dfrac{8ab^2}{12ab}$

10. $\dfrac{7a^2b}{35ab^2}$

11. $\dfrac{(2a)^2}{4a}$

12. $\dfrac{7yx}{8xy}$

13. $\dfrac{5x+2x^2}{3x}$

14. $\dfrac{9x+3}{3x}$

15. $\dfrac{25+7}{25}$

16. $\dfrac{4a+5a^2}{5a}$

17. $\dfrac{3x}{4x-x^2}$

18. $\dfrac{5ab}{15a+10a^2}$

19. $\dfrac{5x+4}{8x}$

20. $\dfrac{12x+6}{6y}$

21. $\dfrac{5x+10y}{15xy}$

22. $\dfrac{18a-3ab}{6a^2}$

23. $\dfrac{4ab+8a^2}{2ab}$

24. $\dfrac{(2x)^2-8x}{4x}$

Example 2

Simplify:

(a) $\dfrac{x^2 + x - 6}{x^2 + 2x - 3} = \dfrac{(x-2)(x+3)}{(x+3)(x-1)} = \dfrac{x-2}{x-1}$

(b) $\dfrac{x^2 + 3x - 10}{x^2 - 4} = \dfrac{(x-2)(x+5)}{(x-2)(x+2)} = \dfrac{x+5}{x+2}$

Example 3

Write $\dfrac{x + \frac{1}{2}}{x + \frac{1}{3}}$ without the fractions.

Multiply 'top' and 'bottom' by 6.

i.e. $\dfrac{6(x + \frac{1}{2})}{6(x + \frac{1}{3})} = \dfrac{6x + 3}{6x + 2}$

Exercise 2

Simplify as far as possible.

1. $\dfrac{x^2 + 2x}{x^2 - 3x}$

2. $\dfrac{x^2 - 3x}{x^2 - 2x - 3}$

3. $\dfrac{x^2 + 4x}{2x^2 \quad 10x}$

4. $\dfrac{x^2 + 6x + 5}{x^2 - x - 2}$

5. $\dfrac{x^2 - 4x - 21}{x^2 - 5x - 14}$

6. $\dfrac{x^2 + 7x + 10}{x^2 - 4}$

7. $\dfrac{x + \frac{1}{x}}{x}$

8. $\dfrac{\frac{1}{4} + x}{\frac{1}{2}}$

9. $\dfrac{2x - \frac{1}{x}}{\frac{1}{x}}$

10. $\dfrac{x - \frac{1}{2}}{\frac{1}{2}}$

11. $\dfrac{3x + \frac{1}{4}}{\frac{1}{4}}$

12. $\dfrac{\frac{1}{4} - x}{\frac{1}{2}}$

13. $\dfrac{3x - \frac{1}{x}}{2}$

14. $\dfrac{x + \frac{1}{3}}{2x - \frac{1}{2}}$

15. $\dfrac{3x + \frac{1}{x}}{x + \frac{2}{x}}$

16. $\dfrac{x - \frac{4}{x}}{x - 2}$

Addition and subtraction of algebraic fractions

Example 4

Write as a single fraction:

(a) $\dfrac{2}{3} + \dfrac{3}{4}$; (b) $\dfrac{2}{x} + \dfrac{3}{y}$

Compare these two workings line for line

(a) $\dfrac{2}{3} + \dfrac{3}{4}$; the L.C.M. of 3 and 4 is 12.

$\therefore \quad \dfrac{2}{3} + \dfrac{3}{4} = \dfrac{8}{12} + \dfrac{9}{12}$

$= \dfrac{17}{12}$

(b) $\dfrac{2}{x} + \dfrac{3}{y}$; the L.C.M. of x and y is xy.

$\therefore \quad \dfrac{2}{x} + \dfrac{3}{y} = \dfrac{2y}{xy} + \dfrac{3x}{xy}$

$= \dfrac{2y + 3x}{xy}$

Exercise 3

Simplify the following.

1. $\dfrac{2}{5} + \dfrac{1}{5}$

2. $\dfrac{2x}{5} + \dfrac{x}{5}$

3. $\dfrac{2}{x} + \dfrac{1}{x}$

4. $\dfrac{1}{7} + \dfrac{3}{7}$

5. $\dfrac{x}{7} + \dfrac{3x}{7}$

6. $\dfrac{1}{7x} + \dfrac{3}{7x}$

7. $\dfrac{5}{8} + \dfrac{1}{4}$

8. $\dfrac{5x}{8} + \dfrac{x}{4}$

9. $\dfrac{5}{8x} + \dfrac{1}{4x}$

10. $\dfrac{2}{3} + \dfrac{1}{6}$

11. $\dfrac{2x}{3} + \dfrac{x}{6}$

12. $\dfrac{2}{3x} + \dfrac{1}{6x}$

13. $\dfrac{3}{4} + \dfrac{2}{5}$

14. $\dfrac{3x}{4} + \dfrac{2x}{5}$

15. $\dfrac{3}{4x} + \dfrac{2}{5x}$

16. $\dfrac{3}{4} - \dfrac{2}{3}$

17. $\dfrac{3x}{4} - \dfrac{2x}{3}$

18. $\dfrac{3}{4x} - \dfrac{2}{3x}$

19. $\dfrac{x}{2} + \dfrac{x+1}{3}$

20. $\dfrac{x-1}{3} + \dfrac{x+2}{4}$

21. $\dfrac{2x-1}{5} + \dfrac{x+3}{2}$

22. $\dfrac{x+1}{3} - \dfrac{(2x+1)}{4}$

23. $\dfrac{x-3}{3} - \dfrac{(x-2)}{5}$

24. $\dfrac{2x+1}{7} - \dfrac{(x+2)}{2}$

25. $\dfrac{1}{x} + \dfrac{2}{x+1}$

26. $\dfrac{3}{x-2} + \dfrac{4}{x}$

27. $\dfrac{5}{x-2} + \dfrac{3}{x+3}$

28. $\dfrac{7}{x+1} - \dfrac{3}{x+2}$

29. $\dfrac{2}{x+3} - \dfrac{5}{x-1}$

30. $\dfrac{3}{x-2} - \dfrac{4}{x+1}$

5.2 CHANGING THE SUBJECT OF A FORMULA

The operations involved in solving ordinary linear equations are exactly the same as the operations required in changing the subject of a formula.

Example 1

(a) Solve the equation $3x + 1 = 12$.

(b) Make x the subject of the formula
$Mx + B = A$.

(a) $3x + 1 = 12$
$$3x = 12 - 1$$
$$x = \frac{12 - 1}{3} = \frac{11}{3}$$

(b) $Mx + B = A$
$$Mx = A - B$$
$$x = \frac{A - B}{M}$$

Example 2

(a) Solve the equation $3(y - 2) = 5$.

(b) Make y the subject of the formula
$x(y - a) = e$.

(a) $3(y - 2) = 5$
$$3y - 6 = 5$$
$$3y = 11$$
$$y = \frac{11}{3}$$

(b) $x(y - a) = e$
$$xy - xa = e$$
$$xy = e + xa$$
$$y = \frac{e + xa}{x}$$

Exercise 4

Make x the subject of the following:

1. $2x = 5$
2. $7x = 21$
3. $Ax = B$
4. $Nx = T$
5. $Mx = K$
6. $xy = 4$
7. $Bx = C$
8. $4x = D$
9. $9x = T + N$
10. $Ax = B - R$
11. $Cx = R + T$
12. $Lx = N - R^2$
13. $R - S^2 = Nx$
14. $x + 5 = 7$

15. $x + 10 = 3$
16. $x + A = T$
17. $x + B = S$
18. $N = x + D$
19. $M = x + B$
20. $L = x + D^2$
21. $N^2 + x = T$
22. $L + x = N + M$
23. $Z + x = R - S$
24. $x - 5 = 2$
25. $x - R = A$
26. $x - A = E$
27. $F = x - B$
28. $F^2 = x - B^2$
29. $x - D = A + B$
30. $x - E = A^2$

Make y the subject of the following:

31. $L = y - B$
32. $N = y - T$
33. $3y + 1 = 7$
34. $2y - 4 = 5$
35. $Ay + C = N$
36. $By + D = L$
37. $Dy + E = F$
38. $Ny - F = H$
39. $Yy - Z = T$
40. $Ry - L = B$

41. $Vy + m = Q$
42. $ty - m = n + a$
43. $qy + n = s - t$
44. $ny - s^2 = t$
45. $V^2y + b = c$
46. $r = ny - 6$
47. $s = my + d$
48. $t = my - b$
49. $j = my + c$
50. $2(y + 1) = 6$

51. $3(y - 1) = 5$
52. $A(y + B) = C$
53. $D(y + E) = F$
54. $h(y + n) = a$
55. $b(y - d) = q$
56. $n = r(y + t)$
57. $t(y - 4) = b$
58. $z = S(y + t)$
59. $s = v(y - d)$
60. $g = m(y + n)$

Example 3

(a) Solve the equation $\dfrac{3a + 1}{2} = 4$.

(b) Make a the subject of the formula
$\dfrac{na + b}{m} = n$.

(a) $\dfrac{3a + 1}{2} = 4$
$$3a + 1 = 8$$
$$3a = 7$$
$$a = \frac{7}{3}$$

(b) $\dfrac{na + b}{m} = n$
$$na + b = mn$$
$$na = mn - b$$
$$a = \frac{mn - b}{n}$$

Example 4

Make a the subject of the formula
$x - na = y$.

Make the 'a' term positive
$$x = y + na$$
$$x - y = na$$
$$\frac{x - y}{n} = a$$

Exercise 5

Make a the subject.

1. $\dfrac{a}{4} = 3$

2. $\dfrac{a}{5} = 2$

3. $\dfrac{a}{D} = B$

4. $\dfrac{a}{B} = T$

5. $\dfrac{a}{N} = R$

6. $b = \dfrac{a}{m}$

7. $\dfrac{a-2}{4} = 6$

8. $\dfrac{a-A}{B} = T$

9. $\dfrac{a-D}{N} = A$

10. $\dfrac{a+Q}{N} = B^2$

11. $g = \dfrac{a-r}{e}$

12. $\dfrac{2a+1}{5} = 2$

13. $\dfrac{A\,a+B}{C} = D$

14. $\dfrac{na+m}{P} = q$

15. $\dfrac{ra-t}{S} = v$

16. $\dfrac{za-m}{q} = t$

17. $\dfrac{m+Aa}{b} = c$

18. $A = \dfrac{Ba+D}{E}$

19. $n = \dfrac{ea-f}{h}$

20. $q = \dfrac{ga+b}{r}$

21. $6 - a = 2$

22. $7 - a = 9$

23. $5 = 7 - a$

24. $A - a = B$

25. $C - a = E$

26. $D - a = H$

27. $n - a = m$

28. $t = q - a$

29. $b = s - a$

30. $v = r - a$

31. $t = m - a$

32. $5 - 2a = 1$

33. $T - Xa = B$

34. $M - Na = Q$

35. $V - Ma = T$

36. $L = N - Ra$

37. $r = v^2 - ra$

38. $t^2 = w - na$

39. $n - qa = 2$

40. $\dfrac{3-4a}{2} = 1$

41. $\dfrac{5-7a}{3} = 2$

42. $\dfrac{B-Aa}{D} = E$

43. $\dfrac{D-Ea}{N} = B$

44. $\dfrac{h-fa}{b} = x$

45. $\dfrac{v^2-ha}{C} = d$

46. $\dfrac{M(a+B)}{N} = T$

47. $\dfrac{f(Na-e)}{m} = B$

48. $\dfrac{T(M-a)}{E} = F$

49. $\dfrac{y(x-a)}{z} = t$

50. $\dfrac{k^2(m-a)}{x} = x$

Example 5

(a) Solve the equation $\dfrac{4}{z} = 7$.

(b) Make z the subject of the formula $\dfrac{n}{z} = k$.

(a) $\dfrac{4}{z} = 7$

$4 = 7z$

$\dfrac{4}{7} = z$

(b) $\dfrac{n}{z} = k$

$n = kz$

$\dfrac{n}{k} = z$

Example 6

Make t the subject of the formula $\dfrac{x}{t} + m = a$.

$\dfrac{x}{t} = a - m$

$x = (a-m)t$

$\dfrac{x}{(a-m)} = t$

Exercise 6

Make a the subject.

1. $\dfrac{7}{a} = 14$

2. $\dfrac{5}{a} = 3$

3. $\dfrac{B}{a} = C$

4. $\dfrac{T}{a} = X$

5. $\dfrac{M}{a} = R$

6. $m = \dfrac{n}{a}$

7. $t = \dfrac{v}{a}$

8. $\dfrac{n}{a} = \sin 20°$

9. $\dfrac{7}{a} = \cos 30°$

10. $\dfrac{B}{a} = x$

11. $\dfrac{5}{a} = \dfrac{3}{4}$

12. $\dfrac{N}{a} = \dfrac{B}{D}$

13. $\dfrac{H}{a} = \dfrac{N}{M}$

14. $\dfrac{t}{a} = \dfrac{b}{e}$

15. $\dfrac{v}{a} = \dfrac{m}{s}$

16. $\dfrac{t}{b} = \dfrac{m}{a}$

17. $\dfrac{5}{a+1} = 2$

18. $\dfrac{7}{a-1} = 3$

19. $\dfrac{B}{a+D} = C$

20. $\dfrac{Q}{a-C} = T$

21. $\dfrac{V}{a-T} = D$

22. $\dfrac{L}{Ma} = B$

23. $\dfrac{N}{Ba} = C$

24. $\dfrac{m}{ca} = d$

25. $t = \dfrac{b}{c-a}$

26. $x = \dfrac{z}{y-a}$

Make x the subject.

27. $\dfrac{2}{x} + 1 = 3$

28. $\dfrac{5}{x} - 2 = 4$

29. $\dfrac{A}{x} + B = C$

30. $\dfrac{V}{x} + G = H$

31. $\dfrac{r}{x} - t = n$

32. $q = \dfrac{b}{x} + d$

33. $t = \dfrac{m}{x} - n$

34. $h = d - \dfrac{b}{x}$

35. $C - \dfrac{d}{x} = e$

36. $r - \dfrac{m}{x} = e^2$

37. $t^2 = b - \dfrac{n}{x}$

38. $\dfrac{d}{x} + b = mn$

39. $\dfrac{M}{x+q} - N = 0$

40. $\dfrac{Y}{x-c} - T = 0$

41. $3M = M + \dfrac{N}{P+x}$

42. $A = \dfrac{B}{c+x} - 5A$

43. $\dfrac{K}{Mx} + B = C$

44. $\dfrac{z}{xy} - z = y$

45. $\dfrac{m^2}{x} - n = -p$

46. $t = w - \dfrac{q}{x}$

15. $g = \sqrt{(c-x)}$

16. $\sqrt{(M-Nx)} = P$

17. $\sqrt{(Ax+B)} = \sqrt{D}$

18. $\sqrt{(x-D)} = A^2$

19. $x^2 = g$

20. $x^2 + 1 = 17$

21. $x^2 = B$

22. $x^2 + A = B$

23. $x^2 - A = M$

24. $b = a + x^2$

25. $C - x^2 = m$

26. $n = d - x^2$

27. $mx^2 = n$

28. $b = ax^2$

Make k the subject.

29. $\dfrac{kz}{a} = t$

30. $ak^2 - t = m$

31. $n = a - k^2$

32. $\sqrt{(k^2 - 4)} = 6$

33. $\sqrt{(k^2 - A)} = B$

34. $\sqrt{(k^2 + y)} = x$

35. $t = \sqrt{(m + k^2)}$

36. $2\sqrt{(k+1)} = 6$

37. $A\sqrt{(k+B)} = M$

38. $\sqrt{\left(\dfrac{M}{k}\right)} = N$

39. $\sqrt{\left(\dfrac{N}{k}\right)} = B$

40. $\sqrt{(a-k)} = b$

41. $\sqrt{(a^2 - k^2)} = t$

42. $\sqrt{(m - k^2)} = x$

43. $2\pi\sqrt{(k+t)} = 4$

44. $A\sqrt{(k+1)} = B$

45. $\sqrt{(ak^2 - b)} = C$

46. $a\sqrt{(k^2 - x)} = b$

47. $k^2 + b = x^2$

48. $\dfrac{k^2}{a} + b = c$

49. $\sqrt{(c^2 - ak)} = b$

50. $\dfrac{m}{k^2} = a + b$

Example 7

Make x the subject of the formulae.

(a) $\sqrt{(x^2 + A)} = B$

$\qquad x^2 + A = B^2$ (square both sides)

$\qquad x^2 = B^2 - A$

$\qquad x = \pm\sqrt{(B^2 - A)}$

(b) $(Ax - B)^2 = M$

$\qquad Ax - B = \pm\sqrt{M}$ (square root both sides)

$\qquad Ax = B \pm \sqrt{M}$

$\qquad x = \dfrac{B \pm \sqrt{M}}{A}$

(c) $\sqrt{(R - x)} = T$

$\qquad R - x = T^2$

$\qquad R = T^2 + x$

$\qquad R - T^2 = x$

Exercise 7

Make x the subject.

1. $\sqrt{x} = 2$

2. $\sqrt{(x+1)} = 5$

3. $\sqrt{(x-2)} = 3$

4. $\sqrt{(x+A)} = B$

5. $\sqrt{(x+C)} = D$

6. $\sqrt{(x-E)} = H$

7. $\sqrt{(ax+b)} = c$

8. $\sqrt{(x-m)} = a$

9. $b = \sqrt{(gx-t)}$

10. $r = \sqrt{(b-x)}$

11. $\sqrt{(d-x)} = t$

12. $b = \sqrt{(x-d)}$

13. $c = \sqrt{(n-x)}$

14. $f = \sqrt{(b-x)}$

Example 8

Make x the subject of the formulae.

(a) $\quad Ax - B = Cx + D$

$\qquad Ax - Cx = D + B$

$\qquad x(A - C) = D + B$ (factorise)

$$x = \dfrac{D + B}{A - C}$$

(b) $\quad x + a = \dfrac{x + b}{c}$

$\qquad c(x + a) = x + b$

$\qquad cx + ca = x + b$

$\qquad cx - x = b - ca$

$\qquad x(c - 1) = b - ca$ (factorise)

$$x = \dfrac{b - ca}{c - 1}$$

Exercise 8

Make y the subject.

1. $5(y-1)=2(y+3)$
2. $7(y-3)=4(3-y)$
3. $Ny+B=D-Ny$
4. $My-D=E-2My$
5. $ay+b=3b+by$
6. $my-c=e-ny$
7. $xy+4=7-ky$
8. $Ry+D=Ty+C$
9. $ay-x=z+by$
10. $m(y+a)=n(y+b)$
11. $x(y-b)=y+d$
12. $\dfrac{a-y}{a+y}=b$
13. $\dfrac{1-y}{1+y}=\dfrac{c}{d}$
14. $\dfrac{M-y}{M+y}=\dfrac{a}{b}$
15. $m(y+n)=n(n-y)$
16. $y+m=\dfrac{2y-5}{m}$
17. $y-n=\dfrac{y+2}{n}$
18. $y+b=\dfrac{ay+e}{b}$
19. $\dfrac{ay+x}{x}=4-y$
20. $c-dy=e-ay$
21. $y(a-c)=by+d$
22. $y(m+n)=a(y+b)$
23. $t-ay=s-by$
24. $\dfrac{y+x}{y-x}=3$
25. $\dfrac{v-y}{v+y}=\dfrac{1}{2}$
26. $y(b-a)=a(y+b+c)$
27. $\sqrt{\left(\dfrac{y+x}{y-x}\right)}=2$
28. $\sqrt{\left(\dfrac{z+y}{z-y}\right)}=\dfrac{1}{3}$
29. $\sqrt{\left[\dfrac{m(y+n)}{y}\right]}=p$
30. $n-y=\dfrac{4y-n}{m}$

Example 9

Make w the subject of the formula

$$\sqrt{\left(\dfrac{w}{w+a}\right)}=c.$$

Squaring both sides, $\quad \dfrac{w}{w+a}=c^2$

Multiplying by $(w+a)$, $\quad w=c^2(w+a)$

$$w=c^2w+c^2a$$

$$w-c^2w=c^2a$$

$$w(1-c^2)=c^2a$$

$$w=\dfrac{c^2a}{1-c^2}$$

Exercise 9

Make the letter in square brackets the subject.

1. $ax+by+c=0$ $[x]$
2. $\sqrt{\{a(y^2-b)\}}=e$ $[y]$
3. $\dfrac{\sqrt{(k-m)}}{n}=\dfrac{1}{m}$ $[k]$
4. $a-bz=z+b$ $[z]$
5. $\dfrac{x+y}{x-y}=2$ $[x]$
6. $\sqrt{\left(\dfrac{a}{z}-c\right)}=e$ $[z]$
7. $lm+mn+a=0$ $[n]$
8. $t=2\pi\sqrt{\left(\dfrac{d}{g}\right)}$ $[d]$
9. $t=2\pi\sqrt{\left(\dfrac{d}{g}\right)}$ $[g]$
10. $\sqrt{(x^2+a)}=2x$ $[x]$
11. $\sqrt{\left\{\dfrac{b(m^2+a)}{e}\right\}}=t$ $[m]$
12. $\sqrt{\left(\dfrac{x+1}{x}\right)}=a$ $[x]$
13. $a+h-mx=0$ $[m]$
14. $\sqrt{(a^2+b^2)}=x^2$ $[a]$
15. $\dfrac{a}{k}+b=\dfrac{c}{k}$ $[k]$
16. $a-y=\dfrac{b+y}{a}$ $[y]$
17. $G=4\pi\sqrt{(x^2+T^2)}$ $[x]$
18. $M(ax+by+c)=0$ $[y]$
19. $x=\sqrt{\left(\dfrac{y-1}{y+1}\right)}$ $[y]$
20. $a\sqrt{\left(\dfrac{x^2-n}{m}\right)}=\dfrac{a^2}{b}$ $[x]$
21. $\dfrac{M}{N}+E=\dfrac{P}{N}$ $[N]$
22. $\dfrac{Q}{P-x}=R$ $[x]$
23. $\sqrt{(z-ax)}=t$ $[a]$
24. $e+\sqrt{(x+f)}=g$ $[x]$
25. $\dfrac{m(ny-e^2)}{p}+n=5n$ $[y]$

5.3 VARIATION

Direct variation

There are several ways of expressing a relationship between two quantities x and y. Here are some examples.

  x varies as y

  x varies directly as y

  x is proportional to y

These three all mean the same and they are written in symbols as follows.

  $x \propto y$

The '$\propto$' sign can always be replaced by '$=k$' where k is a constant:

  $x = ky$

Suppose $x = 3$ when $y = 12$;
then $\quad 3 = k \times 12$
and $\quad k = \frac{1}{4}$
We can then write $x = \frac{1}{4}y$, and this allows us to find the value of x for any value of y, and *vice versa*.

Example 1

y varies as z, and $y = 2$ when $z = 5$; find
(a) the value of y when $z = 6$
(b) the value of z when $y = 5$

Because $y \propto z$, then $y = kz$ where k is a constant.
$\quad y = 2$ when $z = 5$
$\therefore \quad 2 = k \times 5$
$\quad k = \frac{2}{5}$
So $\quad y = \frac{2}{5}z$.

(a) When $z = 6$, $y = \frac{2}{5} \times 6 = 2\frac{2}{5}$.

(b) When $y = 5$, $5 = \frac{2}{5}z$
$\qquad\qquad\qquad z = \frac{25}{2} = 12\frac{1}{2}$.

Example 2

The value V of a diamond is proportional to the square of its weight W. If a diamond weighing 10 grams is worth £200, find
(a) the value of a diamond weighing 30 grams
(b) the weight of a diamond worth £5000.

$\qquad V \propto W^2$
or $\qquad V = kW^2$ where k is a constant.

$\qquad V = 200$ when $W = 10$
$\therefore \quad 200 = k \times 10^2$
$\qquad k = 2$
So $\quad V = 2W^2$

(a) When $W = 30$,
$\qquad V = 2 \times 30^2 = 2 \times 900$
$\qquad V = £1800$
So a diamond of weight 30 grams is worth £1800.

(b) When $\quad V = 5000$,
$\qquad 5000 = 2 \times W^2$
$\qquad W^2 = \dfrac{5000}{2} = 2500$
$\qquad W = \sqrt{2500} = 50$
So a diamond of value £5000 weighs 50 grams.

Exercise 10

1. Rewrite the statement connecting each pair of variables using a constant k instead of '$\propto$'.

   (a) $S \propto e$ $\qquad$ (b) $v \propto t$ $\qquad$ (c) $x \propto z^2$
   (d) $y \propto \sqrt{x}$ $\qquad$ (e) $T \propto \sqrt{L}$ $\qquad$ (f) $C \propto r$
   (g) $A \propto r^2$ $\qquad$ (h) $V \propto r^3$

2. y varies as t. If $y = 6$ when $t = 4$, calculate
   (a) the value of y, when $t = 6$
   (b) the value of t, when $y = 4$.

3. z is proportional to m, If $z = 20$ when $m = 4$, calculate
   (a) the value of z, when $m = 7$
   (b) the value of m, when $z = 55$.

4. A varies directly as r^2. If $A = 12$, when $r = 2$, calculate
   (a) the value of A, when $r = 5$
   (b) the value of r, when $A = 48$.

5. Given that $z \propto x$, copy and complete the table.

x	1	3		$5\frac{1}{2}$
z	4		16	

6. Given that $V \propto r^3$, copy and complete the table.

r	1	2		$1\frac{1}{2}$
V	4		256	

7. Given that $w \propto \sqrt{h}$, copy and complete the table.

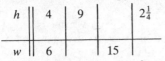

h	4	9		$2\frac{1}{4}$
w	6		15	

8. s is proportional to $(v-1)^2$. If $s = 8$, when $v = 3$, calculate
   (a) the value of s, when $v = 4$
   (b) the value of v, when $s = 2$.

9. m varies as $(d + 3)$. If $m = 28$ when $d = 1$, calculate
   (a) the value of m, when $d = 3$
   (b) the value of d, when $m = 49$.

10. The pressure of the water P at any point below the surface of the sea varies as the depth of the point below the surface d. If the pressure is 200 newtons/cm² at a depth of 3 m, calculate the pressure at a depth of 5 m.

11. The distance d through which a stone falls from rest is proportional to the square of the time taken t. If the stone falls 45 m in 3 seconds, how far will it fall in 6 seconds?
    How long will it take to fall 20 m?

12. The energy E stored in an elastic band varies as the square of the extension x. When the elastic is extended by 3 cm, the energy stored is 243 joules. What is the energy stored when the extension is 5 cm?
    What is the extension when the stored energy is 36 joules?

13. In the first few days of its life, the length of an earthworm l is thought to be proportional to the square root of the number of hours n which have elapsed since its birth. If a worm is 2 cm long after 1 hour, how long will it be after 4 hours?
    How long will it take to grow to a length of 14 cm?

14. It is well known that the number of golden eggs which a goose lays in a week varies as the cube root of the average number of hours of sleep she has. When she has 8 hours sleep, she lays 4 golden eggs. How long does she sleep when she lays 5 golden eggs?

15. The resistance to motion of a car is proportional to the square of the speed of the car. If the resistance is 4000 newtons at a speed of 20 m/s, what is the resistance at a speed of 30 m/s?
    At what speed is the resistance 6250 newtons?

16. A road research organisation recently claimed that the damage to road surfaces was proportional to the fourth power of the axle load. The axle load of a 44-ton HGV is about 15 times that of a car. Calculate the ratio of the damage to road surfaces made by a 44-ton HGV and a car.

Inverse variation

There are several ways of expressing an inverse relationship between two variables,

   x varies inversely as y
   x is inversely proportional to y.

We write $x \propto \dfrac{1}{y}$ for both statements and proceed using the method outlined in the previous section.

Example 3

z is inversely proportional to t^2 and $z = 4$ when $t = 1$. Calculate
(a) z when $t = 2$
(b) t when $z = 16$.

We have $\quad z \propto \dfrac{1}{t^2}$

or $\qquad z = k \times \dfrac{1}{t^2}$ (k is a constant)

$z = 4$ when $t = 1$,

$\therefore \quad 4 = k\left(\dfrac{1}{1^2}\right)$

so $\quad k = 4$

$\therefore \quad z = 4 \times \dfrac{1}{t^2}$

(a) when $t = 2$, $z = 4 \times \dfrac{1}{2^2} = 1$.

(b) when $z = 16$, $16 = 4 \times \dfrac{1}{t^2}$

$$16t^2 = 4$$
$$t^2 = \tfrac{1}{4}$$
$$t = \pm\tfrac{1}{2}.$$

Exercise 11

1. Rewrite the statements connecting the variables using a constant of variation, k.

   (a) $x \propto \dfrac{1}{y}$ (b) $s \propto \dfrac{1}{t^2}$ (c) $t \propto \dfrac{1}{\sqrt{q}}$

   (d) m varies inversely as w
   (e) z is inversely proportional to t^2.

2. b varies inversely as e. If $b = 6$ when $e = 2$, calculate
   (a) the value of b when $e = 12$
   (b) the value of e when $b = 3$.

3. q varies inversely as r. If $q = 5$ when $r = 2$, calculate
(a) the value of q when $r = 4$
(b) the value of r when $q = 20$.

4. x is inversely proportional to y^2. If $x = 4$ when $y = 3$, calculate
(a) the value of x when $y = 1$
(b) the value of y when $x = 2\frac{1}{4}$.

5. R varies inversely as v^2. If $R = 120$ when $v = 1$, calculate
(a) the value of R when $v = 10$
(b) the value of v when $R = 30$.

6. T is inversely proportional to x^2. If $T = 36$ when $x = 2$, calculate
(a) the value of T when $x = 3$
(b) the value of x when $T = 1\cdot44$.

7. p is inversely proportional to $\sqrt{y}$. If $p = 1\cdot2$ when $y = 100$, calculate
(a) the value of p when $y = 4$
(b) the value of y when $p = 3$.

8. y varies inversely as z. If $y = \frac{1}{8}$ when $z = 4$, calculate
(a) the value of y when $z = 1$
(b) the value of z when $y = 10$.

9. Given that $z \propto \dfrac{1}{y}$, copy and complete the table:

y	2	4		$\frac{1}{4}$
z	8		16	

10. Given that $v \propto \dfrac{1}{t^2}$, copy and complete the table:

t	2	5		10
v	25		$\frac{1}{4}$	

11. Given that $r \propto \dfrac{1}{\sqrt{x}}$, copy and complete the table:

x	1	4		
r	12		$\frac{3}{4}$	2

12. e varies inversely as $(y - 2)$. If $e = 12$ when $y = 4$, find
(a) e when $y = 6$ (b) y when $e = \frac{1}{2}$.

13. M is inversely proportional to the square of l. If $M = 9$ when $l = 2$, find
(a) M when $l = 10$ (b) l when $M = 1$.

14. Given $z = \dfrac{k}{x^n}$, find k and n, then copy and complete the table.

x	1	2	4	
z	100	$12\frac{1}{2}$		$\frac{1}{10}$

15. Given $y = \dfrac{k}{\sqrt[n]{v}}$, find k and n, then copy and complete the table.

v	1	4	36	
y	12	6		$\frac{3}{25}$

16. The volume V of a given mass of gas varies inversely as the pressure P. When $V = 2$ m^3, $P = 500$ N/m^2. Find the volume when the pressure is 400 N/m^2. Find the pressure when the volume is 5 m^3.

17. The number of hours N required to dig a certain hole is inversely proportional to the number of men available x. When 6 men are digging, the hole takes 4 hours. Find the time taken when 8 men are available. If it takes $\frac{1}{2}$ hour to dig the hole, how many men are there?

18. The life expectancy L of a rat varies inversely as the square of the density d of poison distributed around his home. When the density of poison is 1 g/m^2 the life expectancy is 50 days. How long will he survive if the density of poison is
(a) 5 g/m^2?   (b) $\frac{1}{2}$ g/m^2?

19. The force of attraction F between two magnets varies inversely as the square of the distance d between them. When the magnets are 2 cm apart, the force of attraction is 18 newtons. How far apart are they if the attractive force is 2 newtons?

5.4 INDICES

Rules of indices

1. $a^n \times a^m = a^{n+m}$ e.g. $7^2 \times 7^4 = 7^6$
2. $a^n \div a^m = a^{n-m}$ e.g. $6^6 \div 6^2 = 6^4$
3. $(a^n)^m = a^{nm}$ e.g. $(3^2)^5 = 3^{10}$

Also, $a^{-n} = \dfrac{1}{a^n}$ e.g. $5^{-2} = \dfrac{1}{5^2}$

$a^{\frac{1}{n}}$ means 'the nth root of a'   e.g. $9^{\frac{1}{2}} = \sqrt[2]{9}$

$a^{\frac{m}{n}}$ means 'the nth root of a raised to the power m'

     e.g. $4^{\frac{3}{2}} = (\sqrt{4})^3 = 8$

Example 1

Simplify (a) $x^7 . x^{13}$ (b) $x^3 \div x^7$
          (c) $(x^4)^3$ (d) $(3x^2)^3$
          (e) $(2x^{-1})^2 \div x^{-5}$ (f) $3y^2 \times 4y^3$

(a) $x^7 . x^{13} = x^{7+13} = x^{20}$

(b) $x^3 \div x^7 = x^{3-7} = x^{-4} = \dfrac{1}{x^4}$

(c) $(x^4)^3 = x^{12}$

(d) $(3x^2)^3 = 3^3 . (x^2)^3 = 27x^6$

(e) $(2x^{-1})^2 \div x^{-5} = 4x^{-2} \div x^{-5}$
               $= 4x^{(-2--5)}$
               $= 4x^3.$

(f) $3y^2 \times 4y^3 = 12y^5.$

Exercise 12

Express in index form

1. $3 \times 3 \times 3 \times 3$ 2. $4 \times 4 \times 5 \times 5 \times 5$

3. $3 \times 7 \times 7 \times 7$ 4. $2 \times 2 \times 2 \times 7$

5. $\dfrac{1}{10 \times 10 \times 10}$ 6. $\dfrac{1}{2 \times 2 \times 3 \times 3 \times 3}$

7. $\sqrt{15}$ 8. $\sqrt[3]{3}$

9. $\sqrt[5]{10}$ 10. $(\sqrt{5})^3$

Simplify

11. $x^3 \times x^4$ 12. $y^6 \times y^7$

13. $z^7 \div z^3$ 14. $z^{50} \times z^{50}$

15. $m^3 \div m^2$ 16. $e^{-3} \times e^{-2}$

17. $y^{-2} \times y^4$ 18. $w^4 \div w^{-2}$

19. $y^{\frac{1}{2}} \times y^{\frac{1}{2}}$ 20. $(x^2)^5$

21. $x^{-2} \div x^{-2}$ 22. $w^{-3} \times w^{-2}$

23. $w^{-7} \times w^2$ 24. $x^3 \div x^{-4}$

25. $(a^2)^4$ 26. $(k^{\frac{1}{2}})^6$

27. $e^{-4} \times e^4$ 28. $x^{-1} \times x^{30}$

29. $(y^4)^{\frac{1}{2}}$ 30. $(x^{-3})^{-2}$

31. $z^2 \div z^{-2}$ 32. $t^{-3} \div t$

33. $(2x^3)^2$ 34. $(4y^5)^2$

35. $2x^2 \times 3x^2$ 36. $5y^3 \times 2y^2$

37. $5a^3 \times 3a$ 38. $(2a)^3$

39. $3x^3 \div x^3$ 40. $8y^3 \div 2y$

41. $10y^2 \div 4y$ 42. $8a \times 4a^3$

43. $(2x)^2 \times (3x)^3$ 44. $4z^4 \times z^{-7}$

45. $6x^{-2} \div 3x^2$ 46. $5y^3 \div 2y^{-2}$

47. $(x^2)^{\frac{1}{2}} : (x^{\frac{1}{3}})^3$ 48. $7w^{-2} \times 3w^{-1}$

49. $(2n)^4 \div 8n^0$ 50. $4x^{\frac{3}{2}} \div 2x^{\frac{1}{2}}$

Example 2

Evaluate (a) $9^{\frac{1}{2}}$ (b) 5^{-1}
          (c) $4^{-\frac{1}{2}}$ (d) $25^{\frac{3}{2}}$
          (c) $(5^{\frac{1}{2}})^3 . 5^{\frac{1}{2}}$ (f) 7^0

(a) $9^{\frac{1}{2}} = \sqrt{9} = 3$

(b) $5^{-1} = \frac{1}{5}$

(c) $4^{-\frac{1}{2}} = \dfrac{1}{4^{\frac{1}{2}}} = \dfrac{1}{\sqrt{4}} = \dfrac{1}{2}$

(d) $25^{\frac{3}{2}} = (\sqrt{25})^3 = 5^3 = 125$

(e) $(5^{\frac{1}{2}})^3 . 5^{\frac{1}{2}} = 5^{\frac{3}{2}} . 5^{\frac{1}{2}} = 5^2$
                  $= 25$

(f) $7^0 = 1 \left[\text{consider } \dfrac{7^3}{7^3} = 7^{3-3} = 7^0 = 1 \right]$

Remember. $a^0 = 1$ for any non-zero value of a.

Exercise 13

Evaluate the following

1. $3^2 \times 3$
2. 100^0
3. 3^{-2}
4. $(5^{-1})^{-2}$
5. $4^{\frac{1}{2}}$
6. $16^{\frac{1}{2}}$
7. $81^{\frac{1}{2}}$
8. $8^{\frac{1}{3}}$
9. $9^{\frac{3}{2}}$
10. $27^{\frac{1}{3}}$
11. $9^{-\frac{1}{2}}$
12. $8^{-\frac{1}{3}}$
13. $1^{\frac{5}{2}}$
14. $25^{-\frac{1}{2}}$
15. $1000^{\frac{1}{3}}$
16. $2^{-2} \times 2^5$
17. $2^4 \div 2^{-1}$
18. $8^{\frac{2}{3}}$
19. $27^{-\frac{2}{3}}$
20. $4^{-\frac{3}{2}}$
21. $36^{\frac{1}{2}} \times 27^{\frac{1}{3}}$
22. $10\,000^{\frac{1}{4}}$
23. $100^{\frac{3}{2}}$
24. $(100^{\frac{1}{2}})^{-3}$
25. $(9^{\frac{1}{2}})^{-2}$
26. $(-16 \cdot 371)^0$
27. $81^{\frac{1}{4}} \div 16^{\frac{1}{4}}$
28. $(5^{-4})^{\frac{1}{2}}$
29. $1000^{-\frac{1}{3}}$
30. $(4^{-\frac{1}{2}})^2$
31. $8^{-\frac{2}{3}}$
32. $100^{\frac{5}{2}}$
33. $1^{\frac{4}{3}}$
34. 2^{-5}
35. $(0 \cdot 01)^{\frac{1}{2}}$
36. $(0 \cdot 04)^{\frac{1}{2}}$
37. $(2 \cdot 25)^{\frac{1}{2}}$
38. $(7 \cdot 63)^0$
39. $3^5 \times 3^{-3}$
40. $(3\frac{3}{8})^{\frac{1}{3}}$
41. $(11\frac{1}{9})^{-\frac{1}{2}}$
42. $(\frac{1}{8})^{-2}$
43. $(\frac{1}{1000})^{\frac{2}{3}}$
44. $(\frac{9}{25})^{-\frac{1}{2}}$
45. $(10^{-6})^{\frac{1}{3}}$
46. $7^2 \div (7^{\frac{1}{2}})^4$
47. $(0 \cdot 0001)^{-\frac{1}{2}}$
48. $\dfrac{9^{\frac{1}{2}}}{4^{-\frac{1}{2}}}$
49. $\dfrac{25^{\frac{3}{2}} \times 4^{\frac{1}{2}}}{9^{-\frac{1}{2}}}$
50. $(-\frac{1}{7})^2 \div (-\frac{1}{7})^3$

Example 3

Simplify   (a) $(2a)^3 \div (9a^2)^{\frac{1}{2}}$
            (b) $(3ac^2)^3 \times 2a^{-2}$
            (c) $(2x)^2 \div 2x^2$

(a) $(2a)^3 \div (9a^2)^{\frac{1}{2}} = 8a^3 \div 3a$
                           $= \frac{8}{3}a^2$

(b) $(3ac^2)^3 \times 2a^{-2} = 27a^3c^6 \times 2a^{-2}$
                          $= 54ac^6$

(c) $(2x)^2 \div 2x^2 = 4x^2 \div 2x^2$
                     $= 2$

Exercise 14

Rewrite without brackets

1. $(5x^2)^2$
2. $(7y^3)^2$
3. $(10ab)^2$
4. $(2xy^2)^2$
5. $(4x^2)^{\frac{1}{2}}$
6. $(9y)^{-1}$
7. $(x^{-2})^{-1}$
8. $(2x^{-2})^{-1}$
9. $(5x^2y)^0$

10. $(\frac{1}{2}x)^{-1}$
11. $(3x)^2 \times (2x)^2$
12. $(5y)^2 \div y$
13. $(2x^{\frac{1}{2}})^4$
14. $(3y^{\frac{1}{3}})^3$
15. $(5x^0)^2$
16. $[(5x)^0]^2$
17. $(7y^0)^2$
18. $[(7y)^0]^2$
19. $(2x^2y)^3$
20. $(10xy^3)^2$

Simplify the following:

21. $(3x^{-1})^2 \div 6x^{-3}$
22. $(4x)^{\frac{1}{2}} \div x^{\frac{3}{2}}$
23. $x^2y^2 \times xy^3$
24. $4xy \times 3x^2y$
25. $10x^{-1}y^3 \times xy$
26. $(3x)^2 \times (\frac{1}{9}x^2)^{\frac{1}{2}}$
27. $z^3yx \times x^2yz$
28. $(2x)^{-2} \times 4x^3$
29. $(3y)^{-1} \div (9y^2)^{-1}$
30. $(xy)^0 \times (9x)^{\frac{3}{5}}$
31. $(x^2y)(2xy)(5y^3)$
32. $(4x^{\frac{1}{2}}) \times (8x^{\frac{3}{2}})$
33. $5x^{-3} \div 2x^{-5}$
34. $[(3x^{-1})^{-2}]^{-1}$
35. $(2a)^{-2} \times 8a^4$
36. $(abc^2)^3$

37. Write in the form 2^p (e.g. $4 = 2^2$)
   (a) 32   (b) 128   (c) 64   (d) 1

38. Write in the form 3^q
   (a) $\frac{1}{27}$ (b) $\frac{1}{81}$ (c) $\frac{1}{3}$ (d) $9 \times \frac{1}{81}$

Evaluate, with $x = 16$ and $y = 8$.

39. $2x^{\frac{1}{2}} \times y^{\frac{1}{3}}$
40. $x^{\frac{1}{4}} \times y^{-1}$
41. $(y^2)^{\frac{1}{6}} \div (9x)^{\frac{1}{2}}$
42. $(x^2y^3)^0$
43. $x + y^{-1}$
44. $x^{-\frac{1}{2}} + y^{-1}$
45. $y^{\frac{1}{3}} \div x^{\frac{3}{4}}$
46. $(1000y)^{\frac{1}{3}} \times x^{-\frac{5}{2}}$
47. $(x^{\frac{1}{4}} + y^{-1}) \div x^{\frac{1}{4}}$
48. $x^{\frac{1}{2}} - y^{\frac{2}{3}}$
49. $(x^{\frac{3}{4}}y)^{-\frac{1}{3}}$
50. $\left(\dfrac{x}{y}\right)^{-2}$

Solve the equations for x

51. $2^x = 8$
52. $3^x = 81$
53. $5^x = \frac{1}{5}$
54. $10x = \frac{1}{100}$
55. $3^{-x} = \frac{1}{27}$
56. $4^x = 64$
57. $6^{-x} = \frac{1}{6}$
58. $100\,000^x = 10$
59. $12^x = 1$
60. $10^x = 0 \cdot 0001$
61. $2^x + 3^x = 13$
62. $(\frac{1}{2})^x = 32$
63. $5^{2x} = 25$
64. $1\,000\,000^{3x} = 10$
65. These two are more difficult. Use a calculator to find solutions correct to three significant figures.
   (a) $x^x = 100$ (b) $x^x = 10\,000$

5.5 ITERATIVE METHODS

These methods are particularly well suited when a computer or calculator is available.

Example 1

Suppose we want to find $\sqrt{11}$.
Let our first approximation be a_1.

(a) Try $a_1 = 3$.
Since $3^2 = 9$, we deduce that 3 is smaller than $\sqrt{11}$.
We also deduce that $\frac{11}{3}$ is larger than $\sqrt{11}$.

(b) For our second approximation a_2, try the *average* of 3 and $\frac{11}{3}$.
i.e. $a_2 = \frac{1}{2}(3 + \frac{11}{3}) \approx 3 \cdot 333$.

(c) Now, $3 \cdot 333^2 > 11$
$\therefore \quad \dfrac{11}{3 \cdot 333} < \sqrt{11}$ by the argument above.

(d) Find the average of $3 \cdot 333$ and $\dfrac{11}{3 \cdot 333}$

i.e. $u_3 = \dfrac{1}{2}\left(3 \cdot 333 + \dfrac{11}{3 \cdot 333}\right)$
$= 3 \cdot 317$

This is already $\sqrt{11}$ correct to four significant figures. This is an example of an iterative method.

The iterative formula for finding the square root of a number N is

$$a_{n+1} = \frac{1}{2}\left(a_n + \frac{N}{a_n}\right)$$

where a_n is the approximation after n repetitions of the formula and a_{n+1} is the next approximation.

Exercise 15

1. Start by making a suitable 'guess' and then use the formula above to find the square root of the following numbers, correct to four significant figures.
   (a) 8   (b) 18   (c) 39
   (d) 7·63   (e) 111   (f) 413

2. The iterative formula for finding the cube root of a number N is
   $$a_{n+1} = \frac{1}{3}\left(2a_n + \frac{N}{a_n^2}\right)$$

Use the formula to find the cube root of the following numbers, correct to four significant figures.
   (a) 9   (b) 28   (c) 70
   (d) 0·962   (e) 1050

3. The iterative formula for finding the fifth root of a number N is
   $$a_{n+1} = \frac{1}{5}\left(4a_n + \frac{N}{a_n^4}\right)$$
   Find the fifth root of the following numbers correct to four significant figures.
   (a) 35   (b) 260   (c) 115 000
   (d) 27

4. It is possible to find a solution to the equation $x^3 - 4x^2 - 20 = 0$ by using the iterative formula
   $$x_{n+1} = \frac{20}{x_n^2} + 4$$
   Find a solution, correct to three significant figures. (Try $x_1 = 1$.)

5. The equation $x^4 - 3x^3 - 10 = 0$ may be solved approximately by using the formula
   $$x_{n+1} = \frac{10}{x_n^3} + 3$$
   Find a solution, correct to three significant figures. (Try $x_1 = 1$.)

6. Find the value of $\frac{1}{13}$ correct to six decimal places by using the iterative formula
   $$x_{n+1} = x_n(2 - 13x_n).$$

7. One solution of the quadratic equation $x^2 - 10x + 5 = 0$ may be found by using the iterative formula
   $$x_{n+1} = 10 - \frac{5}{x_n}$$
   Find this solution, correct to four significant figures by using as an initial approximation:
   (a) $x_1 = 8$ (b) $x_1 = -7$ (c) $x_1 = 100$
   You will notice that each time the formula gives the same solution but fails to find the other solution. This illustrates a weakness of this type of simple 'first order' formula.

8. Use the formula
   $$x_{n+1} = \frac{1}{2}\left(3 + \frac{1}{x_n}\right)$$
   to find a solution of the equation $2x^2 - 3x - 1 = 0$.
   Again try a wide variety of initial approximations.

9. Consider the equation $x^2 + x - 5 = 0$.
Rearranging,
$$x^2 + x = 5$$
$$x(x + 1) = 5$$
$$x = \frac{5}{x + 1} \quad \text{or} \quad x + 1 = \frac{5}{x}$$
$$x = \frac{5}{x} - 1$$

Use the iteration formulae $x_{n+1} = \frac{5}{x_n + 1}$ and

$x_{n+1} = \frac{5}{x_n} - 1$ with

(a) an initial value $x_1 = 2$,
(b) an initial value $x_1 = -3$.

What do you notice?

Flow charts

Exercise 16

1. Copy the flow chart and put each of the numbers 1, 2, 3, 4, 5, 6, 7 in at the box marked N. Work out what number would be printed in each case and record the results in a table:

Input	1	2	3	4	5	6	7
Output							

(a)

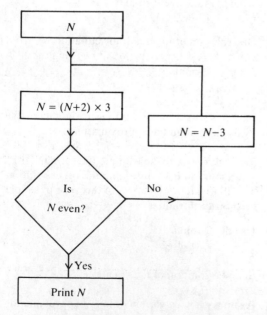

(b)

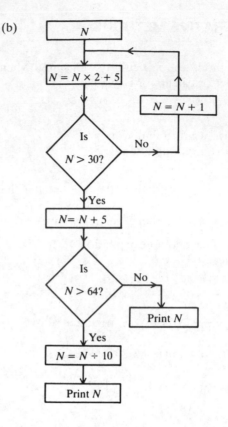

2. Take any number (we have taken 7) and work through the flow charts several times.
What do you notice?
What happens if you change the first number (say $x = 11$ or 197)?
(a)

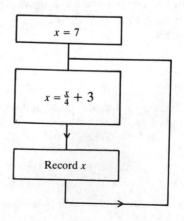

Can you explain what is happening?

(b)

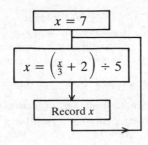

(c)

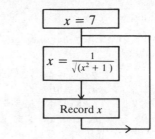

5.6   INEQUALITIES

$x < 4$ means 'x is *less than* 4'
$y > 7$ means 'y is *greater than* 7'
$z \leqslant 10$ means 'z is *less than or equal to* 10'
$t \geqslant -3$ means 't is *greater than or equal to* -3'

Solving inequalities

We follow the same procedure used for solving equations except that when we multiply or divide by a *negative* number the inequality is *reversed*.

c.g. $4 > -2$
but multiplying by -2,
$-8 < 4$

Example 1

Solve the inequalities

(a) $2x - 1 > 5$
$\qquad 2x > 5 + 1$
$\qquad x > \dfrac{6}{2}$
$\qquad x > 3$

(b) $5 - 3x \leqslant 1$
$\qquad 5 \leqslant 1 + 3x$
$\qquad 5 - 1 \leqslant 3x$
$\qquad \dfrac{4}{3} \leqslant x$

Exercise 17

Introduce one of the symbols $<$, $>$ or $=$ between each pair of numbers.

1. -2, 1
2. $(-2)^7$, 1
3. $\frac{1}{4}$, $\frac{1}{5}$
4. $0 \cdot 2$, $\frac{1}{5}$
5. 10^2, 2^{10}
6. $\frac{1}{4}$, $0 \cdot 4$
7. 40%, $0 \cdot 4$
8. $(-1)^2$, $(-\frac{1}{2})^2$
9. 5^2, 2^5
10. $3\frac{1}{3}$, $\sqrt{10}$
11. π^2, 10
12. $-\frac{1}{3}$, $-\frac{1}{2}$
13. 2^{-1}, 3^{-1}
14. 50%, $\frac{1}{5}$
15. 1%, 100^{-1}

State whether the following are true or false:

16. $0 \cdot 7^2 > \frac{1}{2}$
17. $10^3 = 30$
18. $\frac{1}{8} > 12\%$
19. $(0 \cdot 1)^3 = 0 \cdot 0001$
20. $(-\frac{1}{5})^0 = -1$
21. $\dfrac{1}{5^2} > \dfrac{1}{2^5}$
22. $(0 \cdot 2)^2 < (0 \cdot 3)^2$
23. $\frac{6}{7} > \frac{7}{8}$
24. $0 \cdot 1^2 > 0 \cdot 1$

Solve the following inequalities:

25. $x - 3 > 10$
26. $x + 1 < 0$
27. $5 > x - 7$
28. $2x + 1 \leqslant 6$
29. $3x - 4 > 5$
30. $10 \leqslant 2x - 6$
31. $5x < x + 1$
32. $2x \geqslant x - 3$
33. $4 + x < -4$
34. $3x + 1 < 2x + 5$
35. $2(x + 1) > x - 7$
36. $7 < 15 - x$
37. $9 > 12 - x$
38. $4 - 2x \leqslant 2$
39. $3(x - 1) < 2(1 - x)$
40. $7 - 3x < 0$

The number line

The inequality $x < 4$ is represented on the number line as

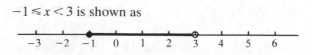

$x \geqslant -2$ is shown as

In the first case, 4 is *not* included so we have ○.
In the second case, -2 *is* included so we have ●.

$-1 \leqslant x < 3$ is shown as

Exercise 18

For questions **1** to **25**, solve each inequality and show the result on a number line.

1. $2x + 1 > 11$
2. $3x - 4 \leqslant 5$
3. $2 < x - 4$
4. $6 \geqslant 10 - x$
5. $8 < 9 - x$
6. $8x - 1 < 5x - 10$
7. $2x > 0$
8. $1 < 3x - 11$
9. $4 - x > 6 - 2x$
10. $\dfrac{x}{3} < -1$
11. $1 < x < 4$
12. $-2 \leqslant x \leqslant 5$
13. $1 \leqslant x < 6$
14. $0 \leqslant 2x < 10$
15. $-3 \leqslant 3x \leqslant 21$
16. $x^2 < 4$
17. $x^2 \leqslant 16$
18. $x^2 > 1$
19. $x^2 \geqslant 9$
20. $1 < 2x + 1 < 9$
21. $10 \leqslant 2x \leqslant x + 9$
22. $x < 3x + 2 < 2x + 6$
23. $10 \leqslant 2x - 1 \leqslant x + 5$
24. $x < 3x - 1 < 2x + 7$
25. $x - 10 < 2(x - 1) < x$

(Hint: in questions **20** to **25**, solve the two inequalities separately.)

For questions **26** to **35**, find the solution set, subject to the given condition.

26. $3a + 1 < 20$; a is a positive integer
27. $b - 1 \geqslant 6$; b is a prime number less than 20
28. $2e - 3 < 21$; e is a positive even number
29. $1 < z < 50$; z is a square number
30. $0 < 3x < 40$; x is divisible by 5

31. $2x > -10$; x is a negative integer
32. $x + 1 < 2x < x + 13$; x is an integer
33. $x^2 < 100$; x is a positive square number
34. $0 \leqslant 2z - 3 \leqslant z + 8$; z is a prime number
35. $\dfrac{a}{2} + 10 > a$; a is a positive even number

36. State the smallest integer n for which $4n > 19$.
37. Find an integer value of x such that
$2x - 7 < 8 < 3x - 11$.
38. Find an integer value of y such that
$3y - 4 < 12 < 4y - 5$.
39. Find any value of z such that $9 < z + 5 < 10$.
40. Find any value of p such that $9 < 2p + 1 < 11$.
41. Find a simple fraction q such that $\frac{4}{9} < q < \frac{5}{9}$.
42. Find an integer value of a such that
$a - 3 \leqslant 11 \leqslant 2a + 10$.
43. State the largest prime number z for which
$3z < 66$.
44. Find a simple fraction r such that $\frac{1}{3} < r < \frac{2}{3}$.
45. Find the largest prime number p such that
$p^2 < 400$.
46. Illustrate on a number line the solution set of each pair of simultaneous inequalities:
(a) $x < 6$; $-3 \leqslant x \leqslant 8$
(b) $x > -2$; $-4 < x < 2$
(c) $2x + 1 \leqslant 5$; $-12 \leqslant 3x - 3$
(d) $3x - 2 < 19$; $2x \geqslant -6$
47. Find the integer n such that $n < \sqrt{300} < n + 1$

Graphical display

It is useful to represent inequalities on a graph, particularly where two variables are involved.

Example 3

Draw a sketch graph and shade the area which represents the set of points that satisfy each of these inequalities.
(a) $x > 2$
(b) $1 \leqslant y \leqslant 5$
(c) $x + y \leqslant 8$

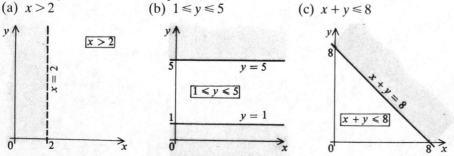

In each graph, the unwanted region is shaded so that the region representing the set of points is left clearly visible.

In (a), the line $x = 2$ is shown as a broken line to indicate that the points on the line are *not* included.

In (b) and (c), the lines $y = 1$, $y = 5$ and $x + y = 8$ are shown as solid lines because points on the line *are* included in the solution set.

An inequality can thus be regarded as a set of points, for example, the unshaded region in (c) may be described as

$$\{(x, y) : x + y \leqslant 8\}$$

i.e. the set of points (x, y) such that $x + y \leqslant 8$.

Exercise 19

In questions **1** to **9** describe the region left unshaded.

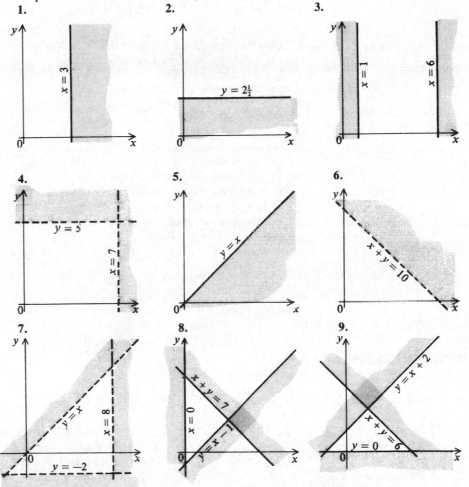

For questions **10** to **27**, draw a sketch graph similar to those above and indicate the set of points which satisfy the inequalities by shading the unwanted regions.

10. $2 \leqslant x \leqslant 7$

11. $0 \leqslant y \leqslant 3\frac{1}{2}$

12. $-2 < x < 2$

13. $x < 6$ and $y \leqslant 4$

14. $0 < x < 5$ and $y < 3$

15. $1 \leqslant x \leqslant 6$ and $2 \leqslant y \leqslant 8$

16. $-3 < x < 0$ and $-4 < y < 2$

17. $y \leqslant x$

18. $x + y < 5$

19. $y > x + 2$ and $y < 7$

20. $x \geqslant 0$ and $y \geqslant 0$ and $x + y \leqslant 7$

21. $x \geqslant 0$ and $x + y < 10$ and $y > x$

22. $8 \geqslant y \geqslant 0$ and $x + y > 3$

23. $x + 2y < 10$ and $x \geqslant 0$ and $y \geqslant 0$

24. $3x + 2y \leqslant 18$ and $x \geqslant 0$ and $y \geqslant 0$

25. $x \geqslant 0$, $y \geqslant x - 2$, $x + y \leqslant 10$

26. $3x + 5y \leqslant 30$ and $y > \dfrac{x}{2}$

27. $y \geqslant \dfrac{x}{2}$, $y \leqslant 2x$ and $x + y \leqslant 8$

For questions **28** to **30** draw an accurate graph to represent the inequalities listed, using shading to show the unwanted regions.

28. $x + y \leqslant 11$; $y \geqslant 3$; $y \leqslant x$.
Find the point having whole number coordinates and satisfying these inequalities which gives
(a) the maximum value of $x + 4y$
(b) the minimum value of $3x + y$

29. $3x + 2y > 24$; $x + y < 12$; $y < \frac{1}{2}x$; $y > 1$.
Find the point having whole number coordinates and satisfying these inequalities which gives
(a) the maximum value of $2x + 3y$
(b) the minimum value of $x + y$

30. $3x + 2y \leqslant 60$; $x + 2y \leqslant 30$; $x \geqslant 10$; $y \geqslant 0$.
Find the point having whole number coordinates and satisfying these inequalities which gives
(a) the maximum value of $2x + y$
(b) the maximum value of xy

REVISION EXERCISE 5A

1. Express the following as single fractions:

(a) $\dfrac{x}{4} + \dfrac{x}{5}$

(b) $\dfrac{1}{2x} + \dfrac{2}{3x}$

(c) $\dfrac{x+2}{2} + \dfrac{x-4}{3}$

(d) $\dfrac{7}{x-1} - \dfrac{2}{x+3}$

2. (a) Factorise $x^2 - 4$

(b) Simplify $\dfrac{3x - 6}{x^2 - 4}$

3. Given that $s - 3t = rt$, express
(a) s in terms of r and t
(b) r in terms of s and t
(c) t in terms of s and r.

4. (a) Given that $x - z = 5y$, express z in terms of x and y.
(b) Given that $mk + 3m = 11$, express m in terms of k.
(c) For the formula $T = C\sqrt{z}$, express z in terms of T and C.

5. It is given that $y = \dfrac{k}{x}$ and that $1 \leqslant x \leqslant 10$.
(a) If the smallest possible value of y is 5, find the value of the constant k.
(b) Find the largest possible value of y.

6. Given that y varies as x^2 and that $y = 36$ when $x = 3$, find
  (a) the value of y when $x = 2$
  (b) the value of x when $y = 64$.

7. (a) Evaluate (i) $9^{\frac{1}{2}}$ (ii) $8^{\frac{2}{3}}$ (iii) $16^{-\frac{1}{2}}$
  (b) Find x, given that
    (i) $3^x = 81$ (ii) $7^x = 1$.

8. List the integer values of x which satisfy.
  (a) $2x - 1 < 20 < 3x - 5$
  (b) $5 < 3x + 1 < 17$.

9. Given that $t = k\sqrt{(x + 5)}$, express x in terms of t and k.

10. Given that $z = \dfrac{3y + 2}{y - 1}$, express y in terms of z.

11. Given that $y = \dfrac{k}{k + w}$
  (a) Find the value of y when $k = \frac{1}{2}$ and $w = \frac{1}{3}$.
  (b) Express w in terms of y and k.

12. On a suitable sketch graph, identify clearly the region A defined by $x \geqslant 0$, $x + y \leqslant 8$ and $y \geqslant x$.

13. Without using tables, calculate the value of
  (a) $9^{-\frac{1}{2}} + (\frac{1}{8})^{\frac{1}{3}} + (-3)^0$
  (b) $(1000)^{-\frac{1}{3}} - (0 \cdot 1)^2$

14. It is given that $10^x = 3$ and $10^y = 7$. What is the value of 10^{x+y}?

15. Make x the subject of the following formulae
  (a) $x + a = \dfrac{2x - 5}{a}$
  (b) $cz + ax + b = 0$
  (c) $a = \sqrt{\left(\dfrac{x + 1}{x - 1}\right)}$

16. Write the following as single fractions
  (a) $\dfrac{3}{x} + \dfrac{1}{2x}$
  (b) $\dfrac{3}{a - 2} + \dfrac{1}{a^2 - 4}$
  (c) $\dfrac{3}{x(x + 1)} - \dfrac{2}{x(x - 2)}$

17. p varies jointly as the square of t and inversely as s. Given that $p = 5$ when $t = 1$ and $s = 2$, find a formula for p in terms of t and s.

18. Use the iteration formula $x_{n+1} = \dfrac{7}{x_n + 1}$ to find a solution of the equation $x^2 + x - 7 = 0$. Choose different values for the initial approximation x_1.

19. The shaded region A is formed by the lines $y = 2$, $y = 3x$ and $x + y = 6$. Write down the three inequalities which define A.

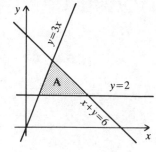

20. The shaded region B is formed by the lines $x = 0$, $y = x - 2$ and $x + y = 7$.

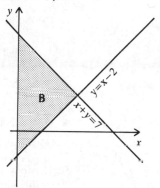

Write down the three inequalities which define B.

21. In the diagram below, the solution set $-1 \leqslant x < 2$ is shown on a number line.

Illustrate, on similar diagrams, the solution set of the following pairs of simultaneous inequalities.
  (a) $x > 2$, $x \leqslant 7$
  (b) $4 + x \geqslant 2$, $x + 4 < 10$
  (c) $2x + 1 \geqslant 3$, $x - 3 \leqslant 3$.

22. In a laboratory we start with 2 cells in a dish. The number of cells in the dish doubles every 30 minutes.
  (a) How many cells are in the dish after four hours?
  (b) After what time are there 2^{13} cells in the dish?
  (c) After $10\frac{1}{2}$ hours there are 2^{22} cells in the dish and an experimental fluid is added which eliminates half of the cells. How many cells are left?

EXAMINATION EXERCISE 5B

1. Given that $T = 2\pi \sqrt{\dfrac{\ell}{g}}$,
   (a) estimate the value of T, correct to 1 significant figure, when $\ell = 243$, and $g = 981$,
   (b) express g in terms of T, π and ℓ.   [M]

2. Jasbir is doing a simple experiment to test the relationship between two quantities m and ℓ. His theory is that m varies as the square of ℓ. He has already found that $m = 18$ when $\ell = 6$. Assuming that his theory is correct, what value of m should he obtain when $\ell = 4$?   [L]

3. (a) Make h the subject of the formula $d = \sqrt{\dfrac{3h}{2}}$
   (b) Solve the inequalities
       (i) $9 - 2x \leqslant 5$
       (ii) $\frac{1}{2}(2 - x) > -2$
       (iii) List the integers which satisfy both these inequalities simultaneously.
   (c) A positive integer r is such that $pr^2 = 168$, where p lies between 3 and 5. List the possible values of r.   [MEI]

4. Given that $f(x) = 8^x$, calculate
   (i) $f(\frac{2}{3})$,   (ii) $f(-\frac{1}{3})$.   [N]

5. The following table shows some values of the variables x and y which are linked by the equation $y = 8x^n$.

x	$\frac{1}{4}$	1	16
y	16	8	2

   (i) Find the value of n.
   (ii) Find y when $x = 36$.
   (iii) Find x when $y = 20$.
   (iv) Express x in terms of y.   [W]

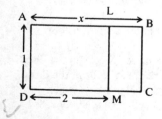

6. The rectangle ABCD is divided into two smaller rectangles by the line LM. $AD = 1$, $DM = 2$, $AB = x$.
   (a) Write down MC in terms of x.
   (b) Rectangle ABCD and BCML are similar. Use this fact to write down two expressions that must necessarily be equal to each other and show that $x^2 - 2x - 1 = 0$.
   (c) Show that the equation $x^2 - 2x - 1 = 0$ can be written in the form
       $$x = 2 + \frac{1}{x}.$$
       With $x_0 = 3$ use $x_{n+1} = 2 + \dfrac{1}{x_n}$ to calculate x, and x_2. Write down all the figures shown by your calculator.
   (d) Continue the iteration and find a solution of the equation $x^2 - 2x - 1 = 0$ correct to three significant figures.   [L]

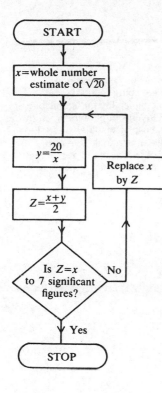

START

x=whole number estimate of $\sqrt{20}$

$y=\dfrac{20}{x}$

Replace x by Z

$Z=\dfrac{x+y}{2}$

Is $Z=x$ to 7 significant figures? No

Yes

STOP

7. (a) If p is the largest whole number which is a perfect square and which is also less than or equal to n, find p when
   (i) $n = 4 \cdot 9$, (ii) $n = 49$, (iii) $n = 490$.
   (b) The following flow diagram can be used to find the approximate square root of 20, after first choosing an appropriate whole number x as an estimate of $\sqrt{20}$.
   Start with $x = 4$ and make out a table like the one below to show every stage in finding $\sqrt{20}$ to 7 significant figures. Show all the figures on your calculator at each stage.

x	y	Z	difference between x and Z
4·0	5·0	4·5	0·5
4·5	4·4444444		
4·4722222			
etc.			

[S]

8. A sequence is defined by $u_n = \frac{1}{2}n(n+1)$.
   (a) Work out u_0, u_1, u_2, u_3 and u_4.
   (b) Work out $u_1 - u_0$, $u_2 - u_1$, $u_3 - u_2$ and $u_4 - u_3$.
   (c) Write down u_{n-1}, and simplify $u_n - u_{n-1}$.
   (d) Hence show that $1 + 2 + 3 + \cdots + n = u_n$.
   (e) If the sum of the first n positive integers is 4950, show that $n^2 + n - 9900 = 0$, and hence find the value of n. [M]

9. A business requires containers (made from thin sheet metal), such that the internal measurements are related as shown in the diagram.
   (i) The internal volume of the container is V m^3.
      (a) Calculate the value of V when $x = 0 \cdot 6$.
      (b) Write down a formula for V in terms of x.
   (ii) The value of V is to be as close to 1 as possible.

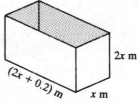

$2x$ m

$(2x + 0.2)$ m x m

   (a) One possible rearrangement of the equation obtained by putting $V = 1$ in the formula is

$$x = \frac{1}{4x^2} - 0 \cdot 1.$$

   Use an iterative method with this form of the equation to find the values of x_1, x_2 and x_3 when $x_0 = 0 \cdot 6$.
   (b) Show that the equation may also be rearranged as

$$x = \frac{1}{2\sqrt{(x + 0 \cdot 1)}}$$

   Use an iterative method with this form of the equation to find the values of x_1, x_2 and x_3 when $x_0 = 0 \cdot 6$.
   (c) State, correct to 3 significant figures, the required value of x. [N]

6 Trigonometry

Leonard Euler (1707–1783) was born near Basel in Switzerland but moved to St Petersburg in Russia and later to Berlin. He had an amazing facility for figures but delighted in speculating in the realms of pure intellect. In trigonometry he introduced the use of small letters for the sides and capitals for the angles of a triangle. He also wrote r, R and s for the radius of the inscribed and of the circumscribed circles and the semi-perimeter, giving the beautiful formula $4rRs = abc$.

6.1 RIGHT-ANGLED TRIANGLES

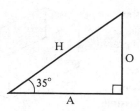

The side opposite the right angle is called the hypotenuse (we will use H). It is the longest side.

The side opposite the marked angle of 35° is called the opposite (we will use O).

The other side is called the adjacent (we will use A).

Consider two triangles, one of which is an enlargement of the other.

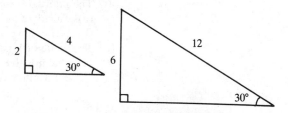

It is clear that the *ratio* $\dfrac{O}{H}$ will be the same in both triangles.

Sine, cosine and tangent

Three important functions are defined as follows:

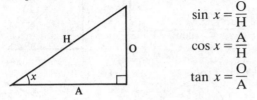

$$\sin x = \frac{O}{H}$$

$$\cos x = \frac{A}{H}$$

$$\tan x = \frac{O}{A}$$

It is important to get the letters in the right order. Some people find a simple sentence helpful where the first letters of each word describe sine, cosine or tangent and Hypotenuse, Opposite and Adjacent. An example is:

Silly Old Harry Caught A Herring Trawling Off Afghanistan.

e.g.   S O H : $\sin = \dfrac{O}{H}$

For any angle x the values for $\sin x$, $\cos x$ and $\tan x$ can be found using either a calculator or tables.

Exercise 1

1. Draw a circle of radius 10 cm and construct a tangent to touch the circle at T.

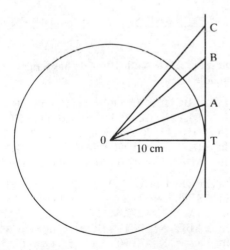

Draw OA, OB and OC where $A\hat{O}T = 20°$
$B\hat{O}T = 40°$
$C\hat{O}T = 50°$.

Measure the length AT and compare it with the value for tan 20° given on a calculator or in tables. Repeat for BT, CT and for other angles of your own choice.

Finding the length of a side

Example 1

Find the side marked x

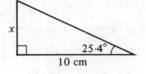

(a) Label the sides of the triangle H, O, A (in brackets)

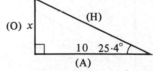

(b) In this example, we know nothing about H so we need the function involving O and A.

$$\tan 25\cdot4° = \frac{O}{A} = \frac{x}{10}$$

(c) Find tan 25·4° from tables

$$0\cdot4748 = \frac{x}{10}$$

(d) Solve for x

$$x = 10 \times 0\cdot4748 = 4\cdot748$$
$$x = 4\cdot75 \text{ cm (3 significant figures)}$$

Example 2

Find the side marked z.

(a) Label H, O, A.

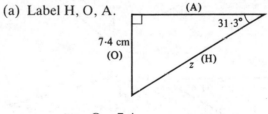

(b) $\sin 31\cdot3° = \dfrac{O}{H} = \dfrac{7\cdot4}{z}$

(c) Multiply by z.

$$z \times (\sin 31\cdot3°) = 7\cdot4$$
$$z = \frac{7\cdot4}{\sin 31\cdot3}$$

(d) On a calculator, press the keys as follows:

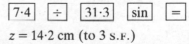

$$z = 14\cdot2 \text{ cm (to 3 s.f.)}$$

Exercise 2

In questions **1** to **22**, find the length of the side marked with a letter. Give your answers to three significant figures.

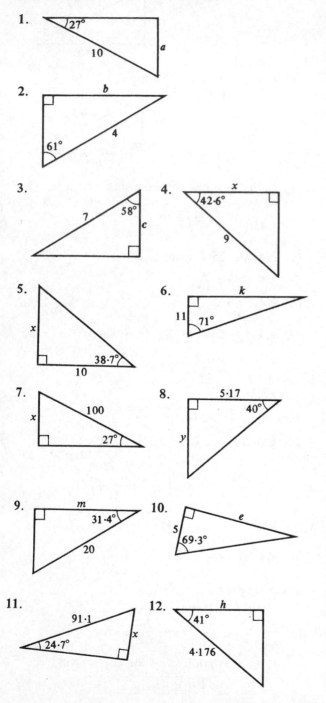

1.

2.

3.

4.

5.

6.

7.

8.

9.

10.

11.

12.

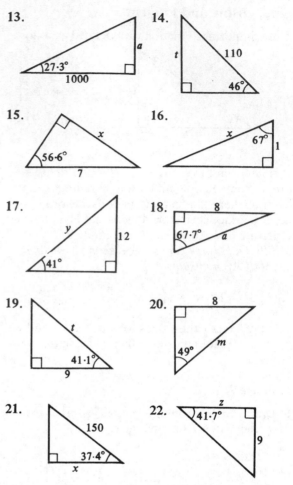

13.

14.

15.

16.

17.

18.

19.

20.

21.

22.

In questions **23** to **34**, the triangle has a right angle at the middle letter.

23. In △ABC, Ĉ = 40°, BC = 4 cm. Find AB
24. In △DEF, F̂ = 35·3°, DF = 7 cm. Find ED
25. In △GHI, Î = 70°, GI = 12 m. Find HI
26. In △JKL, L̂ = 55°, KL = 8·21 m. Find JK
27. In △MNO, M̂ = 42·6°, MO = 14 cm. Find ON

28. In △PQR, P̂ = 28°, PQ = 5·071 m. Find PR
29. In △STU, Ŝ = 39°, TU = 6 cm. Find SU
30. In △VWX, X̂ = 17°, WV = 30·7 m. Find WX
31. In △ABC, Â = 14°17′, BC = 14 m. Find AC
32. In △KLM, K̂ = 72°50′, KL = 5·04 cm. Find LM

33. In △PQR, R̂ = 31°43′, QR = 0·81 cm. Find PR
34. In △YXZ, X̂ = 81°4′, YZ = 52·6 m. Find XY

Example 3

Find the length marked x.

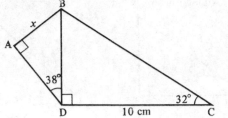

(a) Find BD from triangle BDC

$$\tan 32° = \frac{BD}{10}$$

$\therefore \qquad BD = 10 \times \tan 32° \qquad \dots [1]$

(b) Now find x from triangle ABD

$$\sin 38° = \frac{x}{BD}$$

$\therefore \qquad x = BD \times \sin 38°$
$\qquad\qquad x = 10 \times \tan 32° \times \sin 38 \text{ (from [1])}$
$\qquad\qquad x = 3\cdot85 \text{ cm (to 3 s.f.)}$

Notice that BD was *not* calculated from [1].

It is better to do all the multiplications at one time.

Exercise 3

In questions **1** to **10**, find each side marked with a letter.

1.

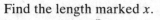

2.

3.   **4.**

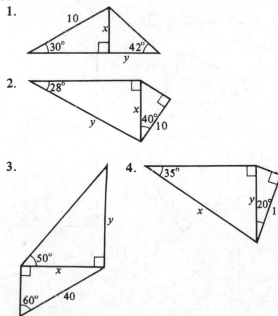

5.   **6.**

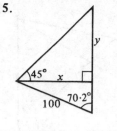
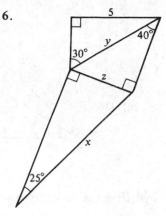

7.   **8.**

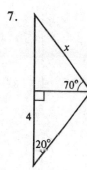
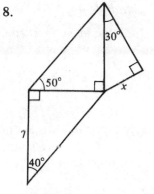

9.   **10.**

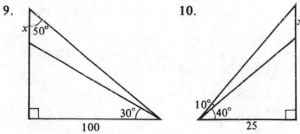

11. $B\hat{A}D = A\hat{C}D = 90°$
$\qquad C\hat{A}D = 35°$
$\qquad B\hat{D}A = 41°$
$\qquad AD = 20 \text{ cm}$
$\qquad$Calculate
$\qquad$(a) AB
$\qquad$(b) DC
$\qquad$(c) BD

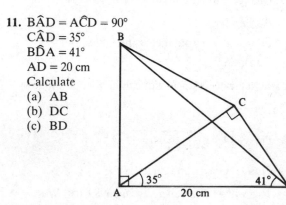

12. AB̂D = AD̂C = 90°
    CÂD = 31°
    BD̂A = 43°
    AD = 10 cm
    Calculate
    (a) AB
    (b) CD
    (c) DB

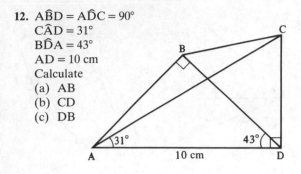

Finding an unknown angle

Example 4

Find the angle marked *m*.

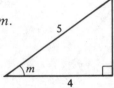

(a) Label the sides of the triangle
     H, O, A in relation
     to angle *m*.

(b) In this example, we do not know 'O' so we
     need the cosine.
$$\cos m = \left(\frac{A}{H}\right) = \frac{4}{5}$$

(c) Change $\frac{4}{5}$ to a decimal: $\frac{4}{5} = 0.8$

(d) $\cos m = 0.8$

     Find angle *m* from the cosine table
     $m = 36.9°$

Note: On a calculator, angles can be found as follows:

If $\cos m = \frac{4}{5}$

(a) Press 4 ÷ 5 =

(b) Press INV and then COS

     This will give the angle as 36·86989765°. We
     require the angle to 1 place of decimals so
     $m = 36.9°$.

Exercise 4

For questions **1** to **15**, find the angle marked with a
letter. All lengths are in cm.

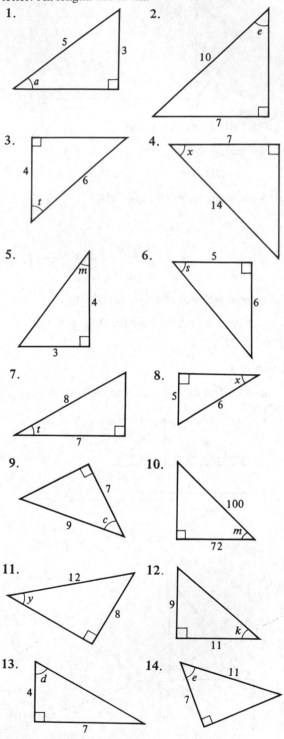

15.

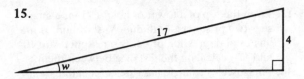

In questions **16** to **20**, the triangle has a right angle at the middle letter.

16. In $\triangle ABC$, $BC = 4$, $AC = 7$. Find $\hat{A}$.
17. In $\triangle DEF$, $EF = 5$, $DF = 10$. Find $\hat{F}$.
18. In $\triangle GHI$, $GH = 9$, $HI = 10$. Find $\hat{I}$.
19. In $\triangle JKL$, $JL = 5$, $KL = 3$. Find $\hat{J}$.
20. In $\triangle MNO$, $MN = 4$, $NO = 5$. Find $\hat{M}$.

In questions **21** to **26**, find the angle x.

21.   **22.**

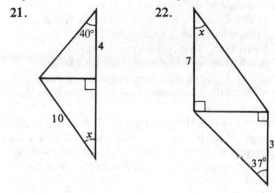

23.   **24.**

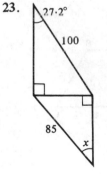

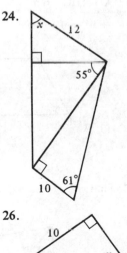

25.   **26.**

Example 5

A ship sails 22 km from A on a bearing of 042°, and a further 30 km on a bearing of 090° to arrive at B. What is the distance and bearing of B from A?

(a) Draw a clear diagram and label extra points as shown.

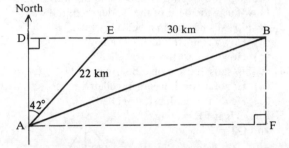

(b) Find DE and AD.

  (i) $\sin 42° = \dfrac{DE}{22}$

     ∴ $DE = 22 \times \sin 42° = 14\cdot72$ km

  (ii) $\cos 42° = \dfrac{AD}{22}$

     ∴ $AD = 22 \times \cos 42° = 16\cdot35$ km

(c) Using triangle ABF,
        $AB^2 = AF^2 + BF^2$ (Pythagoras' Theorem)
and $AF = DE + EB$
         $AF = 14\cdot72 + 30 = 44\cdot72$ km
and $BF = AD = 16\cdot35$ km.

  ∴ $AB^2 = 44\cdot72^2 + 16\cdot35^2$
              $= 2267\cdot2$

       $AB = 47\cdot6$ km (to 3 s.f.)

(d) The bearing of B from A is given by the angle DAB.
But $D\hat{A}B = A\hat{B}F$.

      $\tan A\hat{B}F = \dfrac{AF}{BF} = \dfrac{44\cdot72}{16\cdot35}$
             $= 2\cdot7352$

  ∴ $A\hat{B}F = 69\cdot9°$.

B is 47·6 km from A on a bearing of 069·9°.

Exercise 5

In this exercise, start by drawing a clear diagram.

1. A ladder of length 6 m leans against a vertical wall so that the base of the ladder is 2 m from the wall. Calculate the angle between the ladder and the wall.

2. A ladder of length 8 m rests against a wall so that the angle between the ladder and the wall is 31°. How far is the base of the ladder from the wall?

3. A ship sails 35 km on a bearing of 042°.
   (a) How far north has it travelled?
   (b) How far east has it travelled?

4. A ship sails 200 km on a bearing of 243·7°.
   (a) How far south has it travelled?
   (b) How far west has it travelled?

5. Find TR if PR = 10 m and QT = 7 m.

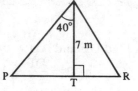

6. Find *d*.

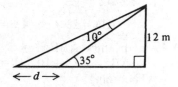

7. An aircraft flies 400 km from a point O on a bearing of 025° and then 700 km on a bearing of 080° to arrive at B.
   (a) How far north of O is B?
   (b) How far east of O is B?
   (c) Find the distance and bearing of B from O.

8. An aircraft flies 500 km on a bearing of 100° and then 600 km on a bearing of 160°.
   Find the distance and bearing of the finishing point from the starting point.

For questions **9** to **12**, plot the points for each question on a sketch graph with *x*- and *y*-axes drawn to the same scale.

9. For the points A(5, 0) and B(7, 3), calculate the angle between AB and the *x*-axis.

10. For the points C(0, 2) and D(5, 9), calculate the angle between CD and the *y*-axis.

11. For the points A(3, 0), B(5, 2) and C(7, −2), calculate the angle BAC.

12. For the points P(2, 5), Q(5, 1) and R(0, −3), calculate the angle PQR.

13. From the top of a tower of height 75 m, a guard sees two prisoners, both due West of him. If the angles of depression of the two prisoners are 10° and 17°, calculate the distance between them.

14. An isosceles triangle has sides of length 8 cm, 8 cm and 5 cm. Find the angle between the two equal sides.

15. The angles of an isosceles triangle are 66°, 66° and 48°. If the shortest side of the triangle is 8·4 cm, find the length of one of the two equal sides.

16. A chord of length 12 cm subtends an angle of 78·2° at the centre of a circle. Find the radius of the circle.

17. Find the acute angle between the diagonals of a rectangle whose sides are 5 cm and 7 cm.

18. A kite flying at a height of 55 m is attached to a string which makes an angle of 55° with the horizontal. What is the length of the string?

19. A boy is flying a kite from a string of length 150 m. If the string is taut and makes an angle of 67° with the horizontal, what is the height of the kite?

20. A rocket flies 10 km vertically, then 20 km at an angle of 15° to the vertical and finally 60 km at an angle of 26° to the vertical. Calculate the vertical height of the rocket at the end of the third stage.

21. Find *x*, given AD = BC = 6 m.

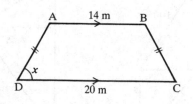

22. Find *x*.

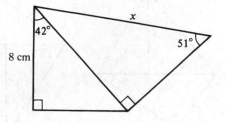

23. Ants can hear each other up to a range of 2 m. An ant A, 1 m from a wall sees her friend B about to be eaten by a spider. If the angle of elevation of B from A is 62°, will the spider have a meal or not? (Assume B escapes if he hears A calling.)

24. A hedgehog wishes to cross a road without being run over. He observes the angle of elevation of a lamp post on the other side of the road to be 27° from the edge of the road and 15° from a point 10 m back from the road. How wide is the road? If he can run at 1 m/s, how long will he take to cross?
If cars are travelling at 20 m/s, how far apart must they be if he is to survive?

25. From a point 10 m from a vertical wall, the angles of elevation of the bottom and the top of a statue of Sir Isaac Newton, set in the wall, are 40° and 52°. Calculate the length of the statue.

26. A rectangular paving stone 3 m by 1 m rests against a vertical wall as shown.

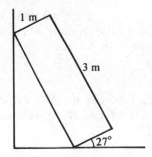

What is the height of the highest point of the stone above the ground?

27. A rectangular piece of paper 30 cm by 21 cm is folded so that opposite corners coincide. How long is the crease?

6.2 SCALE DRAWING

Exercise 6

Make a scale drawing and then answer the questions.

1. A field has four sides as shown below

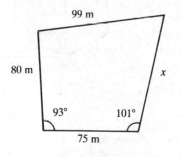

How long is the side *x* in metres?

2. A destroyer and a cruiser leave a port at the same time. The destroyer sails at 38 knots on a bearing of 042° and the cruiser sails at 25 knots on a bearing of 315°. How far apart are the ships two hours later?

3. Two radar stations A and B are 80 km apart and B is due East of A.
One aircraft is on a bearing of 030° from A and 346° from B. A second aircraft is on a bearing of 325° from A and 293° from B.
How far apart are the two aircraft?

4. A ship sails 95 km on a bearing of 140°, then a further 102 km on a bearing of 260° and then returns directly to its starting point. Find the length and bearing of the return journey.

5. A control tower observes the flight of an unidentified flying object.
At 09 23 the U.F.O. is 580 km away on a bearing of 043°.
At 09 25 the U.F.O. is 360 km away on a bearing of 016°.
What is the speed and the course of the U.F.O.? [Use a scale of 1 cm to 50 km]

6. Make a scale drawing of the diagram below and find the length of CD in km.

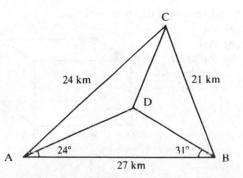

6.3 THREE-DIMENSIONAL PROBLEMS

Always draw a large, clear diagram. It is often helpful to redraw the triangle which contains the length or angle to be found.

Example 1

A rectangular box with top WXYZ and base ABCD has AB = 6 cm, BC = 8 cm and WA = 3 cm.

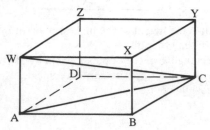

Calculate
(a) the length of AC
(b) the angle between WC and AC.

(a) Redraw triangle ABC

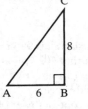

$AC^2 = 6^2 + 8^2 = 100$
$AC = 10$ cm.

(b) Redraw triangle WAC

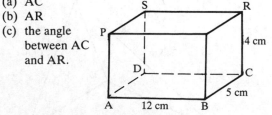

Let $W\widehat{C}A = \theta$
$\tan \theta = \frac{3}{10}$
$\theta = 16.7°$

The angle between WC and AC is 16.7°.

Exercise 7

1. In the rectangular box shown, find
   (a) AC
   (b) AR
   (c) the angle between AC and AR.

2. A vertical pole BP stands at one corner of a horizontal rectangular field as shown.

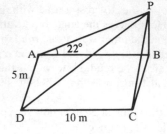

If AB = 10 m, AD = 5 m and the angle of elevation of P from A is 22°, calculate
   (a) the height of the pole
   (b) the angle of elevation of P from C
   (c) the length of a diagonal of the rectangle ABCD
   (d) the angle of elevation of P from D.

3. In the cube shown, find
   (a) BD
   (b) AS
   (c) BS
   (d) the angle SBD
   (e) the angle ASB

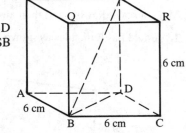

4. In the cuboid shown, find
   (a) WY
   (b) DY
   (c) WD
   (d) the angle WDY

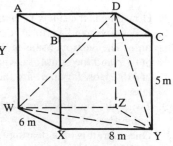

5. In the square-based pyramid, V is vertically above the middle of the base, AB = 10 cm and VC = 20 cm. Find

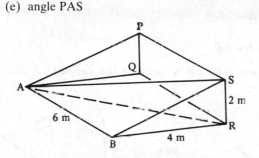

(a) AC
(b) the height of the pyramid
(c) The angle between VC and the base ABCD
(d) the angle AVB
(c) the angle AVC.

6. In the wedge shown, PQRS is perpendicular to ABRQ; PQRS and ABRQ are rectangles with AB = QR = 6 m, BR = 4 m, RS = 2 m. Find

(a) BS   (b) AS
(c) angle BSR   (d) angle ASR
(e) angle PAS

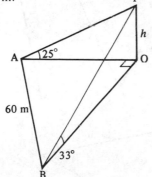

7. The edges of a box are 4 cm, 6 cm and 8 cm. Find the length of a diagonal and the angle it makes with the diagonal on the largest face.

8. In the diagram A, B and O are points in a horizontal plane and P is vertically above O, where OP = h m.

A is due West of O, B is due South of O and AB = 60 m. The angle of elevation of P from A is 25° and the angle of elevation of P from B is 33°.
(a) Find the length AO in terms of h
(b) Find the length BO in terms of h
(c) Find the value of h.

9. The angle of elevation of the top of a tower is 38° from a point A due South of it. The angle of elevation of the top of the tower from another point B, due East of the tower is 29°. Find the height of the tower if the distance AB is 50 m.

10. An observer at the top of a tower of height 15 m sees a man due West of him at an angle of depression 31°. He sees another man due South at an angle of depression 17°. Find the distance between the men.

11. The angle of elevation of the top of a tower is 27° from a point A due East of it. The angle of elevation of the top of the tower is 11° from another point B due South of the tower. Find the height of the tower if the distance AB is 40 m.

12. The figure shows a triangular pyramid on a horizontal base ABC, V is vertically above B where VB = 10 cm, A\hat{B}C = 90° and AB = BC = 15 cm. Point M is the mid point of AC.

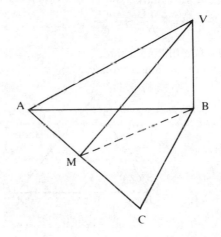

Calculate the size of angle VMB.

6.4 THE SINE RULE

The sine rule enables us to calculate sides and angles in some triangles where there is not a right angle.

In △ABC, we use the convention that
a is the side opposite $\widehat{A}$
b is the side opposite $\widehat{B}$, etc.

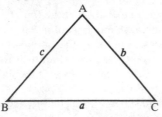

Sine rule

either $\quad \dfrac{a}{\sin A} = \dfrac{b}{\sin B} = \dfrac{c}{\sin C} \qquad \ldots[1]$

or $\quad \dfrac{\sin A}{a} = \dfrac{\sin B}{b} = \dfrac{\sin C}{c} \qquad \ldots[2]$

Use [1] when finding a *side*,
and [2] when finding an *angle*.

Example 1

Find c.

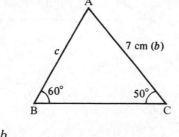

$$\frac{c}{\sin C} = \frac{b}{\sin B}$$
$$\frac{c}{\sin 50°} = \frac{7}{\sin 60°}$$
$$c = \frac{7 \times \sin 50°}{\sin 60°} = 6 \cdot 19 \text{ cm (3 s.f.)}$$

Although we cannot have an angle of more than 90° in a right-angled triangle, it is still useful to define sine, cosine and tangent for these angles. For an obtuse angle x,
we have $\sin x = \sin(180 - x)$
Examples $\quad \sin 130° = \sin 50°$
$\sin 170° = \sin 10°$
$\sin 116° = \sin 64°$
Most people simply use a calculator when finding the sine of an obtuse angle.

Example 2

Find $\widehat{B}$.

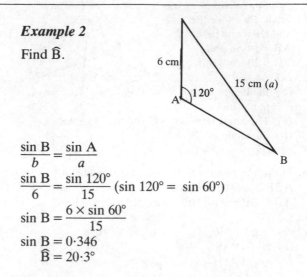

$$\frac{\sin B}{b} = \frac{\sin A}{a}$$
$$\frac{\sin B}{6} = \frac{\sin 120°}{15} \quad (\sin 120° = \sin 60°)$$
$$\sin B = \frac{6 \times \sin 60°}{15}$$
$$\sin B = 0 \cdot 346$$
$$\widehat{B} = 20 \cdot 3°$$

Exercise 8

For questions **1** to **6**, find each side marked with a letter.

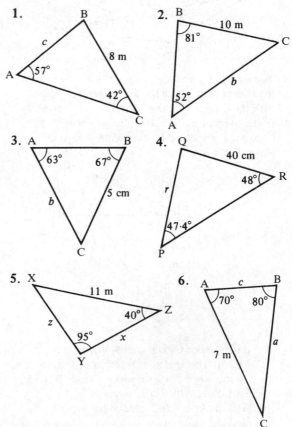

7. In △ABC, Â = 61°, B̂ = 47°, AC = 7·2 cm.
   Find BC.
8. In △XYZ, Ẑ = 32°, Ŷ = 78°, XY = 5·4 cm.
   Find XZ.
9. In △PQR, Q̂ = 100°, R̂ = 21°, PQ = 3·1 cm.
   Find PR.
10. In △LMN, L̂ = 21°, N̂ = 30°, MN = 7 cm.
    Find LN.

In questions **11** to **18**, find each angle marked *. All lengths are in centimetres.

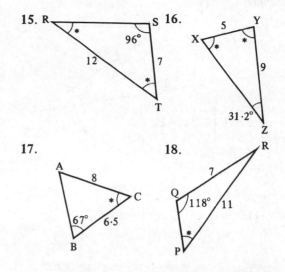

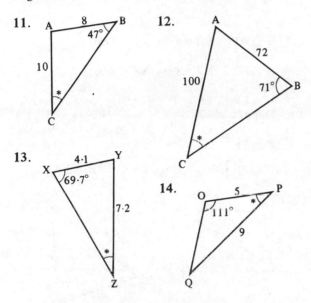

17.   18.

19. In △ABC, Â = 62°, BC = 8, AB = 7.
    Find Ĉ.
20. In △XYZ, Ŷ = 97·3°, XZ = 22, XY = 14.
    Find Ẑ.
21. In △DEF, D̂ = 58°, EF = 7·2, DE = 5·4.
    Find F̂.
22. In △LMN, M̂ = 127·1°, LN = 11·2, LM = 7·3.
    Find L̂.

6.5   THE COSINE RULE

We use the cosine rule when we have either
(a) two sides and the included angle or
(b) all three sides.

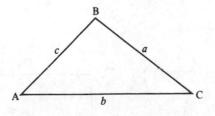

There are two forms.

1. To find the length of a side.

$$a^2 = b^2 + c^2 - (2bc \cos A)$$
$$\text{or} \quad b^2 = c^2 + a^2 - (2ac \cos B)$$
$$\text{or} \quad c^2 = a^2 + b^2 - (2ab \cos C)$$

2. To find an angle when given all three sides.

$$\cos A = \frac{b^2 + c^2 - a^2}{2bc}$$
$$\text{or} \quad \cos B = \frac{a^2 + c^2 - b^2}{2ac}$$
$$\text{or} \quad \cos C = \frac{a^2 + b^2 - c^2}{2ab}$$

For an obtuse angle x we have

$$\cos x = -\cos(180 - x)$$

Examples $\cos 120° = -\cos 60°$
            $\cos 142° = -\cos 38°$

Example 1

Find b.

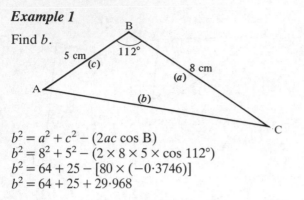

$b^2 = a^2 + c^2 - (2ac \cos B)$
$b^2 = 8^2 + 5^2 - (2 \times 8 \times 5 \times \cos 112°)$
$b^2 = 64 + 25 - [80 \times (-0.3746)]$
$b^2 = 64 + 25 + 29.968$

(Notice the change of sign for the obtuse angle)

$b = \sqrt{(118.968)} = 10.9$ cm (to 3 s.f.)

Example 2

Find angle C.

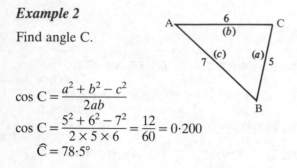

$\cos C = \dfrac{a^2 + b^2 - c^2}{2ab}$

$\cos C = \dfrac{5^2 + 6^2 - 7^2}{2 \times 5 \times 6} = \dfrac{12}{60} = 0.200$

$\widehat{C} = 78.5°$

Exercise 9

Find the sides marked ∗.

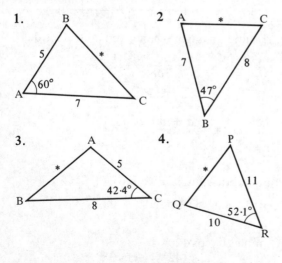

1.

2

3.

4.

5.

6.

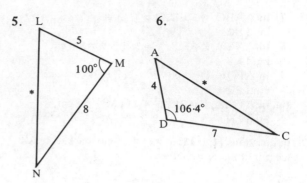

7. In △ABC, AB = 4 cm, AC = 7 cm, $\widehat{A} = 57°$.
Find BC.

8. In △XYZ, XY = 3 cm, YZ = 3 cm, $\widehat{Y} = 90°$.
Find XZ.

9. In △LMN, LM = 5.3 cm, MN = 7.9 cm, $\widehat{M} = 127°$.
Find LN.

10. In △ PQR, $\widehat{Q} = 117°$, PQ = 80 cm, QR = 100 cm.
Find PR.

In questions **11** to **16**, find the angles marked ∗.

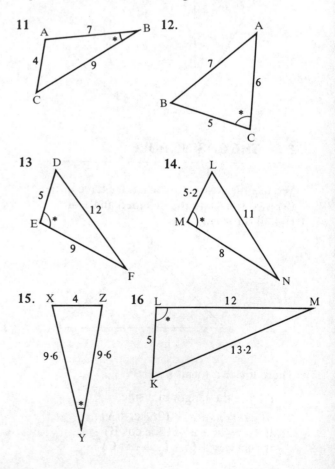

11

12.

13

14.

15.

16

17. In △ABC, $a = 4\cdot3$, $b = 7\cdot2$, $c = 9$. Find $\hat{C}$.
18. In △DEF, $d = 30$, $e = 50$, $f = 70$. Find $\hat{E}$.
19. In △PQR, $p = 8$, $q = 14$, $r = 7$. Find $\hat{Q}$.
20. In △LMN, $l = 7$, $m = 5$, $n = 4$. Find $\hat{N}$.
21. In △XYZ, $x = 5\cdot3$, $y = 6\cdot7$, $z = 6\cdot14$. Find $\hat{Z}$.
22. In △ABC, $a = 4\cdot1$, $c = 6\cdot3$, $\hat{B} = 112°10'$. Find b.
23. In △PQR, $r = 0\cdot72$, $p = 1\cdot14$, $\hat{Q} = 94°33'$. Find q.
24. In △LMN, $n = 7\cdot206$, $l = 6\cdot3$, $\hat{L} = 51°10'$, $\hat{N} = 63°$. Find m.

Example 3

A ship sails from a port P a distance of 7 km on a bearing of 306° and then a further 11 km on a bearing of 070° to arrive at X. Calculate the distance from P to X.

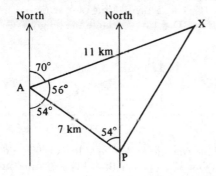

$$PX^2 = 7^2 + 11^2 - (2 \times 7 \times 11 \times \cos 56°)$$
$$= 49 + 121 - (86\cdot12)$$

$$PX^2 = 83\cdot88$$
$$PX = 9\cdot16 \text{ km (to 3 s.f.)}$$

The distance from P to X is 9·16 km.

Exercise 10

Start each question by drawing a large, clear diagram.

1. In triangle PQR $\hat{Q} = 72°$, $\hat{R} = 32°$ and PR = 12 cm. Find PQ.
2. In triangle LMN $\hat{M} = 84°$, LM = 7 m and MN = 9 m. Find LN.
3. A destroyer D and a cruiser C leave port P at the same time. The destroyer sails 25 km on a bearing 040° and the cruiser sails 30 km on a bearing of 320°. How far apart are the ships?
4. Two honeybees A and B leave the hive H at the same time; A flies 27 m due south and B flies 9 m on a bearing of 111°. How far apart.are they?
5. Find all the angles of a triangle in which the sides are in the ratio 5 : 6 : 8.

6. A golfer hits his ball B a distance of 170 m on a hole H which measures 195 m from the tee T to the green. If his shot is directed 10° away from the true line to the hole, find the distance between his ball and the hole.
7. From A, B lies 11 km away on a bearing of 041° and C lies 8 km away on a bearing of 341°. Find
   (a) the distance between B and C
   (b) the bearing of B from C.
8. From a lighthouse L an aircraft carrier A is 15 km away on a bearing of 112° and a submarine S is 26 km away on a bearing of 200°. Find
   (a) the distance between A and S
   (b) the bearing of A from S.
9. Find (a) AE
          (b) EÂC
   If the line BCD is horizontal, find the angle of elevation of E from A.

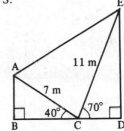

10. An aircraft flies from its base 200 km on a bearing 162°, then 350 km on a bearing 260°, and then returns directly to base. Calculate the length and bearing of the return journey.
11. Town Y is 9 km due north of town Z. Town X is 8 km from Y, 5 km from Z and somewhere to the west of the line YZ.
    (a) Draw triangle XYZ and find angle YZX
    (b) During an earthquake, town X moves due south until it is due west of Z. Find how far it has moved.
12. Calculate WX, given YZ = 15 m.

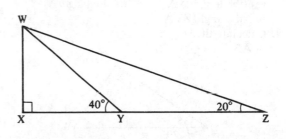

13. A golfer hits her ball a distance of 127 m so that it finishes 31 m from the hole. If the length of the hole is 150 m, calculate the angle between the line of her shot and the direct line to the hole.

14. (a) Find the sine of all the angles 0°, 10°, 20°, 30°,
        ... 350°, 360°.
    (b) Draw a graph of $y = \sin x$ for $0° \leqslant x \leqslant 360°$. Use
        a scale of 1 cm to 20° on the x-axis and
        5 cm to 1 unit on the y-axis.

15. (a) Find the cosine of all the angles 0°, 10°, 20°,
        30°, ... 350°, 360°.
    (b) Draw a graph of $y = \cos x$ for $0° \leqslant x \leqslant 360°$.
        Use a scale of 1 cm to 20° on the x-axis and
        5 cm to 1 unit on the y-axis.

16. Draw a graph of $y = \tan x$ for $0° \leqslant x \leqslant 360°$. Use
    a scale of 1 cm to 20° on the x-axis and 1 cm to
    1 unit on the y-axis.

17. Use the graphs in questions 14, 15, 16 to find
    approximate solutions to the equations
    (a) $\sin x = 0 \cdot 3$
    (b) $\cos x = 0 \cdot 75$
    (c) $\tan x = 2 \cdot 5$

REVISION EXERCISE 6A

1. Calculate the side or angle marked with a letter.

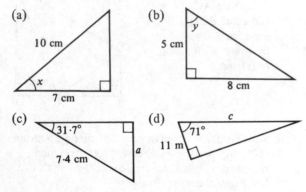

(a)   (b)

(c)   (d)

2. Given that x is an acute angle and that
    $$3 \tan x - 2 = 4 \cos 35 \cdot 3°$$
   calculate
   (a) $\tan x$
   (b) the value of x in degrees correct to 1 D.P.

3. In the triangle XYZ, XY = 14 cm, XZ = 17 cm
   and angle YXZ = 25°. A is the foot of the
   perpendicular from Y to XZ. Calculate
   (a) the length XA   (b) the length YA
   (c) the angle ZYA

4. Calculate the length
   of AB.

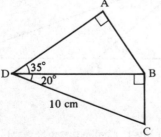

5. (a) A lies on a bearing of 040° from B.
       Calculate the bearing of B from A.
   (b) The bearing of X from Y is 115°.
       Calculate the bearing of Y from X.

6. Given BD = 1 m, calculate the length AC.

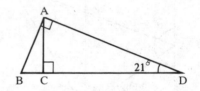

7. In the triangle PQR, angle PQR = 90° and angle
   RPQ = 31°. The length of PQ is 11 cm. Calculate
   (a) the length of QR
   (b) the length of PR
   (c) the length of the perpendicular from Q to PR.

8. $\hat{BAD} = \hat{DCA} = 90°$, $\hat{CAD} = 32 \cdot 4°$, $\hat{BDA} = 41°$ and
   AD = 100 cm.

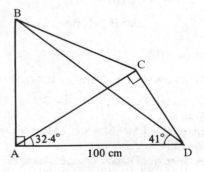

Calculate
(a) the length of AB
(b) the length of DC
(c) the length of BD.

9. An observer at the top of a tower of height 20 m sees a man due East of him at an angle of depression of 27°. He sees another man due South of him at an angle of depression of 30°. Find the distance between the men on the ground.

10. The figure shows a cube of side 10 cm.

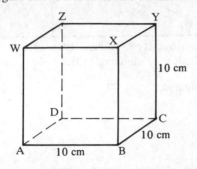

Calculate
(a) the length of AC
(b) the angle YAC
(c) the angle ZBD.

11. The diagram shows a rectangular block.
AY = 12 cm, AB = 8 cm, BC = 6 cm.

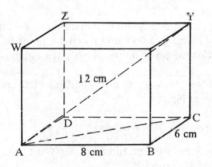

Calculate
(a) the length YC
(b) the angle YÂZ

12. VABCD is a pyramid in which the base ABCD is a square of side 8 cm; V is vertically above the centre of the square and
VA = VB = VC = VD = 10 cm.
Calculate
(a) the length AC
(b) the height of V above the base
(c) the angle VĈA.

Questions 13 to 18 may be answered either by scale drawing or by using the sine and cosine rules.

13. Two lighthouses A and B are 25 km apart and A is due West of B. A submarine S is on a bearing of 137° from A and on a bearing of 170° from B. Find the distance of S from A and the distance of S from B.

14. In triangle PQR, PQ = 7 cm, PR = 8 cm and QR = 9 cm. Find angle QPR.

15. In triangle XYZ, XY = 8 m, $\hat{X}$ = 57° and $\hat{Z}$ = 50°. Find the lengths YZ and XZ.

16. In triangle ABC, $\hat{A}$ = 22° and $\hat{C}$ = 44°.

Find the ratio $\dfrac{BC}{AB}$.

17. Given cos AĈB = 0·6, AC = 4 cm, BC = 5 cm and CD = 7 cm, find the length of AB and AD.

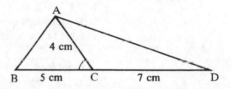

18. Find the smallest angle in a triangle whose sides are of length 3x, 4x and 6x.

EXAMINATION EXERCISE 6B

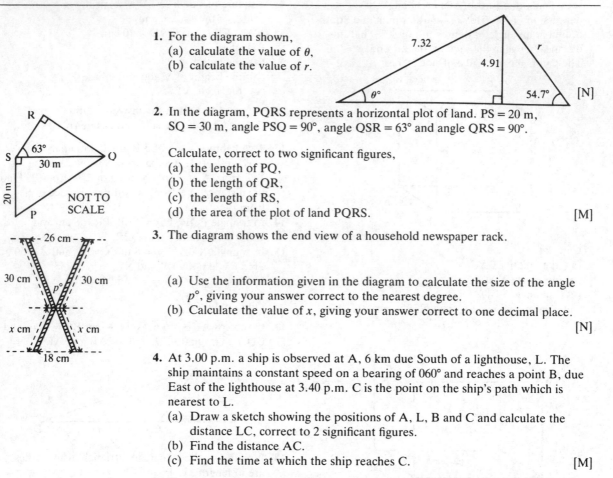

1. For the diagram shown,
   (a) calculate the value of θ,
   (b) calculate the value of r. [N]

2. In the diagram, PQRS represents a horizontal plot of land. PS = 20 m, SQ = 30 m, angle PSQ = 90°, angle QSR = 63° and angle QRS = 90°.

   Calculate, correct to two significant figures,
   (a) the length of PQ,
   (b) the length of QR,
   (c) the length of RS,
   (d) the area of the plot of land PQRS. [M]

3. The diagram shows the end view of a household newspaper rack.

   (a) Use the information given in the diagram to calculate the size of the angle $p°$, giving your answer correct to the nearest degree.
   (b) Calculate the value of x, giving your answer correct to one decimal place. [N]

4. At 3.00 p.m. a ship is observed at A, 6 km due South of a lighthouse, L. The ship maintains a constant speed on a bearing of 060° and reaches a point B, due East of the lighthouse at 3.40 p.m. C is the point on the ship's path which is nearest to L.
   (a) Draw a sketch showing the positions of A, L, B and C and calculate the distance LC, correct to 2 significant figures.
   (b) Find the distance AC.
   (c) Find the time at which the ship reaches C. [M]

5. Two lookout posts, A and B on a straight coastline running East-West sight a ship (S) on a bearing 067° from A and 337° from B.

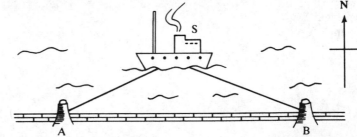

   (a) Explain why angle ASB is 90°.
   The distance from A to B is 5 kilometres.
   (b) Calculate the distance of the ship from A.
   (c) Calculate the distance of the ship from B.
   The ship sails on a course such that angle ASB is always 90°.
   (d) Describe the path the ship must take.
   (e) What is the bearing of the ship from A (to the nearest degree) when it is 3 kilometres from it? [S]

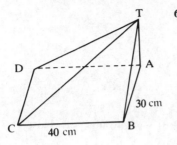

6. The diagram shows a pyramid with a rectangular horizontal base, ABCD, and a vertex, T, which is vertically above the point A. The length of AB is 30 cm and the length of BC is 40 cm. The size of angle ADT is 55·9°.

Calculate, giving your answers correct to 3 significant figures,
(a) the length of AT.   (b) the size of the angle TCA.   [N]

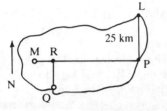

7. The figure shows a sketch map of an island. The lighthouse L is 25 km due North of Pertpearl (P) and the Mount (M) is 96 km due West of Portpearl.

(a) Calculate, to the nearest kilometre, the distance of the lighthouse from the Mount.
(b) Calculate to the nearest degree, the size of angle LMP, and hence find the bearing of L from M.

The Quay (Q) is 20 km from the Mount on a bearing of 125°. The Rooftop Hotel (R) is due North of the Quay and due East of the Mount.

(c) Draw a rough sketch of △MRQ and mark on it the length of MQ and the size of each of its three angles.
(d) Calculate, in km, to 1 decimal place, the distance of the Rooftop Hotel from
   (i) the Mount,   (ii) The Port.   [L]

8. From a point A, a ship sails 200 km due North to a point B. It then changes course and sails 100 km due East to a point C. From C it sails 100 km on a bearing of 210° to a point D.
   (i) On graph paper, and using a scale of 1 cm to represent 10 km, make a scale drawing of the ship's course.
   (ii) Use your scale drawing to find the distance, in km, of D from A in a straight line.
   (iii) Use your protractor to find the bearing of D from A.   [N]

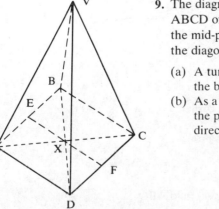

9. The diagram shows an Egyptian pyramid of height 150 m and square base ABCD of side length 240 m. V is the vertex of the right pyramid, E and F are the mid-points of AB and DC respectively and X is the point of intersection of the diagonals of the square base.

(a) A tunnel to the burial chamber runs directly from A to X. Find the length of the base of this tunnel correct to 3 significant figures.
(b) As a tourist attraction, young Egyptians run from the base to the top V of the pyramid. If Abdi runs from F directly to V and Abda runs from A directly to V, what angles to the horizontal do each of their paths make?   [W]

10.

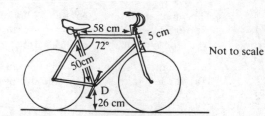

Not to scale

The diagram shows a bicycle. When it stands on horizontal ground, CB is also horizontal and D is 26 cm above the ground.
Angle BCD = 72° and BA is parallel to CD.
AB = 5 cm, BC = 58 cm and CD = 50 cm.
(a) Make an accurate scale drawing of ABCD, one fifth of full size.
(b) Using your drawing, find the total length of tubing needed to construct ABCD.
(c) A tall man requires a cycle with its saddle 1 m above the ground. The saddle of this cycle can be raised to be 22 cm vertically above BC. Is it big enough for him? Show the measurement which you take from your drawing to answer this.   [M]

11. The figure shows a vertical flagpole CT, of height 11·2 m, situated at the corner C of a horizontal square plot ABCD. The angle of elevation of the top of the flagpole T from A is 19°.

(a) Calculate, in metres to one decimal place, the length of AC.
(b) Show that the length of the side of the square is approximately 23 m.
(c) A point E is marked on AD so that the angle ACE is 20°.
     Find:
      (i) the length of EC to 2 decimal places,
     (ii) the angle of elevation of T from E to the nearest degree.   [L]

12. The diagram represents two fields ABD and BCD in a horizontal plane ABCD.
AB = 270 m, BD = 250 m, angle ABD = 57°, angle BCD = 37° and angle CBD = 93°.

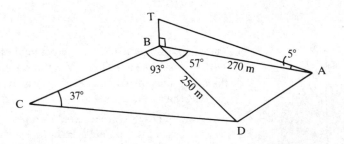

A vertical radio mast BT stands at the corner B of the fields and the angle of elevation of T from A is 5°.
Calculate, correct to three significant figures,
(a) the height of the radio mast,
(b) the length of AD,
(c) the length of BC,
(d) the area, in hectares, of the triangular field ABD.
     (1 hectare = 10^4 square metres.)   [M]

7 Graphs

Rene Descartes (1596–1650) was one of the greatest philosophers of his time. Strangely his restless mind only found peace and quiet as a soldier and he apparently discovered the idea of 'cartesian' geometry in a dream before the battle of Prague. The word 'cartesian' is derived from his name and his work formed the link between geometry and algebra which inevitably led to the discovery of calculus. He finally settled in Holland for ten years, but later moved to Sweden where he soon died of pneumonia.

7.1 DRAWING ACCURATE GRAPHS

Example 1

Draw the graph of $y = 2x - 3$ for values of x from -2 to $+4$.

(a) The coordinates of points on the line are calculated in a table.

x	-2	-1	0	1	2	3	4
$2x$	-4	-2	0	2	4	6	8
-3	-3	-3	-3	-3	-3	-3	-3
y	-7	-5	-3	-1	1	3	5

(b) Draw and label axes using suitable scales.
(c) Plot the points and draw a pencil line through them. Label the line with its equation.

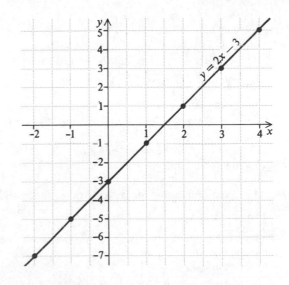

Exercise 1

Draw the following graphs, using a scale of 2 cm to 1 unit on the x-axis and 1 cm to 1 unit on the y-axis.

1. $y = 2x + 1$ for $-3 \leqslant x \leqslant 3$
2. $y = 3x - 4$ for $-3 \leqslant x \leqslant 3$
3. $y = 2x - 1$ for $-3 \leqslant x \leqslant 3$
4. $y = 8 - x$ for $-2 \leqslant x \leqslant 4$
5. $y = 10 - 2x$ for $-2 \leqslant x \leqslant 4$
6. $y = \dfrac{x + 5}{2}$ for $-3 \leqslant x \leqslant 3$
7. $y = 3(x - 2)$ for $-3 \leqslant x \leqslant 3$
8. $y = \frac{1}{2}x + 4$ for $-3 \leqslant x \leqslant 3$
9. $v = 2t - 3$ for $-2 \leqslant t \leqslant 4$
10. $z = 12 - 3t$ for $-2 \leqslant t \leqslant 4$

In each question from **11** to **16**, draw the graphs on the same page and hence find the coordinates of the vertices of the polygon formed. Give the answers as accurately as your graph will allow.

11. (a) $y = x$ (b) $y = 8 - 4x$ (c) $y = 4x$
    Take $-1 \leqslant x \leqslant 3$ and $-4 \leqslant y \leqslant 14$.
12. (a) $y = 2x + 1$ (b) $y = 4x - 8$ (c) $y = 1$
    Take $0 \leqslant x \leqslant 5$ and $-8 \leqslant y \leqslant 12$.
13. (a) $y = 3x$ (b) $y = 5 - x$ (c) $y = x - 4$
    Take $-2 \leqslant x \leqslant 5$ and $-9 \leqslant y \leqslant 8$.
14. (a) $y = -x$ (b) $y = 3x + 6$ (c) $y = 8$
    (d) $x = 3\frac{1}{2}$
    Take $-2 \leqslant x \leqslant 5$ and $-6 \leqslant y \leqslant 10$.
15. (a) $y = \frac{1}{2}(x - 8)$ (b) $2x + y = 6$
    (c) $y = 4(x + 1)$
    Take $-3 \leqslant x \leqslant 4$ and $-7 \leqslant y \leqslant 7$.
16. (a) $y = 2x + 7$ (b) $3x + y = 10$
    (c) $y = x$ (d) $2y + x = 4$
    Take $-2 \leqslant x \leqslant 4$ and $0 \leqslant y \leqslant 13$.
17. The equation connecting the annual mileage, M miles, of a certain car and the annual running cost, £C is $C = \dfrac{M}{20} + 200$.

    Draw the graph for $0 \leqslant M \leqslant 10\,000$ using scales of 1 cm for 1000 miles for M and 2 cm for £100 for C. From the graph find
    (a) the cost when the annual mileage is 7200 miles,
    (b) the annual mileage corresponding to a cost of £320.
18. The equation relating the cooking time t hours and the weight w kg for a joint of meat is $t = \dfrac{3w + 1}{2}$.

    Draw the graph for $0 \leqslant w \leqslant 5$. From the graph find
    (a) the weight of a joint requiring a cooking time of 2·8 hours,
    (b) the cooking time for a joint of weight 4·4 kg.

19. Some drivers try to estimate their annual cost of repairs £c in relation to their average speed of driving s km/h using the equation $c = 6s + 50$. Draw the graph for $0 \leqslant s \leqslant 160$. From the graph find
    (a) the estimated repair bill for a man who drives at an average speed of 23 km/h,
    (b) the average speed at which a motorist drives if his annual repair bill is £1000,
    (c) the annual saving for a man who, on returning from a holiday, reduces his average speed of driving from 100 km/h to 65 km/h.
20. The value of a car £v is related to the number of miles n which it has travelled by the equation
    $$v = 4500 - \frac{n}{20}.$$

    Draw the graph for $0 \leqslant n \leqslant 90\,000$. From the graph find
    (a) the value of a car which has travelled 3700 miles,
    (b) the number of miles travelled by a car valued at £3200,
    (c) the decrease in value of a car when the mileometer reading changes from 28100 to 35700.

7.2 GRADIENTS

The gradient of a straight line is a measure of how steep it is.

Example 1

Find the gradient of the line joining the points A (1, 2) and B (6, 5).

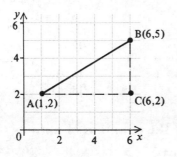

$$\text{gradient of AB} = \frac{BC}{AC} = \frac{3}{5}$$

It is possible to use the formula

$$\text{gradient} = \frac{\text{difference in } y\text{-coordinates}}{\text{difference in } x\text{-coordinates}}.$$

Example 2

Find the gradient of the line joining the points D (1, 5) and E (5, 2).

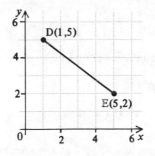

$$\text{gradient of DE} = \frac{5-2}{1-5} = \frac{3}{-4} = -\frac{3}{4}$$

Note

(a) Lines which slope upward to the right have a *positive* gradient.

(b) Lines which slope downward to the right have a *negative* gradient.

Exercise 2

Calculate the gradient of the line joining the following pairs of points.

1. $(3, 1)(5, 4)$ 2. $(1, 1)(3, 5)$
3. $(3, 0)(4, 3)$ 4. $(-1, 3)(1, 6)$
5. $(-2, -1)(0, 0)$ 6. $(7, 5)(1, 6)$
7. $(2, -3)(1, 4)$ 8. $(0, -2)(-2, 0)$
9. $(\frac{1}{2}, 1)(\frac{3}{4}, 2)$ 10. $(-\frac{1}{2}, 1)(0, -1)$
11. $(3\cdot1, 2)(3\cdot2, 2\cdot5)$ 12. $(-7, 10)(0, 0)$
13. $(\frac{1}{3}, 1)(\frac{1}{2}, 2)$ 14. $(3, 4)(-2, 4)$
15. $(2, 5)(1\cdot3, 5)$ 16. $(2, 3)(2, 7)$
17. $(-1, 4)(-1, 7\cdot2)$ 18. $(2\cdot3, -2\cdot2)(1\cdot8, 1\cdot8)$
19. $(0\cdot75, 0)(0\cdot375, -2)$ 20. $(17\cdot6, 1)(1\cdot4, 1)$
21. $(a, b)(c, d)$ 22. $(m, n)(a, -b)$
23. $(2a, f)(a, -f)$ 24. $(2k, -k)(k, 3k)$
25. $(m, 3n)(-3m, 3n)$ 26. $\left(\frac{c}{2}, -d\right)\left(\frac{c}{4}, \frac{d}{2}\right)$

In questions **27** and **28**, find the gradient of each straight line.

27.

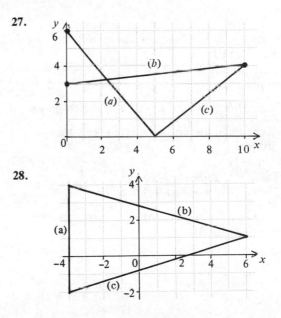

28.

29. Find the value of a if the line joining the points $(3a, 4)$ and $(a, -3)$ has a gradient of 1.

30. (a) Write down the gradient of the line joining the points $(2m, n)$ and $(3, -4)$,

(b) Find the value of n if the line is parallel to the x-axis,

(c) Find the value of m if the line is parallel to the y-axis.

7.3 THE FORM $y = mx + c$

When the equation of a straight line is written in the form $y = mx + c$, the gradient of the line is m and the intercept on the y-axis is c.

Example 1

Draw the line $y = 2x + 3$ on a *sketch* graph.

The word 'sketch' implies that we do not plot a series of points but simply show the position and slope of the line.

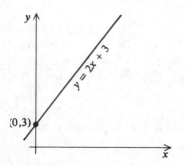

The line $y = 2x + 3$ has a gradient of 2 and cuts the y-axis at $(0, 3)$.

Example 2

Draw the line $x + 2y - 6 = 0$ on a sketch graph.

(a) Rearrange the equation to make y the subject.
$$x + 2y - 6 = 0$$
$$2y = -x + 6$$
$$y = -\tfrac{1}{2}x + 3.$$

(b) The line has a gradient of $-\tfrac{1}{2}$ and cuts the y-axis at $(0, 3)$.

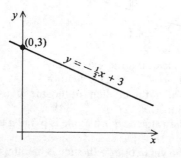

Exercise 3

In questions **1** to **20**, find the gradient of the line and the intercept on the y-axis. Hence draw a small sketch graph of each line.

1. $y = x + 3$
2. $y = x - 2$
3. $y = 2x + 1$
4. $y = 2x - 5$
5. $y = 3x + 4$
6. $y = \tfrac{1}{2}x + 6$
7. $y = 3x - 2$
8. $y = 2x$
9. $y = \tfrac{1}{4}x - 4$
10. $y = -x + 3$
11. $y = 6 - 2x$
12. $y = 2 - x$
13. $y + 2x = 3$
14. $3x + y + 4 = 0$
15. $2y - x = 6$
16. $3y + x - 9 = 0$
17. $4x - y = 5$
18. $3x - 2y = 8$
19. $10x - y = 0$
20. $y - 4 = 0$

Finding the equation of a line

Example 3

Find the equation of the straight line which passes through $(1, 3)$ and $(3, 7)$.

(a) Let the equation of the line take the form $y = mx + c$.

The gradient, $m = \dfrac{7 - 3}{3 - 1} = 2$

so we may write the equation as
$$y = 2x + c \qquad \dots [1]$$

(b) Since the line passes through $(1, 3)$, substitute 3 for y and 1 for x in [1].
$$\therefore \quad 3 = 2 \times 1 + c$$
$$1 = c$$

The equation of the line is $y = 2x + 1$.

Exercise 4

In questions **1** to **11** find the equation of the line which

1. Passes through $(0, 7)$ at a gradient of 3
2. Passes through $(0, -9)$ at a gradient of 2
3. Passes through $(0, 5)$ at a gradient of -1
4. Passes through $(2, 3)$ at a gradient of 2
5. Passes through $(2, 11)$ at a gradient of 3
6. Passes through $(4, 3)$ at a gradient of -1
7. Passes through $(6, 0)$ at a gradient of $\tfrac{1}{2}$
8. Passes through $(2, 1)$ and $(4, 5)$
9. Passes through $(5, 4)$ and $(6, 7)$
10. Passes through $(0, 5)$ and $(3, 2)$
11. Passes through $(3, -3)$ and $(9, -1)$

7.4 SCATTER DIAGRAMS

A scatter diagram is a graph which is plotted to see if there is any relationship between two sets of data.

Example.   Each month the average outdoors temperature was recorded together with the number of therms of gas used to heat the house. The results are plotted on the scatter diagram below.

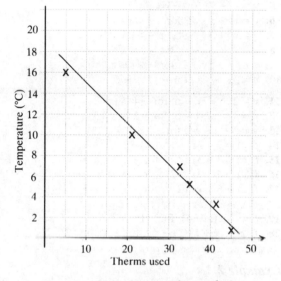

Clearly there is a high degree of *correlation* between these two figures. British Gas do in fact use weather forecasts as their main short-term predictor of future gas consumption over the whole country.
A *line of best fit* has been drawn.
We can estimate that if the outdoor temperature was 12°C then about 17 therms of gas would be used.

Exercise 5

In questions 1 and 2 draw a scatter diagram and put in the line of best fit.

1. The marks of 7 pupils in the two papers of a physics examination were as follows

Paper 1	20	32	40	60	71	80	91
Paper 2	15	25	40	50	64	75	84

A pupil scored a mark of 50 on paper 1. What would you expect her to get on paper 2?

2. The table shows (a) the engine size in litres of various cars and (b) the distance travelled on one litre of petrol.

Engine	0·8	1·6	2·6	1·0	2·1	1·3	1·8
Distance	13	10·2	5·4	12	7·8	11·2	8·5

A car has a 3 litre engine. How far would you expect it to go on one litre of petrol?

3. What sort of pattern would you expect if you took readings of the following and drew a scatter diagram?
   (a) cars on roads; accident rate.
   (b) sales of perfume; advertising costs.
   (c) birth rate; rate of inflation.
   (d) petrol consumption of car; price of petrol.

4. In an experiment, the following measurements of the variables q and t were taken.

q	0·5	1·0	1·5	2·0	2·5	3·0
t	3·85	5·0	6·1	7·0	7·75	9·1

A scientist suspects that q and t are related by an equation of the form $t = mq + c$, (m and c constants). Plot the values obtained from the experiment and draw the line of best fit through the points. Plot q on the horizontal axis with a scale of 4 cm to 1 unit, and t on the vertical axis with a scale of 2 cm to 1 unit. Find the gradient and intercept on the t-axis and hence estimate the values of m and c.

5. In an experiment, the following measurements of p and z were taken:

z	1·2	2·0	2·4	3·2	3·8	4·6
p	11·5	10·2	8·8	7·0	6	3·5

Plot the points on a graph with z on the horizontal axis and draw the line of best fit through the points. Hence estimate the values of n and k if the equation relating p and z is of the form $p = nz + k$.

6. In an experiment the following measurements of t and z were taken:

t	1·41	2·12	2·55	3·0	3·39	3·74
z	3·4	3·85	4·35	4·8	5·3	5·75

Draw a graph, plotting t^2 on the horizontal axis and z on the vertical axis, and hence confirm that the equation connecting t and z is of the form $z = mt^2 + c$. Find approximate values for m and c.

7.5 PLOTTING CURVES

Example 1

Draw the graph of the function
$y = 2x^2 + x - 6$, for $-3 \leqslant x \leqslant 3$.

(a)

x	-3	-2	-1	0	1	2	3
$2x^2$	18	8	2	0	2	8	18
x	-3	-2	-1	0	1	2	3
-6	-6	-6	-6	-6	-6	-6	-6
y	9	0	-5	-6	-3	4	15

(b) Draw and label axes using suitable scales.

(c) Plot the points and draw a smooth curve through them with a pencil.

(d) Check any points which interrupt the smoothness of the curve.

(e) Label the curve with its equation.

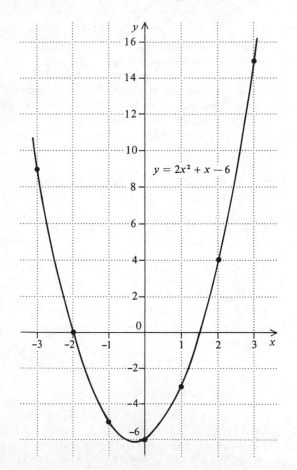

$y = 2x^2 + x - 6$

Exercise 6

Draw the graphs of the following functions using a scale of 2 cm for 1 unit on the x-axis and 1 cm for 1 unit on the y-axis.

1. $y = x^2 + 2x$, for $-3 \leqslant x \leqslant 3$
2. $y = x^2 + 4x$, for $-3 \leqslant x \leqslant 3$
3. $y = x^2 - 3x$, for $-3 \leqslant x \leqslant 3$
4. $y = x^2 + 2$, for $-3 \leqslant x \leqslant 3$
5. $y = x^2 - 7$, for $-3 \leqslant x \leqslant 3$
6. $y = x^2 + x - 2$, for $-3 \leqslant x \leqslant 3$
7. $y = x^2 + 3x - 9$, for $-4 \leqslant x \leqslant 3$
8. $y = x^2 - 3x - 4$, for $-2 \leqslant x \leqslant 4$
9. $y = x^2 - 5x + 7$, for $0 \leqslant x \leqslant 6$
10. $y = 2x^2 - 6x$, for $-1 \leqslant x \leqslant 5$
11. $y = 2x^2 + 3x - 6$, for $-4 \leqslant x \leqslant 2$
12. $y = 3x^2 - 6x + 5$, for $-1 \leqslant x \leqslant 3$
13. $y = 2 + x - x^2$, for $-3 \leqslant x \leqslant 3$
14. $f(x) = 1 - 3x - x^2$, for $-5 \leqslant x \leqslant 2$
15. $f(x) = 3 + 3x - x^2$, for $-2 \leqslant x \leqslant 5$
16. $f(x) = 7 - 3x - 2x^2$, for $-3 \leqslant x \leqslant 3$
17. $f(x) = 6 + x - 2x^2$, for $-3 \leqslant x \leqslant 3$
18. $f : x \rightarrow 8 + 2x - 3x^2$, for $-2 \leqslant x \leqslant 3$
19. $f : x \rightarrow x(x - 4)$, for $-1 \leqslant x \leqslant 6$
20. $f : x \rightarrow (x + 1)(2x - 5)$, for $-3 \leqslant x \leqslant 3$.

Example 2

Draw the graph of $y = \dfrac{12}{x} + x - 6$, for $1 \leqslant x \leqslant 8$.

Use the graph to find approximate values for

(a) the minimum value of $\dfrac{12}{x} + x - 6$

(b) the value of $\dfrac{12}{x} + x - 6$, when $x = 2 \cdot 25$.

(c) the gradient of the tangent to the curve drawn at the point where $x = 5$.

Here is the table of values.

x	1	2	3	4	5	6	7	8	1·5
$\dfrac{12}{x}$	12	6	4	3	2·4	2	1·71	1·5	8
x	1	2	3	4	5	6	7	8	1·5
-6	-6	-6	-6	-6	-6	-6	-6	-6	-6
y	7	2	1	1	1·4	2	2·71	3·5	3·5

Notice that an 'extra' value of y has been calculated at $x = 1.5$ because of the large difference between the y-values at $x = 1$ and $x = 2$.

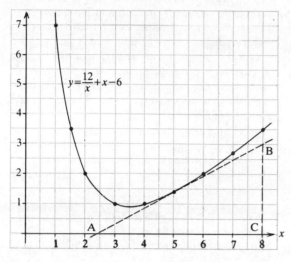

$y = \dfrac{12}{x} + x - 6$

(a) From the graph, the minimum value of $\dfrac{12}{x} + x - 6$ (i.e. y) is approximately 0·9.

(b) At $x = 2·25$, y is approximately 1·6.

(c) The tangent AB is drawn to touch the curve at $x = 5$.

The gradient of $AB = \dfrac{BC}{AC}$.

$\text{gradient} = \dfrac{3}{8 - 2·4} = \dfrac{3}{5·6} \approx 0·54$

It is difficult to obtain an accurate value for the gradient of a tangent so the above result is more realistically 'approximately 0·5'.

Exercise 7

Draw the following curves. The scales given are for one unit of x and y.

1. $y = x^2$, for $0 \leqslant x \leqslant 6$.
   (Scales: 2 cm for x, $\frac{1}{2}$ cm for y)
   Find
   (a) the gradient of the tangent to the curve at $x = 2$,
   (b) the gradient of the tangent to the curve at $x = 4$,
   (c) the y-value at $x = 3·25$.

2. $y = x^2 - 3x$, for $-2 \leqslant x \leqslant 5$.
   (Scales: 2 cm for x, 1 cm for y)
   Find
   (a) the gradient of the tangent to the curve at $x = 3$,
   (b) the gradient of the tangent to the curve at $x = -1$,
   (c) the value of x where the gradient of the curve is zero.

3. $y = 5 + 3x - x^2$, for $-2 \leqslant x \leqslant 5$.
   (Scales: 2 cm for x, 1 cm for y)
   Find
   (a) the maximum value of the function $5 + 3x - x^2$,
   (b) the gradient of the tangent to the curve at $x = 2·5$,
   (c) the two values of x for which $y = 2$.

4. $y = \dfrac{12}{x}$, for $1 \leqslant x \leqslant 10$.
   (Scales: 1 cm for x and y)

5. $y = \dfrac{9}{x}$, for $1 \leqslant x \leqslant 10$.
   (Scales: 1 cm for x and y)

6. $y = \dfrac{12}{x + 1}$, for $0 \leqslant x \leqslant 8$.
   (Scales: 2 cm for x, 1 cm for y)

7. $y = \dfrac{8}{x - 4}$, for $-4 \leqslant x \leqslant 3·5$.
   (Scales: 2 cm for x, 1 cm for y)

8. $y = \dfrac{15}{3 - x}$, for $-4 \leqslant x \leqslant 2$.
   (Scales: 2 cm for x, 1 cm for y)

9. $y = \dfrac{x}{x + 4}$, for $-3·5 \leqslant x \leqslant 4$.
   (Scales: 2 cm for x and y)

10. $y = \dfrac{3x}{5 - x}$, for $-3 \leqslant x \leqslant 4$.
    (Scales: 2 cm for x, 1 cm for y)

11. $y = \dfrac{x + 8}{x + 1}$, for $0 \leqslant x \leqslant 8$.
    (Scales: 2 cm for x and y)

12. $y = \dfrac{x - 3}{x + 2}$, for $-1 \leqslant x \leqslant 6$.
    (Scales: 2 cm for x and y)

13. $y = \dfrac{10}{x} + x$, for $1 \leqslant x \leqslant 7$.
    (Scales: 2 cm for x, 1 cm for y)

14. $y = \dfrac{12}{x} - x$, for $1 \leqslant x \leqslant 7$.
    (Scales: 2 cm for x, 1 cm for y)

15. $y = \dfrac{15}{x} + x - 7$, for $1 \leqslant x \leqslant 7$.

(Scales: 2 cm for x and y)
Find (a) the minimum value of y,
    (b) the y value when $x = 5 \cdot 5$.

16. $y = x^3 - 2x^2$, for $0 \leqslant x \leqslant 4$.
(Scales: 2 cm for x, $\frac{1}{2}$ cm for y)
Find (a) the y value at $x = 2 \cdot 5$
    (b) the x value at $y = 15$

17. $y = \frac{1}{10}(x^3 + 2x + 20)$, for $-3 \leqslant x \leqslant 3$.
(Scales: 2 cm for x and y)
Find
(a) the x-value where $x^3 + 2x + 20 = 0$
(b) the gradient of the tangent to the curve at
    $x = 2$.

18. Copy and complete the table for the function
$y = 7 - 5x - 2x^2$, giving values of y correct to one
decimal place.

x	-4	$-3 \cdot 5$	-3	$-2 \cdot 5$	-2	$-1 \cdot 5$
7	7	7		7		7
$-5x$	20	17·5		12·5		7·5
$-2x^2$	-32	$-24 \cdot 5$		$-12 \cdot 5$		$-4 \cdot 5$
y	5	0		7		10

x	-1	$-0 \cdot 5$	0	$0 \cdot 5$	1	$1 \cdot 5$	2
7		7		7		7	
$-5x$		2·5		$-2 \cdot 5$		$-7 \cdot 5$	
$-2x^2$		$-0 \cdot 5$		$-0 \cdot 5$		$-4 \cdot 5$	
y		9		4		-5	

Draw the graph, using a scale of 2 cm for x and
1 cm for y. Find
(a) the gradient of the tangent to the curve at
    $x = -2 \cdot 5$,
(b) the maximum value of y,
(c) the value of x at which this maximum value
    occurs.

19. Draw the graph of $y = \dfrac{x}{x^2 + 1}$, for $-6 \leqslant x \leqslant 6$.

(Scales: 1 cm for x, 10 cm for y)

20. Draw the graph of $E = \dfrac{5000}{x} + 3x$ for

$10 \leqslant x \leqslant 80$. (Scales: 1 cm to 5 units for x and 1 cm
to 25 units for E)
From the graph find,
(a) the minimum value of E,
(b) the value of x corresponding to this minimum
    value.
(c) the range of values of x for which E is less
    than 275.

Exercise 8

1. A rectangle has a perimeter of 14 cm and length
x cm. Show that the width of the rectangle is
$(7 - x)$ cm and hence that the area A of the
rectangle is given by the formula $A = x(7 - x)$.
Draw the graph, plotting x on the horizontal axis
with a scale of 2 cm to 1 unit, and A on the
vertical axis with a scale of 1 cm to 1 unit. Take x
from 0 to 7. From the graph find,
(a) the area of the rectangle when $x = 2 \cdot 25$ cm,
(b) the dimensions of the rectangle when its area
    is 9 cm^2,
(c) the maximum area of the rectangle,
(d) the length and width of the rectangle
    corresponding to the maximum area.
(e) what shape of rectangle has the largest area.

2. A farmer has 60 m of wire fencing which he uses
to make a rectangular pen for his sheep. He uses a
stone wall as one side of the pen so the wire is
used for only 3 sides of the pen.

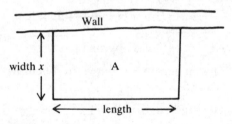

If the width of the pen is x m, what is the length
(in terms of x)?
What is the area A of the pen?
Draw a graph with area A on the vertical axis and
the width x on the horizontal axis. Take values of
x from 0 to 30.
What dimensions should the pen have if the
farmer wants to enclose the largest possible area?

3. A ball is thrown in the air so that t seconds after it
is thrown, its height h metres above its starting
point is given by the function $h = 25t - 5t^2$. Draw
the graph of the function for $0 \leqslant t \leqslant 6$, plotting t
on the horizontal axis with a scale of
2 cm to 1 second, and h on the vertical axis with a
scale of 2 cm for 10 metres. Use the graph to find,
(a) the time when the ball is at its greatest height,
(b) the greatest height reached by the ball,
(c) the interval of time during which the ball is at
    a height of more than 30 m.

4. The velocity v m/s of a missile t seconds after launching is given by the equation $v = 54t - 2t^3$. Draw a graph, plotting t on the horizontal axis with a scale of 2 cm to 1 second, and v on the vertical axis with a scale of 1 cm for 10 m/s. Take values of t from 0 to 5.
Use the graph to find
(a) the maximum velocity reached,
(b) the time taken to accelerate to a velocity of 70 m/s,
(c) the interval of time during which the missile is travelling at more than 100 m/s.

5. Draw the graph of $y = 2^x$, for $-4 \leqslant x \leqslant 4$. (Scales: 2 cm for x, 1 cm for y)

6. Draw the graph of $y = 3^x$, for $-3 \leqslant x \leqslant 3$. (Scales: 2 cm for x, $\frac{1}{2}$ cm for y)
Find the gradient of the tangent to the curve at $x = 1$.

7. Consider the equation $y = \dfrac{1}{x}$.

When $x = \dfrac{1}{2}$, $y = \dfrac{1}{\frac{1}{2}} = 2$.

When $x = \dfrac{1}{100}$, $y = \dfrac{1}{\frac{1}{100}} = 100$.

As the denominator of $\dfrac{1}{x}$ gets smaller, the answer gets larger. An 'infinitely small' denominator gives an 'infinitely large' answer.
We write $\dfrac{1}{0} \to \infty$. '$\dfrac{1}{0}$ tends to an infinitely large number.'

Draw the graph of $y = \dfrac{1}{x}$ for $x = -4, -3, -2,$

$-1, -0.5, -0.25, 0.25, 0.5, 1, 2, 3, 4$
(Scales: 2 cm for x and y)

8. Draw the graph of $y = x + \dfrac{1}{x}$ for $x = -4, -3,$

$-2, -1, -0.5, -0.25, 0.25, 0.5, 1, 2, 3, 4$
(Scales: 2 cm for x and y)

9. Draw the graph of $y = x + \dfrac{1}{x^2}$ for $x = -4, -3,$

$-2, -1, -0.5, -0.25, 0.25, 0.5, 1, 2, 3, 4$
(Scales: 2 cm for x, 1 cm for y)

10. Draw the graph of $y = \dfrac{2^x}{x}$, for $-4 \leqslant x \leqslant 7$,

including $x = -0.5$, $x = 0.5$.
(Scales: 1 cm to 1 unit for x and y)

11. Draw the graph of $y = \dfrac{x^3}{3^x}$, for $x = -1 \leqslant x \leqslant 7$,

including $x = -\dfrac{1}{2}$, $x = \dfrac{1}{2}$.

(Scales: 2 cm to 1 unit for x, 5 cm to 1 unit for y)
(a) What is the maximum value of y?
(b) At what value of x does y have its maximum value?

12. Draw the graph of $y = \dfrac{x^4}{4^x}$, *for* $x = -1, -\frac{3}{4}, -\frac{1}{2},$

$-\frac{1}{4}, 0, \frac{1}{4}, \frac{1}{2}, \frac{3}{4}, 1, 1.5, 2, 2.5, 3, 4, 5, 6, 7.$
(Scales: 2 cm to 1 unit for x, 5 cm to 1 unit for y)
(a) For what values of x is the gradient of the function zero?
(b) For what values of x is $y = 0.5$?

13. Draw the graph of $y = \dfrac{x^5}{5^x}$.

Choose the values of x which best illustrate the interesting behaviour of the function in the region of the origin

7.6   INTERPRETING GRAPHS

Exercise 9

1. Kendal Motors hire out vans at a basic charge of £35 plus a charge of 20p per mile travelled. Copy and complete the table where x is the number of miles travelled and C is the total cost in pounds

x	0	50	100	150	200	250	300
C	35			65			95

Draw a graph of C against x, using scales of 2 cm for 50 miles on the x-axis and 1 cm for £10 on the C-axis.
(a) Use the graph to find the number of miles travelled when the total cost was £71.

(b) What is the formula connecting C and x?

2. A car travels along a motorway and the amount of petrol in its tank is monitored as shown on the graph below.

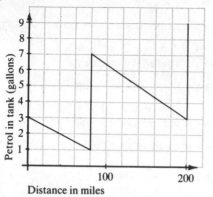

Distance in miles

(a) How much petrol was bought at the first stop?
(b) What was the petrol consumption in miles per gallon:
   (i) before the first stop,
   (ii) between the two stops?
(c) What was the average petrol consumption over the 200 miles?

After it leaves the second service station the car encounters road works and slow traffic for the next 20 miles. Its petrol consumption is reduced to 20 m.p.g. After that, the road clears and the car travels a further 75 miles during which time the consumption is 30 m.p.g. Draw the graph above and extend it to show the next 95 miles. How much petrol is in the tank at the end of the journey?

3. A firm makes a profit of P thousand pounds from producing x thousand tiles.
Corresponding values of P and x are given below

x	0	0·5	1·0	1·5	2·0	2·5	3·0
P	−1.0	0·75	2·0	2·75	3·0	2·75	2·0

Using a scale of 4 cm to one unit on each axis, draw the graph of P against x. [Plot x on the horizontal axis]. Use your graph to find
(a) the number of tiles the firm should produce in order to make the maximum profit.
(b) the minimum number of tiles that should be produced to cover the cost of production.
(c) the range of values of x for which the profit is more than £2850.

4. A small firm increases its monthly expenditure on advertising and records its monthly income from sales.

Month	1	2	3	4	5	6	7
Expenditure (£)	100	200	300	400	500	600	700
Income (£)	280	450	560	630	680	720	740

Draw a graph to display this information.
(a) Is it wise to spend £100 per month on advertising?
(b) Is it wise to spend £700 per month?
(c) What is the most sensible level of expenditure on advertising?

7.7   GRAPHICAL SOLUTION OF EQUATIONS

Accurately drawn graphs enable us to find approximate solutions to a wide range of equations, many of which are impossible to solve exactly by 'conventional' methods.

Example 1

Draw the graph of the function

$$y = 2x^2 - x - 3$$

for $-2 \leqslant x \leqslant 3$. Use the graph to find approximate solutions to the following equations.

(a) $2x^2 - x - 3 = 6$
(b) $2x^2 - x = x + 5$

The table of values for $y = 2x^2 - x - 3$ is found. Note the 'extra' value at $x = \frac{1}{2}$.

x	−2	−1	0	1	2	3	$\frac{1}{2}$
$2x^2$	8	2	0	2	8	18	$\frac{1}{2}$
$-x$	2	1	0	−1	−2	−3	$-\frac{1}{2}$
-3	−3	−3	−3	−3	−3	−3	−3
y	7	0	−3	−2	3	12	−3

The graph drawn from this table is shown on the next page.

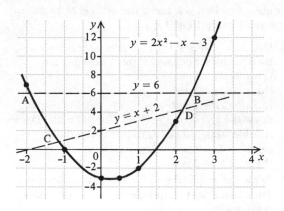

(a) To solve the equation $2x^2 - x - 3 = 6$,
the line $y = 6$ is drawn. At the points of
intersection (A and B), y simultaneously
equals both 6 and $(2x^2 - x - 3)$.
So we may write

$$2x^2 - x - 3 = 6$$

The solutions are the x-values of the points
A and B.
i.e. $x = -1\cdot9$ and $x = 2\cdot4$ approx.

(b) To solve the equation $2x^2 - x = x + 5$, we
rearrange the equation to obtain the function
$(2x^2 - x - 3)$ on the left-hand side.
In this case, subtract 3 from both sides.

$$2x^2 - x - 3 = x + 5 - 3$$
$$2x^2 - x - 3 = x + 2$$

If we now draw the line $y = x + 2$, the
solutions of the equation are given by the
x-values of C and D, the points of
intersection.
i.e. $x = -1\cdot2$ and $x = 2\cdot2$ approx.

It is important to rearrange the equation to be
solved so that the function already plotted is on
one side.

Rearrange $x^2 - 4x + 3 = 0$ in order to obtain
$(x^2 - 3x + 1)$ on the left-hand side.

$$
\begin{array}{ll}
 & x^2 - 4x + 3 = 0 \\
\text{add } x & x^2 - 3x + 3 = x \\
\text{subtract 2} & x^2 - 3x + 1 = x - 2
\end{array}
$$

Therefore draw the line $y = x - 2$ to solve
the equation.

Exercise 10

1. In the diagram below, the graphs of
$y = x^2 - 2x - 3$, $y = -2$ and $y = x$ have been
drawn.

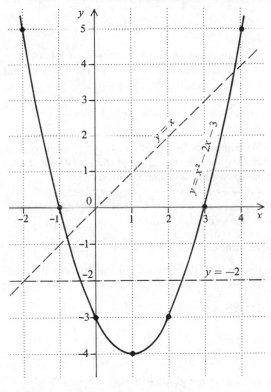

Use the graphs to find approximate solutions to
the following equations.
(a) $x^2 - 2x - 3 = -2$
(b) $x^2 - 2x - 3 = x$
(c) $x^2 - 2x - 3 = 0$
(d) $x^2 - 2x - 1 = 0$

Example 2

Assuming that the graph of $y = x^2 - 3x + 1$ has
been drawn, find the equation of the line which
should be drawn to solve the equation:

$$x^2 - 4x + 3 = 0$$

In questions **2** to **4**, use a scale of 2 cm to 1 unit for x and 1 cm to 1 unit for y.

2. Draw the graphs of the functions $y = x^2 - 2x$ and $y = x + 1$ for $-1 \leqslant x \leqslant 4$. Hence find approximate solutions of the equation $x^2 - 2x = x + 1$.

3. Draw the graphs of the functions $y = x^2 - 3x + 5$ and $y = x + 3$ for $-1 \leqslant x \leqslant 5$. Hence find approximate solutions of the equation $x^2 - 3x + 5 = x + 3$.

4. Draw the graphs of the functions $y = 6x - x^2$ and $y = 2x + 1$ for $0 \leqslant x \leqslant 5$. Hence find approximate solutions of the equation $6x - x^2 = 2x + 1$.

In questions **5** to **9**, do *not* draw any graphs.

5. Assuming the graph of $y = x^2 - 5x$ has been drawn, find the equation of the line which should be drawn to solve the equations
   (a) $x^2 - 5x = 3$ (b) $x^2 - 5x = -2$
   (c) $x^2 - 5x = x + 4$ (d) $x^2 - 6x = 0$
   (e) $x^2 - 5x - 6 = 0$

6. Assuming the graph of $y = x^2 + x + 1$ has been drawn, find the equation of the line which should be drawn to solve the equations
   (a) $x^2 + x + 1 = 6$ (b) $x^2 + x + 1 = 0$
   (c) $x^2 + x - 3 = 0$ (d) $x^2 - x + 1 = 0$
   (e) $x^2 - x - 3 = 0$

7. Assuming the graph of $y = 6x - x^2$ has been drawn, find the equation of the line which should be drawn to solve the equations
   (a) $4 + 6x - x^2 = 0$ (b) $4x - x^2 = 0$
   (c) $2 + 5x - x^2 = 0$ (d) $x^2 - 6x = 3$
   (e) $x^2 - 6x = -2$

8. Assuming the graph of $y = x + \dfrac{4}{x}$ has been drawn, find the equation of the line which should be drawn to solve the equations
   (a) $x + \dfrac{4}{x} - 5 = 0$ (b) $\dfrac{4}{x} - x = 0$
   (c) $x + \dfrac{4}{x} = 0 \cdot 2$ (d) $2x + \dfrac{4}{x} - 3 = 0$
   (e) $x^2 + 4 = 3x$

9. Assuming the graph of $y = x^2 - 8x - 7$ has been drawn, find the equation of the line which should be drawn to solve the equations
   (a) $x = 8 + \dfrac{7}{x}$ (b) $2x^2 = 16x + 9$
   (c) $x^2 = 7$ (d) $x = \dfrac{4}{x - 8}$
   (e) $2x - 5 = \dfrac{14}{x}$.

For questions **10** to **14**, use scales of 2 cm to 1 unit for x and 1 cm to 1 unit for y.

10. Draw the graph of $y = x^2 - 2x + 2$ for $-2 \leqslant x \leqslant 4$. By drawing other graphs, solve the equations
    (a) $x^2 - 2x + 2 = 8$
    (b) $x^2 - 2x + 2 = 5 - x$
    (c) $x^2 - 2x - 5 = 0$

11. Draw the graph of $y = x^2 - 7x$ for $0 \leqslant x \leqslant 7$. Draw suitable straight lines to solve the equations
    (a) $x^2 - 7x + 9 = 0$
    (b) $x^2 - 5x + 1 = 0$

12. Draw the graph of $y = x^2 + 4x + 5$ for $-6 \leqslant x \leqslant 1$. Draw suitable straight lines to find approximate solutions of the equations
    (a) $x^2 + 3x - 1 = 0$ (b) $x^2 + 5x + 2 = 0$

13. Draw the graph of $y = 2x^2 + 3x - 9$ for $-3 \leqslant x \leqslant 2$. Draw suitable straight lines to find approximate solutions of the equations
    (a) $2x^2 + 3x - 4 = 0$
    (b) $2x^2 + 2x - 9 = 1$

14. Draw the graph of $y = 2 + 3x - 2x^2$ for $-2 \leqslant x \leqslant 4$.
    (a) Draw suitable straight lines to find approximate solutions of the equations,
       (i) $2 + 4x - 2x^2 = 0$
       (ii) $2x^2 - 3x - 2 = 0$
    (b) Find the range of values of x for which $2 + 3x - 2x^2 \geqslant -5$.

15. Draw the graph of $y = \dfrac{18}{x}$ for $1 \leqslant x \leqslant 10$, using scales of 1 cm to one unit on both axes. Use the graph to solve approximately
    (a) $\dfrac{18}{x} = x + 2$ (b) $\dfrac{18}{x} + x = 10$
    (c) $x^2 = 18$

16. Draw the graph of $y = \frac{1}{2}x^2 - 6$ for $-4 \leqslant x \leqslant 4$, taking 2 cm to 1 unit on each axis.
    (a) Use your graph to solve approximately the equation $\frac{1}{2}x^2 - 6 = 1$.
    (b) Using tables or a calculator confirm that your solutions are approximately $\pm\sqrt{14}$ and explain why this is so.
    (c) Use your graph to find the square roots of 8.

17. Draw the graph of $y = 6 - 2x - \frac{1}{2}x^3$ for $x = \pm 2$, $\pm 1\frac{1}{2}$, ± 1, $\pm\frac{1}{2}$, 0. Take 4 cm to 1 unit for x and 1 cm to 1 unit for y.
    Use your graph to find approximate solutions of the equations
    (a) $\frac{1}{2}x^3 + 2x - 6 = 0$ (b) $x - \frac{1}{2}x^3 = 0$
    Using tables confirm that two of the solutions to the equation in part (b) are $\pm\sqrt{2}$ and explain why this is so.

18. Draw the graph of $y = x + \dfrac{12}{x} - 5$ for $x = 1, 1\frac{1}{2}$,

2, 3, 4, 5, 6, 7, 8, taking 2 cm to 1 unit on each axis.

(a) From your graph find the range of values of

x for which $x + \dfrac{12}{x} \leqslant 9$

(b) Find an approximate solution of the equation

$2x - \dfrac{12}{x} - 12 = 0$.

19. Draw the graph of $y = 2 \sin x° + 1$ for $0 \leqslant x \leqslant 180°$, taking 1 cm to 10° for x and 5 cm to 1 unit for y. Find approximate solutions to the equations

(a) $2 \sin x + 1 = 2\cdot3$

(b) $\dfrac{1}{(2 \sin x + 1)} = 0\cdot5$

20. Draw the graph of $y = 2 \sin x° + \cos x°$ for $0 \leqslant x \leqslant 180°$, taking 1 cm to 10° for x and 5 cm to 1 unit for y.

(a) Solve approximately the equations

(i) $2 \sin x + \cos x = 1\cdot5$

(ii) $2 \sin x + \cos x = 0$

(b) Estimate the maximum value of y

(c) Find the value of x at which the maximum occurs.

21. Draw the graph of $y = 2^x$ for $-4 \leqslant x \leqslant 4$, taking 2 cm to one unit for x and 1 cm to one unit for y. Find approximate solutions to the equations

(a) $2^x = 6$ (b) $2^x = 3x$ (c) $x2^x = 1$

Find also the approximate value of $2^{2\cdot5}$.

22. Draw the graph of $y = \dfrac{1}{x}$ for $-4 \leqslant x \leqslant 4$, taking

2 cm to one unit on each axis. Find approximate solutions to the equations

(a) $\dfrac{1}{x} = x + 1$

(b) $2x^2 - x - 1 = 0$

23. Draw the graph of $y = \dfrac{2^x}{x}$ for $-2 \leqslant x \leqslant 6$

(including $x = \pm\frac{1}{2}$), taking 2 cm to one unit for x and 1 cm to one unit for y.

(a) Find approximate solutions to the equation

$\dfrac{2^x}{x} = x$

(b) Find the range of values of x for which

$\dfrac{2^x}{x} > 2\frac{1}{2}$.

7.8   AREA UNDER A CURVE

It is sometimes useful to know the area under a curve. The method below, using trapeziums, gives an approximate value for the area but when a computer is used the answer can be found to a high degree of accuracy.

Example 1

Find an approximate value for the area under

the curve $y = \dfrac{12}{x}$ between $x = 2$ and $x = 5$.

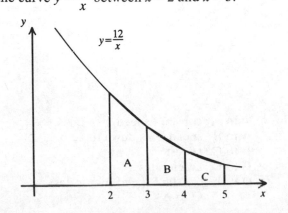

(a) Draw lines parallel to the y-axis at $x = 2, 3, 4, 5$ to form three trapeziums. The area under the curve is approximately equal to the area of the three trapeziums A, B and C.

(b) Calculate the values of y at $x = 2, 3, 4, 5$

at $x = 2$, $y = \frac{12}{2} = 6$

at $x = 3$, $y = \frac{12}{3} = 4$

at $x = 4$, $y = \frac{12}{4} = 3$

at $x = 5$, $y = \frac{12}{5} = 2\cdot4$

(c) Calculate the area of each of the trapeziums using the formula area $= \frac{1}{2}(a + b)h$, where a and b are the parallel sides and h is the distance in between.

Area A $= \frac{1}{2}(6 + 4) \times 1$ $= 5$ sq. units

Area B $= \frac{1}{2}(4 + 3) \times 1$ $= 3\cdot5$ sq. units

Area C $= \frac{1}{2}(3 + 2\cdot4) \times 1 = 2\cdot7$ sq. units

Total area under the curve $\approx (5 + 3\cdot5 + 2\cdot7)$

$\approx 11\cdot2$ sq. units

Exercise 11

1. Find an approximate value for the area under the curve $y = \dfrac{8}{x}$ between $x = 2$ and $x = 5$.

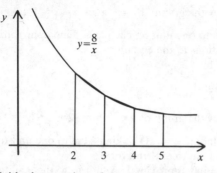

Divide the area into three trapeziums as shown.

2. Find an approximate value for the area under the curve $y = \dfrac{12}{(x+2)}$ between $x = 0$ and $x = 4$.

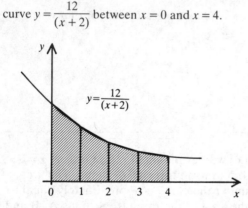

Divide the area into four trapeziums as shown.

3. Find an approximate value for the area under the curve $y = 6x - x^2$ between $x = 1$ and $x = 4$. Divide the area into three trapeziums of equal width.

4. (a) Sketch the curve $y = 16 - x^2$ for $-4 \leqslant x \leqslant 4$.
   (b) Find an approximate value for the area under the curve $y = 16 - x^2$ between $x = 0$ and $x = 3$. (Divide the area into three trapeziums of equal width.)
   (c) State whether your approximate value is greater than or less than the actual value for the area.

5. (a) Find an approximate value for the area under the curve $y = x^2 + 2x + 5$ between $x = 0$ and $x = 4$. Divide the area into four trapeziums.

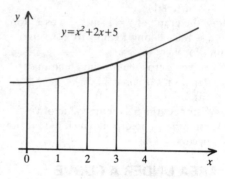

   (b) State whether your approximate value is greater than or less than the actual value for the area.

7.9 DISTANCE-TIME GRAPHS

When a distance-time graph is drawn the *gradient* of the graph gives the *speed* of the object.

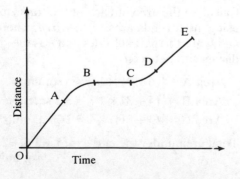

From O to A : constant speed
A to B : speed goes down to zero
B to C : at rest
C to D : accelerates
D to E : constant speed (not as fast as
O to A)

Exercise 12

1. The graph shows the journeys made by a van and a car starting at York, travelling to Durham and returning to York.

   (a) For how long was the van stationary during the journey?
   (b) At what time did the car first overtake the van?
   (c) At what speed was the van travelling between 09 30 and 10 00?
   (d) What was the greatest speed attained by the car during the entire journey?
   (e) What was the average speed of the car over its entire journey?

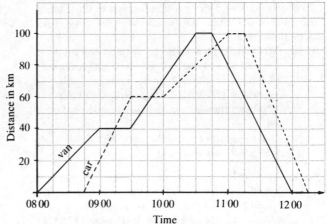

2. The graph shows the journeys of a bus and a car along the same road. The bus goes from Leeds to Darlington and back to Leeds. The car goes from Darlington to Leeds and back to Darlington.

   (a) When did the bus and the car meet for the second time?
   (b) At what speed did the car travel from Darlington to Leeds?
   (c) What was the average speed of the bus over its entire journey?
   (d) Approximately how far apart were the bus and the car at 09 45?
   (e) What was the greatest speed attained by the car during its entire journey?

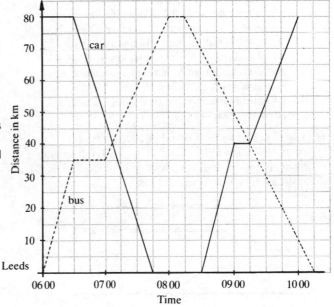

In questions **3**, **4**, **5** draw a travel graph to illustrate the journey described. Draw axes with the same scales as in question **2**.

3. (a) Mrs Chuong leaves home at 08 00 and drives at a speed of 50 km/h. After $\frac{1}{2}$ hour she reduces her speed to 40 km/h and continues at this speed until 09 30. She stops from 09 30 until 10 00 and then returns home at a speed of 60 km/h.
    (b) Use the graph to find the approximate time at which she arrives home.

4. (a) Mr Coe leaves home at 09 00 and drives at a speed of 20 km/h. After $\frac{3}{4}$ hour he increases his speed to 45 km/h and continues at this speed until 10 45. He stops from 10 45 until 11 30 and then returns home at a speed of 50 km/h.
    (b) Use the graph to find the approximate time at which he arrives home.

5. (a) At 10 00 Akram leaves home and cycles to his grandparents' house which is 70 km away. He cycles at a speed of 20 km/h until 11 15, at which time he stops for $\frac{1}{2}$ hour. He then completes the journey at a speed of 30 km/h. At 11 45 Akram's sister, Hameeda, leaves home and drives her car at 60 km/h. Hameeda also goes to her grandparents' house and uses the same road as Akram.
    (b) At approximately what time does Hameeda overtake Akram?

6. A boat can travel at a speed of 20 km/h in still water. The current in a river flows at 5 km/h so that downstream the boat travels at 25 km/h and upstream it travels at only 15 km/h.

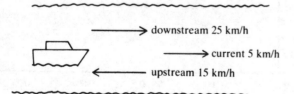

The boat has only enough fuel to last 3 hours. The boat leaves its base and travels downsteam. Draw a distance-time graph and draw lines to indicate the outward and return journeys. After what time must the boat turn round so that it can get back to base without running out of fuel?

7. The boat in question **6** sails in a river where the current is 10 km/h and it has fuel for four hours. At what time must the boat turn round this time if it is not to run out of fuel?

8. The graph shows the motion of three cars A, B and C along the same road.

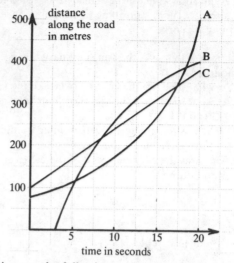

Answer the following questions giving estimates where necessary.

(a) Which car is in front after
    (i) 10s,   (ii) 20s?
(b) When is B in the front?
(c) When are B and C going at the same speed?
(d) When are A and C going at the same speed?
(e) Which car is going fastest after 5s?
(f) Which car starts slowly and then goes faster and faster?

9. Three girls Hanna, Fateema and Carine took part in an egg and spoon race. Describe what happened, giving as many details as possible.

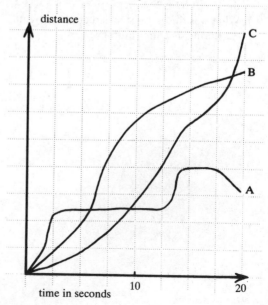

7.10 SPEED-TIME GRAPHS

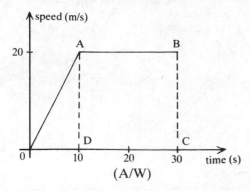

(A/W)

The diagram is the speed-time graph of the first 30 seconds of a car journey. Two quantities are obtained from such graphs:
(a) acceleration = gradient of speed-time graph,
(b) distance travelled = area under graph.

In this example,

(a) The gradient of line OA $= \frac{20}{10} = 2$.

∴ The acceleration in the first 10 seconds is 2 m/s^2.

(b) The distance travelled in the first 30 seconds is given by the area of OAD plus the area of ABCD.

Distance $= (\frac{1}{2} \times 10 \times 20) + (20 \times 20)$
$= 500$ m.

Exercise 13

The graphs show speed v in m/s and time t in seconds.
1. Find
   (a) the acceleration when $t = 4$,
   (b) the total distance travelled,
   (c) the average speed for the whole journey.

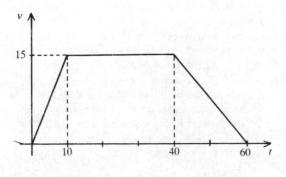

2. Find
   (a) the total distance travelled,
   (b) the average speed for the whole journey,
   (c) the distance travelled in the first 10 seconds.
   (d) the acceleration when $t = 20$.

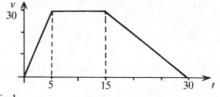

3. Find
   (a) the total distance travelled,
   (b) the distance travelled in the first 40 seconds,
   (c) the acceleration when $t = 15$.

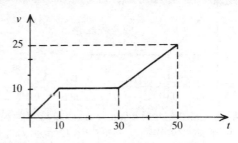

4. Find
   (a) V if the total distance travelled is 900 m,
   (b) the distance travelled in the first 60 seconds.

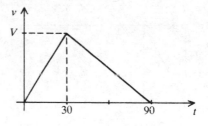

5. Find
   (a) T if the initial acceleration is 2 m/s^2,
   (b) the total distance travelled.
   (c) the average speed for the whole journey.

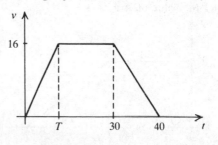

6. Given that the total distance travelled = 810 m, find
   (a) the value of V,
   (b) the rate of change of the speed when $t = 30$,
   (c) the time taken to travel the first 420 m of the journey.

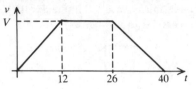

7. Given that the total distance travelled is 1·5 km, find
   (a) the value of V,
   (b) the rate of deceleration after 10 seconds.

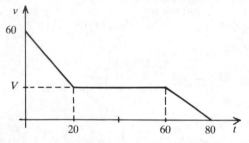

8. Given that the total distance travelled is 1·4 km, and the acceleration is 4 m/s² for the first T seconds, find
   (a) the value of V,
   (b) the value of T.

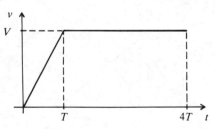

9. Given that the average speed for the whole journey is 37·5 m/s and that the deceleration between T and $2T$ is 2·5 m/s², find
   (a) the value of V,
   (b) the value of T.

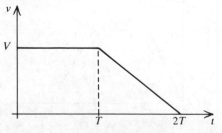

10. Given that the total distance travelled is 4 km and that the initial deceleration is 4 m/s², find
   (a) the value of V,
   (b) the value of T.

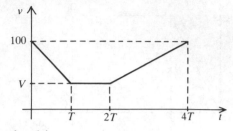

Exercise 14

Sketch a speed-time graph for each question.
All accelerations are taken to be uniform.

1. A car accelerated from 0 to 50 m/s in 9s. How far did it travel in this time?

2. A motor cycle accelerated from 10 m/s to 30 m/s in 6s. How far did it travel is this time?

3. A train slowed down from 50 km/h to 10 km/h in 2 minutes. How far did it travel in this time?

4. When taking off, an aircraft accelerates from 0 to 100 m/s in a distance of 500 m. How long did it take to travel this distance?

5. An earthworm accelerates from a speed of 0·01 m/s to 0·02 m/s over a distance of 0·9 m. How long did it take?

6. A car travelling at 60 m.p.h. is stopped in 6 seconds. How far does it travel in this time? Give the answer in yards. [1 mile = 1760 yards].

7. A car accelerates from 15 m.p.h. to 60 m.p.h. in 3 seconds. How many yards does it travel in this time?

8. At lift-off a rocket accelerates from 0 to 1000 km/h in just 10s. How far does it travel in this time?

9. A coach accelerated from 0 to 60 km/h in 30s. How many metres did it travel in this time?

10. Mr Wheeler was driving a car at 30 m/s when he saw an obstacle 45 m in front of him. It took a reaction time of 0·3 seconds before he could press the brakes and a further 2·5 seconds to stop the car. Did he hit the obstacle?

11. An aircraft is cruising at a speed of 200 m/s. When it lands it must be travelling at a speed of 50 m/s. In the air it can slow down at a rate of 0·2 m/s². On the ground it slows down at a rate of 2 m/s². Draw a velocity time graph for the aircraft as it reduces its speed from 200 m/s to 50 m/s and then to 0 m/s.
How far does it travel in this time?

12. The speed of a train is measured at regular intervals of time from $t = 0$ to $t = 60$s, as shown below.

t s	0	10	20	30	40	50	60
v m/s	0	10	16	19·7	22·2	23·8	24·7

Draw a speed-time graph to illustrate the motion. Plot t on the horizontal axis with a scale of 1 cm to 5s and plot v on the vertical axis with a scale of 2 cm to 5 m/s.
Use the graph to estimate:
(a) the acceleration at $t = 10$,
(b) the distance travelled by the train from $t = 30$ to $t = 60$.

13. The speed of a car is measured at regular intervals of time from $t = 0$ to $t = 60$s, as shown below.

t s	0	10	20	30	40	50	60
v m/s	0	1·3	3·2	6	10·1	16·5	30

Draw a speed-time graph using the same scales as in question **11**. Use the graph to estimate:
(a) the acceleration at $t = 30$.
(b) the distance travelled by the car from $t = 20$ to $t = 50$.

7.11 SKETCH GRAPHS

With a sketch graph we are concerned mainly with the *shape* of the graph. The sketch graph below illustrates the statement: 'Inflation, which has been rising steadily, is now beginning to fall'

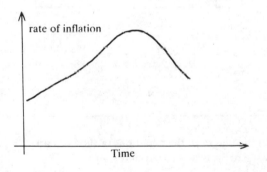

Exercise 15

1. Draw sketch graphs to illustrate the following statements:
(a) Unemployment is still rising but by less each month.
(b) The price of oil was rising more rapidly in 1983 than at any time in the previous ten years.
(c) The birthrate was falling but is now steady.
(d) Unemployment, which rose slowly until 1980, is now rising rapidly.
(e) The price of gold has fallen steadily over the last year.

2. A car hire firm charges £10 per day plus 10p per mile travelled. Draw a sketch graph showing 'miles travelled' across the page and 'total cost' up the page.

3. (a) A car in a Grand Prix goes around the circuit shown. Draw a sketch graph showing speed against time for one lap of the circuit. [Not the first lap. Why not?]

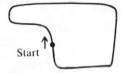

(b) Draw similar sketches for these circuits:

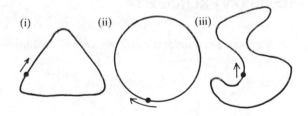

4. (a) An ant crawls at a constant speed along the line through two points A and B. Draw a sketch graph showing 'distance from A' against 'distance from B'.

(b) Sketch a similar graph when the ant walks along a different path.

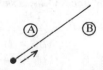

5. Mr Gibson organises a trip for pupils from his
school.
He hires a coach for £100 and decides to divide the
cost equally between the pupils on the trip.
Sketch a graph showing the 'number of pupils'
across the page and the 'cost per pupil' up the
page.
Is it a linear graph?

6. In a video recorder the tape passes the 'heads' at a
constant speed.

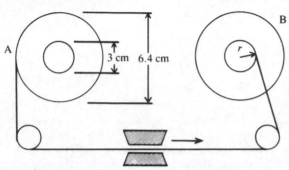

Sketch a graph showing how the radius of the tape
on reel B varies during the 3 hours of a film being
shown on the video.

7. (a) Water flows at a constant rate into the tank
shown. Sketch a graph to show how the depth
of water d varies with time.

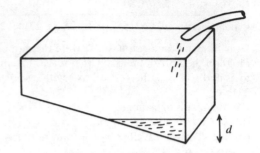

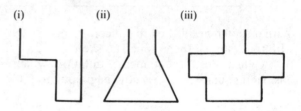

(b) Draw sketch graphs of depth against time for
tanks with the cross sections shown below.
Water flows from the same pipe in each case.

(i)   (ii)   (iii)

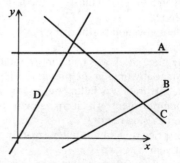

REVISION EXERCISE 7A

1. Find the equation of the straight line satisfied by
the following points.

(a)

x	2	7	10
y	−5	0	3

(b)

x	1	2	3
y	7	9	11

(c)

x	1	2	3
y	8	6	4

(d)

x	3	4	5
y	2	$2\frac{1}{2}$	3

2. Find the gradient of the line joining each pair of
points
(a) $(3, 3)(5, 7)$
(b) $(3, -1)(7, 3)$
(c) $(-1, 4)(1, -3)$
(d) $(2, 4)(-3, 4)$
(e) $(0.5, -3)(0.4, -4)$

3. Find the gradient and the intercept on the y-axis
for the following lines. Draw a *sketch* graph of
each line.
(a) $y = 2x - 7$
(b) $y = 5 - 4x$
(c) $2y = x + 8$
(d) $2y = 10 - x$
(e) $y + 2x = 12$
(f) $2x + 3y = 24$

4. In the diagram, the equations of the lines are
$y = 3x$, $y = 6$, $y = 10 - x$ and $y = \frac{1}{2}x - 3$.

Find the equation corresponding to each line.

5. In the diagram, the equations of the lines are
$2y = x - 8$, $2y + x = 8$, $4y = 3x - 16$ and
$4y + 3x = 16$.

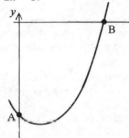

Find the equation corresponding to each line.

6. Find the equations of the lines which pass through
the following pairs of points,
  (a) (2, 1)(4, 5)   (b) (0, 4), (−1, 1)
  (c) (2, 8)(−2, 12)   (d) (0, 7)(−3, 7)

7. The sketch represents a section of the curve
$y = x^2 - 2x - 8$.

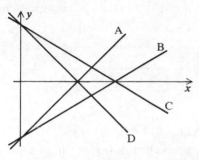

Calculate
  (a) the coordinates of A and of B,
  (b) the gradient of the line AB,
  (c) the equation of the straight line AB.

8. Find the area of the triangle formed by the
intersection of the lines $y = x$, $x + y = 10$ and
$x = 0$.

9. Draw the graph of $y = 7 - 3x - 2x^2$ for $-4 \leqslant x \leqslant 2$.
Find the gradient of the tangent to the curve at the
point where the curve cuts the y-axis.

10. Draw the graph of $y = \dfrac{4000}{x} + 3x$ for $10 \leqslant x \leqslant 80$.

Find the minimum value of y.

11. Draw the graph of $y = \dfrac{1}{x} + 2^x$ for $x = \frac{1}{4}, \frac{1}{2}, \frac{3}{4}, 1, 1\frac{1}{2}, 2, 3$.

12. Assuming that the graph of $y = 4 - x^2$ has been
drawn, find the equation of the straight line which
should be drawn in order to solve the following
equations
  (a) $4 - 3x - x^2 = 0$ (b) $\frac{1}{2}(4 - x^2) = 0$
  (c) $x^2 - x + 7 = 0$ (d) $\dfrac{4}{x} - x = 5$.

13. Draw the graph of $y = 5 - x^2$ for $-3 \leqslant x \leqslant 3$,
taking 2 cm to one unit for x and 1 cm to one unit
for y.
Use the graph to find
  (a) approximate solutions to the equation
       $4 - x - x^2 = 0$,
  (b) the square roots of 5,
  (c) the square roots of 7.

14. Draw the graph of $y = \dfrac{5}{x} + 2x - 3$, for $\frac{1}{2} \leqslant x \leqslant 7$,

taking 2 cm to one unit for x and 1 cm to one unit
for y.
Use the graph to find
  (a) approximate solutions to the equation
       $2x^2 - 10x + 7 = 0$,
  (b) the range of values of x for which
       $\dfrac{5}{x} + 2x - 3 < 6$.
  (c) the minimum value of y.

15. Draw the graph of $y = 4^x$ for $-2 \leqslant x \leqslant 2$.
Use the graph to find
  (a) the approximate value of $4^{1 \cdot 6}$,
  (b) the approximate value of $4^{-\frac{1}{3}}$,
  (c) the gradient of the curve at $x = 0$,
  (d) an approximate solution to the equation
       $4^x = 10$.

16. The diagram is the speed-time graph of a bus.

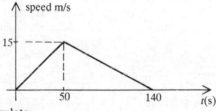

Calculate
  (a) the acceleration during the first 50 seconds,
  (b) the total distance travelled,
  (c) how long it takes before it is moving at 12 m/s
       for the first time.

17. The diagram is the speed-time graph of a car.

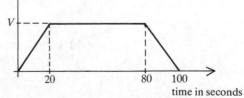

Given that the total distance travelled is 2·4 km,
calculate
  (a) the value of the maximum speed V,
  (b) the distance travelled in the first 30 seconds of
       the motion.

EXAMINATION EXERCISE 7B

1. Given that $y = \dfrac{9}{x}$, complete the following table of values, stating the values, where appropriate, to two decimal places.

x	1	2	3	4	5	6	7	8	9
y	9		3				1·29	1·13	

(a) Draw the graph of $y = \dfrac{9}{x}$ for $1 \leqslant x \leqslant 9$.

Use a scale of 1 cm to 1 unit on both axes.

(b) Using your graph and showing your method clearly, obtain an approximate solution of the equation $\dfrac{9}{x} = 2\cdot4$.   [N]

2. Answer the whole of this question on graph paper.
   (a) Given that $y = 4x^2 - x^3$, copy and complete the following table.

x	0	0·5	1	1·5	2	2·5	3	3·5	4
y	0		3		8	9·375		6·125	0

Using a scale of 4 cm to represent 1 unit on the x-axis and 2 cm to represent 1 unit on the y-axis, draw the graph of $y = 4x^2 - x^3$ for values of x from 0 to 4 inclusive.

   (b) By drawing appropriate straight lines on your graph
   (i) estimate the gradient of the curve $y = 4x^2 - x^3$ at the point $(3\cdot5, 6\cdot125)$,
   (ii) find two solutions of the equation
   $$4x^2 - x^3 = x + 2.$$ [M]

3. (a) Copy and complete the table given below for $y = \dfrac{x^2}{4} + \dfrac{4}{x} - 2$.

x	1	1·5	2	2·5	3	3·5	4
$\dfrac{x^2}{4}$		0·56				3·06	
$\dfrac{4}{x}$		2·67				1·14	
-2		-2				-2	
y		1·23				2·20	

(b) Using a scale of 4 cm to 1 unit on the y-axis and 2 cm to 1 unit on the x-axis, draw the graphs of
   $$y = \frac{x^2}{4} + \frac{4}{x} - 2 \quad \text{for} \quad 1 \leqslant x \leqslant 4.$$

(c) Using the same axes and scales, draw the graph of $y = \dfrac{x}{4} + 1$.

(d) Write down the values of x where the two graphs intersect.

(e) Show that these graphs enable you to find approximate solutions to the equation $x^3 - x^2 - 12x + 16 = 0$.   [L]

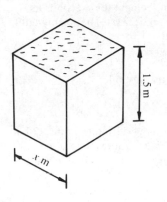

4. This sketch shows a water tank with a square base. It is 1·5 m high, and the length of the base is x metres.
(a) Explain why the volume of the tank is given by the formula $V = 1 \cdot 5x^2$.
(b) Complete the table to show the volume for various values of x.

x	0·1	0·2	0·3	0·4	0·5	0·6	0·7	0·8	0·9	1·0	1·1	1·2	1·3	1·4	1·5
V	0·02	0·06	0·14	0·24		0·54	0·74	0·96	1·2		1·82		2·54		3·38

(c) Draw the graph of $V = 1 \cdot 5x^2$ for values of x from 0 to 1·5.
(d) What value of x will give a volume of 3 m³?
(e) A guest house needs a tank with a volume at least 2 m³. To fit the tank into the loft, it must not be more than 1·3 m wide. Write down the range of values for x which will satisfy these conditions. [S]

5. The petrol consumption (y miles per gallon) of a car is related to its speed (x miles per hour) by the formula
$$y = 200 - \frac{3x}{2} - \frac{4200}{x}$$
(a) Copy and complete the following table of values:

x	40	45	50	55	60	70	80
$\dfrac{3x}{2}$		67·5		82·5			
$\dfrac{4200}{x}$		93·3		76·4			
y		39·2		41·1			

(b) Draw the graph of this relation, taking 1 cm to represent 5 miles per hour on the x-axis and 1 cm to represent 1 mile per gallon on the y-axis. Label the x-axis from 40 to 80 and the y-axis from 27 to 42.
(c) From your graph, estimate
 (i) the petrol consumption at 75 miles per hour,
 (ii) the most economical speed at which to travel in order to conserve petrol. [MEI]

6. The diagram shows the distance-time graph of a particle.

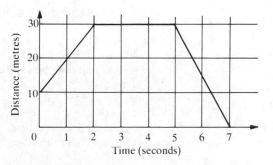

(a) Find the speed of the particle at time 1 second.
(b) Find the average speed of the particle for the first five seconds. [N]

7. A particle is moving with an initial speed of 4 m/s. In the next four seconds its speed increases uniformly to 10 m/s and then the speed decreases uniformly until the particle stops moving after a further eight seconds.
(a) Show this information on a speed-time graph.
(b) Find
    (i) the acceleration in the last eight seconds of the motion,
    (ii) the total distance travelled by the particle.   [L]

8. Answer the whole of this question on graph paper
At 10·00 am George set out from home to ride his cycle to a friend's home 12 km away. He rode at a steady speed of 18 km/h but, after 20 minutes, one of his tyres was punctured. After spending 5 minutes trying to repair it, George walked the rest of the way at 6 km/h.
(a) Using a scale of 2 cm to represent 10 minutes on the time axis and 2 cm to represent 2 km on the distance axis, draw a distance-time graph for George's journey.
(b) Find the time at which George arrived at his friend's home.   [M]

9. A stone thrown from the top of a cliff has a height (h) above the sea after time (t) given by the formula
    $h = 50 + 45t - 5t^2$
where h is in metres and t in seconds.
(a) Find the height of the stone above the sea after 2·2 seconds.
(b) How long does the stone take to reach the water?
(c) Complete this table to show the height of the stone above the sea for the first 5 seconds.

t	0	0·5	1	1·5	2	2·5	3	3·5	4	4·5	5
$5t^2$	0	1·25	5	11·25	20		45		80	101·25	125
$45t$	0	22·5	45	67·5	90		135		180	202·5	225
h	50	71·25	90	106·25	120		140		150	151·25	150

(d) On the graph sheet for this question, draw a graph showing h against t.
(e) Use your graph to find the speed of the stone after 3 seconds.   [S]

10.

Speed (m.p.h.)	0	20	30	40	50	60	70
Stopping distance (feet)	0	40	75	120	175	240	315

The table shows the shortest stopping distance for a car travelling at various speeds.
(a) (i) Using scales of 2 cm to 10 m.p.h. and 2 cm to 50 feet, draw a graph to show stopping distance against speed.
    (ii) From your graph, find the shortest stopping distance for a car travelling at 65 m.p.h.
(b) An approximate rule for estimating stopping distances is 1 yard for each m.p.h. of the car's speed.
    (i) Draw a straight line on your graph to represent this rule.
    (ii) For what speed does this rule give the same stopping distance as the table?
    (iii) For what speeds is the stopping distance estimated from the rule more than 25 ft less than the 'true' stopping distance given by the curve?   [L]

11. The table below shows the prices of some paper-back novels and the number of pages in them.

Price	85p	£1·00	95p	£1·25	£1·50	£1·65	95p	£1·00	£1·35	65p	75p
Pages	224	254	170	236	330	380	210	190	320	136	150

On the graph paper provided construct a scatter diagram for this information.
Use scales of 2 cm to represent 50 pages and 2 cm to represent 20p.
 (i) Draw a line of best fit.
 (ii) Use your line to estimate the cost of a book with 300 pages.   [N]

12. In this question state any assumptions you make.
Jason is arrested for drunken driving at 10·30 p.m. on Saturday. At the Police Station his alcohol level is taken every hour (it is measured in milligrams per 100 millilitres of blood). A record of the measurements is kept.

Time after arrest (hours)	1	2	3	4	5
Level (milligrams)	160	152	139	131	120

(a) Plot these results on a graph.
(b) What would you expect Jason's alcohol level to be 7 hours after his arrest?
(c) By considering a line of best fit, find a formula connecting Jason's alcohol level and the time since his arrest.
He is released from the Police Station when he has a level below 80. Predictions are liable to 12% error.
(d) What time can Jason expect to leave the Police Station?   [S]

13. When a cassette tape recorder starts to re-wind a cassette the counter reads 430. Counter readings (y) after x seconds are as shown in the table.

x	0	10	20	30	40	50	60	70	80	90
y	430	408	382	353	317	278	227	174	104	22

(a) Draw axes showing values of x from 0 to 100 and y from 0 to 450. On these axes draw a graph to show the readings in the table above.
(b) Estimate the time taken to re-wind until the counter reads
    (i) 300   (ii) 0
(c) An approximation to the graph is to be chosen in the form
    $y = 430 + bx + cx^2$ so that it goes through (30, 353) and (60, 227).
    Form simultaneous equations for b and c, hence show that $b = -1·75$ and find c.
(d) Use the equation in (c) to work out the values of y when $x = 0, 50, 80$.   [M]

8 Matrices and transformations

Albert Einstein (1879–1955) working as a patent office clerk in Berne, was responsible for the greatest advance in mathematical physics of this century. His theories of relativity, put forward in 1905 and 1915 were based on the postulate that the velocity of light is absolute: mass, length and even time can only be measured relative to the observer and undergo transformation when studied by another observer. His formula $E = mc^2$ laid the foundations of nuclear physics, a fact that he came to deplore in its application to warfare. In 1933 he moved from Nazi Germany and settled in America.

8.1 MATRIX OPERATIONS

Addition and subtraction

Matrices of the same order are added (or subtracted) by adding (or subtracting) the corresponding elements in each matrix.

Example 1

$$\begin{pmatrix} 2 & -4 \\ 3 & 0 \end{pmatrix} + \begin{pmatrix} 3 & 5 \\ -1 & 7 \end{pmatrix} = \begin{pmatrix} 5 & 1 \\ 2 & 7 \end{pmatrix}$$

Multiplication by a number

Each element of the matrix is multiplied by the multiplying number.

Example 2

$$3 \times \begin{pmatrix} 2 & -1 \\ 1 & 4 \end{pmatrix} = \begin{pmatrix} 6 & -3 \\ 3 & 12 \end{pmatrix}$$

Multiplication by another matrix

For 2×2 matrices,

$$\begin{pmatrix} a & b \\ c & d \end{pmatrix} \begin{pmatrix} w & x \\ y & z \end{pmatrix} = \begin{pmatrix} aw + by & ax + bz \\ cw + dy & cx + dz \end{pmatrix}$$

The same process is used for matrices of other orders.

Example 3

Perform the following multiplications.

(a) $\begin{pmatrix} 3 & 2 \\ 4 & 1 \end{pmatrix} \begin{pmatrix} 2 & 1 \\ 1 & 5 \end{pmatrix} = \begin{pmatrix} 6+2 & 3+10 \\ 8+1 & 4+5 \end{pmatrix}$

$$= \begin{pmatrix} 8 & 13 \\ 9 & 9 \end{pmatrix}$$

(b) $\begin{pmatrix} 2 & 1 & -2 \\ 0 & 1 & 3 \end{pmatrix} \begin{pmatrix} 1 & 0 \\ 1 & -2 \\ 4 & 3 \end{pmatrix}$

$= \begin{pmatrix} 2+1-8 & 0-2-6 \\ 0+1+12 & 0-2+9 \end{pmatrix}$

$= \begin{pmatrix} -5 & -8 \\ 13 & 7 \end{pmatrix}$

Matrices may be multiplied only if they are *compatible*. The number of *columns* in the left-hand matrix must equal the number of *rows* in the right-hand matrix.

Matrix multiplication is not commutative, i.e. for square matrices **A** and **B**, the product **AB** does not necessarily equal the product **BA**.

Exercise 1

In questions **1** to **36**, the matrices have the following values.

$A = \begin{pmatrix} 2 & -1 \\ 3 & 1 \end{pmatrix}$; $B = \begin{pmatrix} 0 & 5 \\ 1 & -2 \end{pmatrix}$; $C = \begin{pmatrix} 4 & 3 \\ 1 & -2 \end{pmatrix}$;

$D = \begin{pmatrix} 1 & 5 & 1 \\ 4 & -6 & 1 \end{pmatrix}$; $E = \begin{pmatrix} 1 & 0 \\ -1 & 1 \\ 2 & 5 \end{pmatrix}$; $F = (4\ \ 5)$;

$G = \begin{pmatrix} 4 \\ 1 \\ 3 \end{pmatrix}$; $H = \begin{pmatrix} 0 & 1 & -2 \\ 3 & -4 & 5 \end{pmatrix}$;

$J = \begin{pmatrix} 3 \\ 1 \end{pmatrix}$; $K = \begin{pmatrix} 1 & -3 \\ 0 & 1 \\ -7 & 0 \end{pmatrix}$

Calculate the resultant value for each question where possible.

1. A + B
2. D + H
3. J + F
4. B − C
5. 2F
6. 3B
7. K − E
8. 2A + B
9. G − J
10. C + B + A
11. 2E − 3K
12. $\frac{1}{2}$A − B
13. AB
14. BA
15. BC
16. CB
17. DG
18. AJ
19. HK
20. (AB)C
21. A(BC)
22. AF
23. CK
24. GF
25. B(2A)
26. (D + H)G
27. JF
28. FJ

29. (A − C)D
30. A²
31. A⁴
32. E²
33. KH
34. (CA)J
35. ED
36. B⁴

In questions **37** to **46**, find the value of the letters.

37. $\begin{pmatrix} 2 & x \\ y & 7 \end{pmatrix} + \begin{pmatrix} 4 & y \\ -3 & 2 \end{pmatrix} = \begin{pmatrix} x & 9 \\ z & 9 \end{pmatrix}$

38. $\begin{pmatrix} x & 2 \\ -1 & -2 \\ w & 3 \end{pmatrix} + \begin{pmatrix} x & y \\ y & -3 \\ v & 5 \end{pmatrix} = \begin{pmatrix} 8 & z \\ x & w \\ w & 8 \end{pmatrix}$

39. $\begin{pmatrix} a & b \\ c & 0 \end{pmatrix} - \begin{pmatrix} 2 & 5 \\ -3 & d \end{pmatrix} = 2 \begin{pmatrix} 1 & a \\ b & -1 \end{pmatrix}$

40. $\begin{pmatrix} x & 3 \\ -2 & y \end{pmatrix} \begin{pmatrix} 2 \\ 1 \end{pmatrix} = \begin{pmatrix} 5 \\ 0 \end{pmatrix}$

41. $\begin{pmatrix} 2 & 0 \\ 0 & -3 \end{pmatrix} \begin{pmatrix} m \\ n \end{pmatrix} = \begin{pmatrix} 10 \\ 1 \end{pmatrix}$

42. $\begin{pmatrix} p & 2 & -1 \\ q & -2 & 2q \end{pmatrix} \begin{pmatrix} 2 \\ 1 \\ 3 \end{pmatrix} = \begin{pmatrix} 5 \\ -10 \end{pmatrix}$

43. $\begin{pmatrix} 3 & 0 \\ 2 & x \end{pmatrix} \begin{pmatrix} y & z \\ 4 & 0 \end{pmatrix} = \begin{pmatrix} 6 & -3 \\ 8 & w \end{pmatrix}$

44. $\begin{pmatrix} 3y & 3z \\ 2y+4x & 2z \end{pmatrix} = \begin{pmatrix} 6 & -3 \\ 8 & w \end{pmatrix}$

45. $\begin{pmatrix} 2 & e \\ a & 3 \end{pmatrix} + k \begin{pmatrix} 3 & 1 \\ 0 & -2 \end{pmatrix} = \begin{pmatrix} 8 & 6 \\ -3 & -1 \end{pmatrix}$

46. $\begin{pmatrix} 4 & 0 \\ 1 & m \end{pmatrix} \begin{pmatrix} n & p \\ -2 & 0 \end{pmatrix} = \begin{pmatrix} 20 & 12 \\ -1 & q \end{pmatrix}$

47. If $A = \begin{pmatrix} 1 & 0 \\ 3 & 2 \end{pmatrix}$, $B = \begin{pmatrix} x & 0 \\ 1 & 3 \end{pmatrix}$,

and **AB = BA**, find x.

48. If $X = \begin{pmatrix} k & 2 \\ 2 & -k \end{pmatrix}$ and $X^2 = 5 \begin{pmatrix} 1 & 0 \\ 0 & 1 \end{pmatrix}$, find k.

49. $B = \begin{pmatrix} 3 & 3 \\ -1 & -1 \end{pmatrix}$

(a) Find k if $B^2 = kB$
(b) Find m if $B^4 = mB$

50. $A = \begin{pmatrix} 5 & 5 \\ -2 & -2 \end{pmatrix}$

(a) Find n if $A^2 = nA$
(b) Find q if $A^3 = qA$

8.2 ROUTE MATRICES AND DATA STORAGE

Route matrices

The diagram shows a map of the roads linking three towns A, B and C. The corresponding *direct* route matrix is shown.

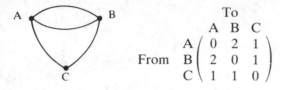

$$\text{From} \quad \begin{array}{c} A \\ B \\ C \end{array} \overset{\text{To}}{\begin{pmatrix} A & B & C \\ 0 & 2 & 1 \\ 2 & 0 & 1 \\ 1 & 1 & 0 \end{pmatrix}}$$

The two-stage route matrix is obtained by squaring the direct route matrix

Two-stage
route matrix

$$\text{From} \quad \begin{array}{c} A \\ B \\ C \end{array} \overset{\text{To}}{\begin{pmatrix} A & B & C \\ 5 & 1 & 2 \\ 1 & 5 & 2 \\ 2 & 2 & 2 \end{pmatrix}}$$

So for example there are two ways of going from A to C in two stages.

Exercise 2

In questions **1** to **4** write down the direct route matrix.

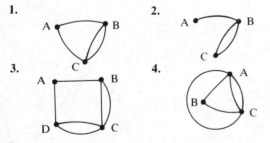

1.

2.

3.

4.

In questions **5** and **6** write down the direct route matrix and work out the two-stage matrix

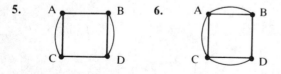

5.

6.

In questions **7** and **8** draw the network.

7.
$$\begin{array}{c} A \\ B \\ C \\ D \end{array} \begin{pmatrix} A & B & C & D \\ 0 & 2 & 0 & 0 \\ 2 & 0 & 1 & 0 \\ 0 & 1 & 0 & 2 \\ 0 & 0 & 0 & 2 \end{pmatrix}$$

8.
$$\begin{array}{c} A \\ B \\ C \\ D \end{array} \begin{pmatrix} A & B & C & D \\ 0 & 1 & 0 & 0 \\ 1 & 0 & 2 & 0 \\ 0 & 2 & 0 & 1 \\ 0 & 0 & 1 & 0 \end{pmatrix}$$

Data storage

The matrix shows the results for four teams.

$$\text{Team} \quad \begin{array}{c} A \\ B \\ C \\ D \end{array} \begin{array}{ccc} \text{win} & \text{draw} & \text{lose} \\ \begin{pmatrix} 1 & 2 & 3 \\ 4 & 0 & 2 \\ 2 & 2 & 2 \\ 3 & 1 & 2 \end{pmatrix} \end{array}$$

The points given are

$$\begin{array}{c} \text{win} \\ \text{draw} \\ \text{lose} \end{array} \begin{pmatrix} 3 \\ 1 \\ 0 \end{pmatrix}$$

By multiplying these two matrices we can work out how many points each team has.

Exercise 3

1. A restaurant sells three meals A, B and C. In two days the sales were as shown in matrix **S**

$$\mathbf{S} = \begin{array}{c} A \\ B \\ C \end{array} \begin{array}{cc} \text{Mon} & \text{Tue} \\ \begin{pmatrix} 10 & 5 \\ 15 & 20 \\ 10 & 10 \end{pmatrix} \end{array}$$

The price in £'s paid for each meal is given by matrix **P**.

$$\mathbf{P} = \begin{array}{ccc} A & B & C \\ (2 & 4 & 7) \end{array}$$

Work out the matrix product **PS** and interpet the result.

2. The results of four soccer teams are shown below together with a matrix showing how points are awarded.

$$\begin{array}{c} A \\ B \\ C \\ D \end{array} \begin{array}{ccc} W & D & L \\ \begin{pmatrix} 7 & 3 & 5 \\ 5 & 4 & 2 \\ 8 & 2 & 5 \\ 6 & 3 & 4 \end{pmatrix} \end{array} \qquad \begin{array}{c} W \\ D \\ L \end{array} \begin{array}{c} \text{Points} \\ \begin{pmatrix} 4 \\ 1 \\ 0 \end{pmatrix} \end{array}$$

Multiply the matrices and hence state which team has most points.

3. Three garages G_1, G_2 and G_3 sell cars of two types A and B. The sales in one week are shown below together with a matrix showing the price paid for each type.

$$\text{Sales.} \quad \begin{array}{c} G_1 \\ G_2 \\ G_3 \end{array} \begin{array}{cc} A & B \\ \begin{pmatrix} 3 & 1 \\ 2 & 0 \\ 4 & 1 \end{pmatrix} \end{array} \qquad \begin{array}{c} \text{Prices.} \\ (\pounds's) \end{array} \begin{array}{c} A \\ B \end{array} \begin{pmatrix} 5000 \\ 7500 \end{pmatrix}$$

Work out the total value of sales for the three garages.

8.3 THE INVERSE OF A MATRIX

The inverse of a matrix **A** is written **A**$^{-1}$, and the inverse exists if

$$\mathbf{AA^{-1} = A^{-1}A = I}$$

where **I** is called the identity matrix.

Only square matrices possess an inverse.

For 2×2 matrices, $\mathbf{I} = \begin{pmatrix} 1 & 0 \\ 0 & 1 \end{pmatrix}$

For 3×3 matrices, $\mathbf{I} = \begin{pmatrix} 1 & 0 & 0 \\ 0 & 1 & 0 \\ 0 & 0 & 1 \end{pmatrix}$, etc.

If $\mathbf{A} = \begin{pmatrix} a & b \\ c & d \end{pmatrix}$, the inverse **A**$^{-1}$ is given by

$$\mathbf{A^{-1}} = \frac{1}{(ad-cb)} \begin{pmatrix} d & -b \\ -c & a \end{pmatrix}$$

Here, the number $(ad - cb)$ is called the *determinant* of the matrix and is written $|\mathbf{A}|$.

If $|\mathbf{A}| = 0$, then the matrix has no inverse.

Example 1

Find the inverse of $\mathbf{A} = \begin{pmatrix} 3 & -4 \\ 1 & -2 \end{pmatrix}$.

$$\mathbf{A^{-1}} = \frac{1}{[3(-2) - 1(-4)]} \begin{pmatrix} -2 & 4 \\ -1 & 3 \end{pmatrix}$$

$$= \frac{1}{-2} \begin{pmatrix} -2 & 4 \\ -1 & 3 \end{pmatrix}$$

Check: $\mathbf{A^{-1}A} = \frac{1}{-2} \begin{pmatrix} -2 & 4 \\ -1 & 3 \end{pmatrix} \begin{pmatrix} 3 & -4 \\ 1 & -2 \end{pmatrix}$

$$= \frac{-1}{2} \begin{pmatrix} -2 & 0 \\ 0 & -2 \end{pmatrix}$$

$$= \begin{pmatrix} 1 & 0 \\ 0 & 1 \end{pmatrix}$$

Multiplying by the inverse of a matrix gives the same result as dividing by the matrix: the effect is similar to ordinary algebraic operations.

e.g. if $\mathbf{AB = C}$
$$\mathbf{A^{-1}AB = A^{-1}C}$$
$$\mathbf{B = A^{-1}C}$$

Exercise 4

In questions **1** to **15**, find the inverse of the matrix.

1. $\begin{pmatrix} 4 & 1 \\ 3 & 1 \end{pmatrix}$ 2. $\begin{pmatrix} 1 & 2 \\ 2 & 5 \end{pmatrix}$ 3. $\begin{pmatrix} 3 & 4 \\ 1 & 2 \end{pmatrix}$

4. $\begin{pmatrix} 5 & 2 \\ 1 & 1 \end{pmatrix}$ 5. $\begin{pmatrix} 2 & -2 \\ -1 & 2 \end{pmatrix}$ 6. $\begin{pmatrix} 4 & -3 \\ -1 & 2 \end{pmatrix}$

7. $\begin{pmatrix} 2 & 1 \\ -2 & 3 \end{pmatrix}$ 8. $\begin{pmatrix} 0 & -3 \\ 2 & 4 \end{pmatrix}$ 9. $\begin{pmatrix} -1 & -2 \\ 1 & -3 \end{pmatrix}$

10. $\begin{pmatrix} 2 & 4 \\ 1 & 2 \end{pmatrix}$ 11. $\begin{pmatrix} 3 & -2 \\ 1 & 4 \end{pmatrix}$ 12. $\begin{pmatrix} -3 & 1 \\ 2 & 1 \end{pmatrix}$

13. $\begin{pmatrix} 2 & -3 \\ 1 & -4 \end{pmatrix}$ 14. $\begin{pmatrix} 7 & 0 \\ -5 & 1 \end{pmatrix}$ 15. $\begin{pmatrix} 2 & 1 \\ -2 & -4 \end{pmatrix}$

16. If $\mathbf{B} = \begin{pmatrix} 2 & 4 \\ 1 & 3 \end{pmatrix}$ and $\mathbf{AB = I}$, find **A**.

17. Find **Y** if $\mathbf{Y} \begin{pmatrix} -2 & 0 \\ 3 & 1 \end{pmatrix} = \begin{pmatrix} 1 & 0 \\ 0 & 1 \end{pmatrix}$.

18. If $\begin{pmatrix} 2 & -3 \\ 0 & 4 \end{pmatrix} + \mathbf{X} = \begin{pmatrix} 1 & 0 \\ 0 & 1 \end{pmatrix}$, find **X**.

19. Find **B** if $\mathbf{A} = \begin{pmatrix} 2 & -2 \\ -1 & 3 \end{pmatrix}$ and $\mathbf{AB} = \begin{pmatrix} 4 & -2 \\ 0 & 7 \end{pmatrix}$.

20. If $\begin{pmatrix} 3 & -3 \\ 2 & 5 \end{pmatrix} - \mathbf{X} = \begin{pmatrix} 1 & 0 \\ 0 & 1 \end{pmatrix}$, find **X**.

21. Find **M** if $\begin{pmatrix} 1 & 1 \\ -2 & 1 \end{pmatrix} \mathbf{M} = 2 \begin{pmatrix} 1 & 0 \\ 0 & 1 \end{pmatrix}$.

22. $\mathbf{A} = \begin{pmatrix} 2 & -3 \\ 0 & 1 \end{pmatrix}$ and $\mathbf{B} = \begin{pmatrix} 1 & -1 \\ -1 & 3 \end{pmatrix}$.
    Find (a) **AB**, (b) **A**$^{-1}$, (c) **B**$^{-1}$.
    Show that $\mathbf{(AB)^{-1} = B^{-1}A^{-1}}$.

23. If $\mathbf{M} = \begin{pmatrix} 3 & 1 \\ 2 & -1 \end{pmatrix}$ and $\mathbf{MN} = \begin{pmatrix} 7 & -9 \\ -2 & -6 \end{pmatrix}$, find **N**.

24. $\mathbf{A} = \begin{pmatrix} 2 & 1 \\ 1 & 1 \end{pmatrix}$; $\mathbf{C} = \begin{pmatrix} 11 \\ 7 \end{pmatrix}$. If **B** is a (2×1) matrix such that $\mathbf{AB = C}$, find **B**.

25. Find x if the determinant of $\begin{pmatrix} x & 3 \\ 1 & 2 \end{pmatrix}$ is
    (a) 5   (b) -1,   (c) 0.

26. If the matrix $\begin{pmatrix} 1 & -2 \\ x & 4 \end{pmatrix}$ has no inverse, what is the value of x?

27. The elements of a (2×2) matrix consist of four different numbers. Find the largest possible value of the determinant of this matrix if the numbers are
    (a) 1, 3, 5, 9   (b) $-1, 2, 3, 4$.

8.4 SIMPLE TRANSFORMATIONS

Reflection

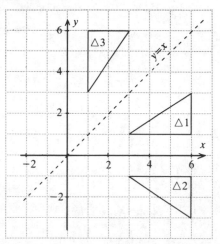

$\triangle 2$ is the image of $\triangle 1$ after reflection in the x-axis.

$\triangle 3$ is the image of $\triangle 1$ after reflection in the line $y = x$.

Exercise 5

In questions **1** to **6** draw the object and its image after reflection in the broken line.

1.

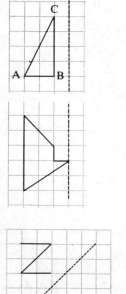

2.

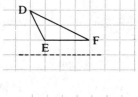

3.

4.

5.

6.

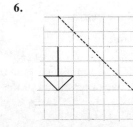

In questions **7**, **8**, **9** draw the image of the given shape after reflection in line M_1 and then reflect this new shape in line M_2.

7.

8.

9.

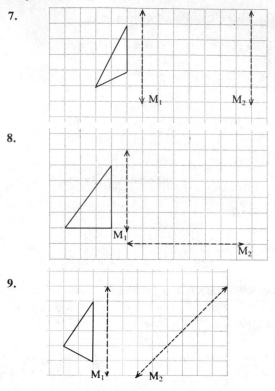

Exercise 6

For each question draw x and y axes with values from -8 to 8.

1. (a) Draw the triangle ABC at A(6, 8), B(2, 8), C(2, 6). Draw the lines $y = 2$ and $y = x$.
   (b) Draw the image of $\triangle ABC$ after reflection in:
       (i) the y-axis. Label it $\triangle 1$.
       (ii) the line $y = 2$. Label it $\triangle 2$.
       (iii) the line $y = x$. Label it $\triangle 3$.
   (c) Write down the coordinates of the image of point A in each case.

2. (a) Draw the triangle DEF at D(−6, 8), E(−2, 8), F(−2, 6). Draw the lines $x = 1$, $y = x$, $y = -x$.
   (b) Draw the image of $\triangle DEF$ after reflection in:
       (i) the line $x = 1$. Label it $\triangle 1$.
       (ii) the line $y = x$. Label it $\triangle 2$.
       (iii) the line $y = -x$. Label it $\triangle 3$.
   (c) Write down the coordinates of the image of point D in each case.

3. (a) Draw the triangle ABC at A(5, 1), B(8, 1), C(8, 3). Draw the lines $x + y = 4$, $y = x - 3$, $x = 2$.
   (b) Draw the image of $\triangle ABC$ after reflection in:
      (i) the line $x + y = 4$. Label it $\triangle 1$.
      (ii) the line $y = x - 3$. Label it $\triangle 2$.
      (iii) the line $x = 2$. Label it $\triangle 3$.
   (c) Write down the coordinates of the image of point A in each case.
4. (a) Draw and label the following triangles:
      (i) $\triangle 1: (3, 3), (3, 6), (1, 6)$
         $\triangle 2: (3, -1), (3, -4), (1, -4)$
         $\triangle 3: (3, 3), (6, 3), (6, 1)$
         $\triangle 4: (-6, -1), (-6, -3), (-3, -3)$
         $\triangle 5: (-6, 5), (-6, 7), (-3, 7)$
   (b) Find the equation of the mirror line for the reflection:
      (i) $\triangle 1$ onto $\triangle 2$ (ii) $\triangle 1$ onto $\triangle 3$
      (iii) $\triangle 1$ onto $\triangle 4$ (iv) $\triangle 4$ onto $\triangle 5$.
5. (a) Draw $\triangle 1$ at (3, 1), (7, 1), (7, 3).
   (b) Reflect $\triangle 1$ in the line $y = x$ onto $\triangle 2$.
   (c) Reflect $\triangle 2$ in the x-axis onto $\triangle 3$.
   (d) Reflect $\triangle 3$ in the line $y = -x$ onto $\triangle 4$.
   (e) Reflect $\triangle 4$ in the line $x = 2$ onto $\triangle 5$.
   (f) Write down the coordinates of $\triangle 5$.
6. (a) Draw $\triangle 1$ at (2, 6), (2, 8), (6, 6).
   (b) Reflect $\triangle 1$ in the line $x + y = 6$ onto $\triangle 2$.
   (c) Reflect $\triangle 2$ in the line $x = 3$ onto $\triangle 3$.
   (d) Reflect $\triangle 3$ in the line $x + y = 6$ onto $\triangle 4$.
   (e) Reflect $\triangle 4$ in the line $y - x = 8$ onto $\triangle 5$.
   (f) Write down the coordinates of $\triangle 5$.

Rotation

Example 2

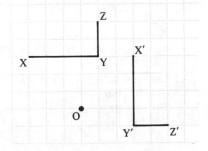

The letter L has been rotated through 90° clockwise about the centre O. The angle, direction, and centre are needed to fully describe a rotation.

We say that the object *maps* onto the image. Here,

X maps onto X′
Y maps onto Y′
Z maps onto Z′

In this work, a clockwise rotation is *negative* and an anticlockwise rotation is *positive*: in Example 2, the letter L has been rotated through −90°. The angle, the direction, and the centre of rotation can be found using tracing paper and a sharp pencil placed where you think the centre of rotation is.

For more accurate work, draw the perpendicular bisector of the line joining two corresponding points, e.g. Y and Y′. Repeat for another pair of corresponding points. The centre of rotation is at the intersection of the two perpendicular bisectors.

Exercise 7

In questions **1** to **4** draw the object and its image under the rotation given. Take O as the centre of rotation in each case.

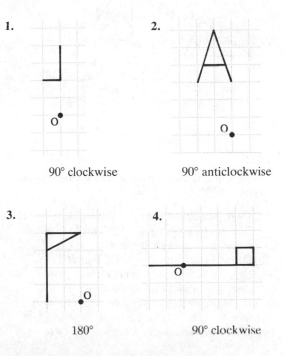

1.

90° clockwise

2.

90° anticlockwise

3.

180°

4.

90° clockwise

In questions **5** to **8**, copy the diagram on squared paper and find the angle, the direction, and the centre of the rotation.

5.

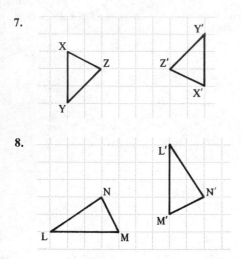

6.

7.

8.

Exercise 8

For all questions draw x- and y-axes for values from −8 to +8.

1. (a) Draw the object triangle ABC at A(1, 3), B(1, 6), C(3, 6), rotate ABC through 90° clockwise about (0, 0), mark A′B′C′.

(b) Draw the object triangle DEF at D(3, 3), E(6, 3), F(6, 1), rotate DEF through 90° clockwise about (0, 0), mark D′E′F′.

(c) Draw the object triangle PQR at P(−4, 7), Q(−4, 5), R(−1, 5), rotate PQR through 90° anticlockwise about (0, 0), mark P′Q′R′.

2. (a) Draw △1 at (1, 4), (1, 7), (3, 7).

(b) Draw the images of △1 under the following reflections:

(i) 90° clockwise, centre (0, 0). Label it △2.

(ii) 180°, centre (0, 0). Label it △3.

(iii) 90° anticlockwise, centre (0, 0). Label it △4.

3. (a) Draw triangle PQR at P(1, 2), Q(3, 5), R(6, 2).

(b) Find the image of PQR under the following rotations:

(i) 90° anticlockwise, centre (0, 0); label the image P′Q′R′

(ii) 90° clockwise, centre (−2, 2); label the image P″Q″R″

(iii) 180°, centre (1, 0); label the image P*Q*R*.

(c) Write down the coordinates of P′, P″, P*.

4. (a) Draw △1 at (1, 2), (1, 6), (3, 5).

(b) Rotate △1 90° clockwise, centre (1, 2) onto △2.

(c) Rotate △2 180°, centre (2, −1) onto △3.

(d) Rotate △3 90° clockwise, centre (2, 3) onto △4.

(e) Write down the coordinates of △4.

5. (a) Draw and label the following triangles:
△1 : (3, 1), (6, 1), (6, 3)
△2 : (−1, 3), (−1, 6), (−3, 6)
△3 : (1, 1), (−2, 1), (−2, −1)
△4 : (3, −1), (3, −4), (5, −4)
△5 : (4, 4), (1, 4), (1, 2)

(b) Describe fully the following rotations:

(i) △1 onto △2   (ii) △1 onto △3
(iii) △1 onto △4   (iv) △1 onto △5
(v) △5 onto △4   (vi) △3 onto △2

6. (a) Draw △1 at (4, 7), (8, 5), (8, 7).

(b) Rotate △1 90° clockwise, centre (4, 3) onto △2.

(c) Rotate △2 180°, centre (5, −1) onto △3.

(d) Rotate △3 90° anticlockwise, centre (0, −8) onto △4

(e) Describe fully the following rotations:

(i) △4 onto △1,
(ii) △4 onto △2.

Translation

The triangle ABC below has been transformed
onto the triangle A'B'C' by a *translation*.

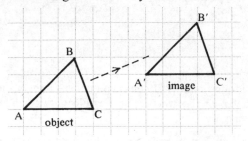

Here the translation is 7 squares to the right and
2 squares up the page. The translation can be
described by a column vector.

In this case the translation is $\begin{pmatrix} 7 \\ 2 \end{pmatrix}$.

Exercise 9

1. Make a copy of the diagram below and write down
   the column vector for each of the following
   translations.
   (a) D onto A   (b) B onto F
   (c) E onto A   (d) A onto C
   (e) F onto C   (f) C onto B
   (g) F onto E   (h) B onto C.

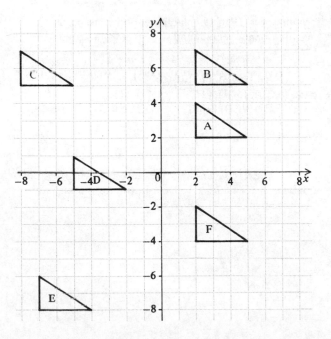

For questions **2** to **11** draw x and y axes with values for
-8 to 8. Draw object triangle ABC at A$(-4, -1)$,
B$(-4, 1)$, C$(-1, -1)$ and shade it.
Draw the image of ABC under the translations
described by the vectors below. For each question,
write down the new coordinates of point C.

2. $\begin{pmatrix} 6 \\ 3 \end{pmatrix}$ **3.** $\begin{pmatrix} 6 \\ 7 \end{pmatrix}$ **4.** $\begin{pmatrix} 9 \\ -4 \end{pmatrix}$ **5.** $\begin{pmatrix} 1 \\ 7 \end{pmatrix}$

6. $\begin{pmatrix} 5 \\ -6 \end{pmatrix}$ **7.** $\begin{pmatrix} -2 \\ 5 \end{pmatrix}$ **8.** $\begin{pmatrix} -2 \\ -4 \end{pmatrix}$ **9.** $\begin{pmatrix} 0 \\ -7 \end{pmatrix}$

10. $\begin{pmatrix} 3 \\ 1 \end{pmatrix}$ followed by $\begin{pmatrix} 3 \\ 2 \end{pmatrix}$

11. $\begin{pmatrix} -2 \\ 0 \end{pmatrix}$ followed by $\begin{pmatrix} 0 \\ 3 \end{pmatrix}$ followed by $\begin{pmatrix} 1 \\ -1 \end{pmatrix}$

Enlargement

In the diagram below, the letter T has been
enlarged by a scale factor of 2 using the point O
as the centre of the enlargement.

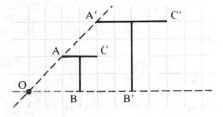

Notice that   OA' = 2 × OA
               OB' = 2 × OB

The scale factor and the centre of enlargement
are both required to describe an enlargement.

Example 3

Draw the image of
triangle ABC under
an enlargement
scale factor of $\frac{1}{2}$
using O as centre
of enlargement.

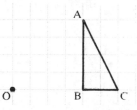

(a) Draw lines through OA, OB and OC.

(b) Mark A' so that OA' = $\frac{1}{2}$OA
     Mark B' so that OB' = $\frac{1}{2}$OB
     Mark C' so that OC' = $\frac{1}{2}$OC.

(c) Join A'B'C as shown overleaf.

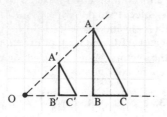

Remember always to measure the lengths from O, not from A, B, or C.

Example 4

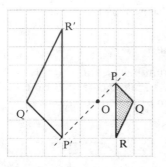

P′Q′R′ is the image of PQR after enlargement with scale factor −2 and centre O.
Notice that P′ and P are on opposite sides of point O. Similarly Q′ and Q, R′ and R.

Exercise 10

In questions **1** to **6** copy the diagram and draw an enlargement using the centre O and the scale factor given.

1. Scale factor 2

2. Scale factor 3

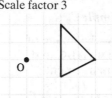

3. Scale factor 3

4. Scale factor −2

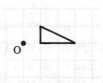

5. Scale factor −3　　　**6.** Scale factor 1½

Answer questions **7** to **19** on graph paper taking x and y from 0 to 15. The vertices of the object are given in coordinate form.

In questions **7** to **10**, enlarge the object with the centre of enlargement and scale factor indicated.

	object	centre	scale factor
7.	(2, 4)(4, 2)(5, 5)	(0, 0)	+2
8.	(2, 4)(4, 2)(5, 5)	(1, 2)	+2
9.	(1, 1)(4, 2)(2, 3)	(1, 1)	+3
10.	(4, 4)(7, 6)(9, 3)	(7, 4)	+2

In questions **11** to **14** plot the object and image and find the centre of enlargement and the scale factor.

11. object　A(2, 1), B(5, 1), C(3, 3)
　　　image　A′(2, 1), B′(11, 1), C′(5, 7)
12. object　A(2, 5), B(9, 3), C(5, 9)
　　　image　A′(6½, 7), B′(10, 6), C′(8, 9)
13. object　A(2, 2), B(4, 4), C(2, 6)
　　　image　A′(11, 8), B′(7, 4), C′(11, 0)
14. object　A(0, 6), B(4, 6), C(3, 0)
　　　image　A′(12, 6), B′(8, 6), C′(9, 12)

In questions **15** to **19** enlarge the object using the centre of enlargement and scale factor indicated.

	object	centre	S.F.
15.	(1, 2), (13, 2), (1, 10)	(0, 0)	+½
16.	(5, 10), (5, 7), (11, 7)	(2, 1)	+⅓
17.	(7, 3), (9, 3), (7, 8)	(5, 5)	−1
18.	(1, 1), (3, 1), (3, 2)	(4, 3)	−2
19	(9, 2), (14, 2), (14, 6)	(7, 4)	−½

The next exercise contains questions involving the four basic transformations: reflection; rotation; translation; enlargement.

Exercise 11

1. (a) Copy the diagram below

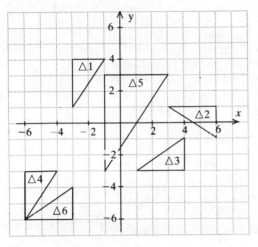

(b) Describe fully the following transformations
   (i) $\triangle 1 \rightarrow \triangle 2$ (ii) $\triangle 1 \rightarrow \triangle 3$
   (iii) $\triangle 4 \rightarrow \triangle 1$ (iv) $\triangle 1 \rightarrow \triangle 5$
   (v) $\triangle 3 \rightarrow \triangle 6$ (vi) $\triangle 6 \rightarrow \triangle 4$

2. Plot and label the following triangles:
   $\triangle 1 : (-5, -5), (-1, -5), (-1, -3)$
   $\triangle 2 : (1, 7), (1, 3), (3, 3)$
   $\triangle 3 : (3, -3), (7, -3), (7, -1)$
   $\triangle 4 : (-5, -5), (-5, -1), (-3, -1)$
   $\triangle 5 : (1, -6), (3, -6), (3, -5)$
   $\triangle 6 : (-3, 3), (-3, 7), (-5, 7)$
   Describe fully the following transformations:
   (a) $\triangle 1 \rightarrow \triangle 2$ (b) $\triangle 1 \rightarrow \triangle 3$
   (c) $\triangle 1 \rightarrow \triangle 4$ (d) $\triangle 1 \rightarrow \triangle 5$
   (e) $\triangle 1 \rightarrow \triangle 6$ (f) $\triangle 5 \rightarrow \triangle 3$
   (g) $\triangle 2 \rightarrow \triangle 3$

3. Plot and label the following triangles:
   $\triangle 1 : (-3, -6), (-3, -2), (-5, -2)$
   $\triangle 2 : (-5, -1), (-5, -7), (-8, -1)$
   $\triangle 3 : (-2, -1), (2, -1), (2, 1)$
   $\triangle 4 : (6, 3), (2, 3), (2, 5)$
   $\triangle 5 : (8, 4), (8, 8), (6, 8)$
   $\triangle 6 : (-3, 1), (-3, 3), (-4, 3)$
   Describe fully the following transformations:
   (a) $\triangle 1 \rightarrow \triangle 2$ (b) $\triangle 1 \rightarrow \triangle 3$
   (c) $\triangle 1 \rightarrow \triangle 4$ (d) $\triangle 1 \rightarrow \triangle 5$
   (e) $\triangle 1 \rightarrow \triangle 6$ (f) $\triangle 3 \rightarrow \triangle 5$
   (g) $\triangle 6 \rightarrow \triangle 2$

8.5   COMBINED TRANSFORMATIONS

It is convenient to denote transformations by a symbol.

   Let **A** denote 'reflection in line $x = 3$' and
   B denote 'translation $\begin{pmatrix} 2 \\ 1 \end{pmatrix}$'.

Perform **A** on $\triangle 1$

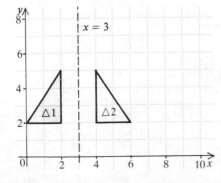

Fig. 1

$\triangle 2$ is the image of $\triangle 1$ under the reflection in $x = 3$
   i.e. $\mathbf{A}(\triangle 1) = \triangle 2$

$\mathbf{A}(\triangle 1)$ means 'perform the transformation **A** on triangle $\triangle 1$'

Perform **B** on $\triangle 2$

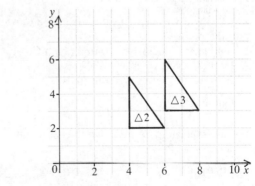

Fig. 2

From Fig. 2 we can see that
   $\mathbf{B}(\triangle 2) = \triangle 3$
The effect of going from $\triangle 1$ to $\triangle 3$ may be written
   $\mathbf{BA}(\triangle 1) = \triangle 3$
It is very important to notice that $\mathbf{BA}(\triangle 1)$ means do **A** first and then **B**.

Repeated transformations

XX(P) means 'perform transformation **X** on P and then perform **X** on the image'.

It may be written $X^2(P)$

Similarly $TTT(P) = T^3(P)$.

Inverse transformations

If translation **T** has vector $\begin{pmatrix} 3 \\ -2 \end{pmatrix}$, the translation which has the opposite effect has vector $\begin{pmatrix} -3 \\ 2 \end{pmatrix}$. This is written T^{-1}.

If rotation **R** denotes 90° clockwise rotation about (0, 0), then R^{-1} denotes 90° *anti*clockwise rotation about (0, 0).

The *inverse* of a transformation is the transformation which takes the *image* back to the object.

Note
For all reflections, the inverse is the same reflection.
e.g. if **X** is reflection in $x = 0$, then X^{-1} is also reflection in $x = 0$.

The symbol T^{-3} means $(T^{-1})^3$ i.e. perform T^{-1} three times.

Exercise 12

Draw x- and y-axes with values from −8 to +8 and plot the point P(3, 2).
R denotes 90° clockwise rotation about (0, 0);
X denotes reflection in $x = 0$.
H denotes 180° rotation about (0, 0);
T denotes translation $\begin{pmatrix} 3 \\ 2 \end{pmatrix}$.

For each question, write down the coordinates of the final image of P.

1. R(P)	2. TR(P)
3. T(P)	4. RT(P)
5. TH(P)	6. XT(P)
7. HX(P)	8. XX(P)
9. R^{-1}(P)	10. T^{-1}(P)
11. X^3(P)	12. T^{-2}(P)
13. R^2(P)	14. $T^{-1}R^2$(P)
15. THX(P)	16. R^3(P)
17. TX^{-1}(P)	18. T^3X(P)
19. T^2H^{-1}(P)	20. XTH(P)

Exercise 13

For questions **1** to **24**, make a copy of the diagram shown; then copy and complete the data and write down the equivalent single transformation if
  M_x denotes reflection in x axis
  M_y denotes reflection in y axis
  M_3 denotes reflection in $y = x$
  M_4 denotes reflection in $y = -x$
  R_a denotes rotation 90° anticlockwise about (0, 0)
  R_b denotes rotation 180° about (0, 0)
  R_c denotes rotation 90° clockwise about (0, 0)
  I is the identity: image is same as object.

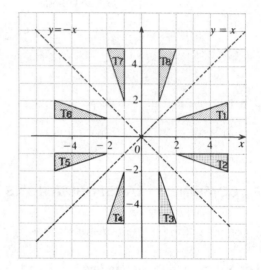

For example,

$$T_1 \xrightarrow{M_y} \xrightarrow{M_x} \text{ becomes } T_1 \xrightarrow{M_y} T_6 \xrightarrow{M_x} T_5$$

The equivalent single transformation $T_1 \rightarrow T_5$ is R_b

1. $T_1 \xrightarrow{M_y} \xrightarrow{M_x}$	2. $T_2 \xrightarrow{M_4} \xrightarrow{R_a}$
3. $T_3 \xrightarrow{M_y} \xrightarrow{R_a}$	4. $T_5 \xrightarrow{M_y} \xrightarrow{R_a}$
5. $T_8 \xrightarrow{M_y} \xrightarrow{R_a}$	6. $T_3 \xrightarrow{R_a} \xrightarrow{M_y}$
7. $T_6 \xrightarrow{R_a} \xrightarrow{M_y}$	8. $T_5 \xrightarrow{R_a} \xrightarrow{M_3}$
9. $T_7 \xrightarrow{M_y} \xrightarrow{M_3}$	10. $T_4 \xrightarrow{R_b} \xrightarrow{R_c}$
11. $T_6 \xrightarrow{R_c} \xrightarrow{M_4}$	12. $T_5 \xrightarrow{M_4} \xrightarrow{M_3}$
13. $T_5 \xrightarrow{M_3} \xrightarrow{M_4}$	14. $T_8 \xrightarrow{M_x} \xrightarrow{M_x}$
15. $T_2 \xrightarrow{R_c} \xrightarrow{M_y}$	16. $T_5 \xrightarrow{M_x} \xrightarrow{M_4} \xrightarrow{R_c}$
17. $T_1 \xrightarrow{M_y} \xrightarrow{R_b} \xrightarrow{M_3}$	18. $T_1 \xrightarrow{R_a} \xrightarrow{M_3} \xrightarrow{R_b}$
19. $T_3 \xrightarrow{R_c} \xrightarrow{M_3} \xrightarrow{R_b}$	20. $T_4 \xrightarrow{M_x} \xrightarrow{M_3} \xrightarrow{M_4}$

21. $T_7 \xrightarrow{M_3} \xrightarrow{R_a} \xrightarrow{M_y}$ **22.** $T_6 \xrightarrow{M_x} \xrightarrow{M_4} \xrightarrow{R_a}$

23. $T_8 \xrightarrow{R_b} \xrightarrow{R_c} \xrightarrow{M_x}$ **24.** $T_2 \xrightarrow{M_y} \xrightarrow{R_a} \xrightarrow{M_x}$

25. Is the equivalent single transformation the same if you change the order of the transformations?

Exercise 14

In this exercise, transformations **A**, **B**, ... **H**, are as follows:

A denotes reflection in $x = 2$

B denotes $180°$ rotation, centre $(1, 1)$

C denotes translation $\begin{pmatrix} -6 \\ 2 \end{pmatrix}$

D denotes reflection in $y = x$

E denotes reflection in $y = 0$

F denotes translation $\begin{pmatrix} 4 \\ 3 \end{pmatrix}$

G denotes $90°$ rotation clockwise, centre $(0, 0)$

H denotes enlargement, scale factor $+\frac{1}{2}$, centre $(0, 0)$

Draw x- and y-axes with values from -8 to $+8$.

1. Draw triangle LMN at L(2, 2), M(6, 2), N(6, 4). Find the image of LMN under the following combinations of transformations. Write down the coordinates of the image of point L in each case:
  (a) **CA**(LMN)   (b) **ED**(LMN)
  (c) **DB**(LMN)   (d) **BE**(LMN)
  (e) **EB**(LMN).

2. Draw triangle PQR at P(2, 2), Q(6, 2), R(6, 4). Find the image of PQR under the following combinations of transformations. Write down the coordinates of the image of point P in each case:
  (a) **AF**(PQR)   (b) **CG**(PQR)
  (c) **AG**(PQR)   (d) **HE**(PQR).

3. Draw triangle XYZ at X(-2, 4), Y(-2, 1), Z(-4, 1). Find the image of XYZ under the following combinations of transformations and state the equivalent single transformation in each case.
  (a) **G²E**(XYZ)   (b) **CB**(XYZ)
  (c) **DA**(XYZ).

4. Draw triangle OPQ at O(0, 0), P(0, 2), Q(3, 2). Find the image of OPQ under the following combinations of transformations and state the equivalent single transformation in each case.
  (a) **DE**(OPQ)   (b) **FC**(OPQ)
  (c) **DEC**(OPQ)   (d) **DFE**(OPQ).

5. Draw triangle RST at R(-4, -1), S(-2$\frac{1}{2}$, -2), T(-4, -4). Find the image of RST under the following combinations of transformations and state the equivalent single transformation in each case.
  (a) **EAG**(RST)   (b) **FH**(RST)
  (c) **GF**(RST).

6. Write down the inverses of the transformations **A**, **B**, ... **H**.

7. Draw triangle JKL at J(-2, 2), K(-2, 5), L(-4, 5). Find the image of JKL under the following transformations. Write down the coordinates of the image of point J in each case.
  (a) $\mathbf{C}^{-1}$ (b) $\mathbf{F}^{-1}$ (c) $\mathbf{G}^{-1}$
  (d) $\mathbf{D}^{-1}$ (e) $\mathbf{A}^{-1}$.

8. Draw triangle PQR at P(-2, 4), Q(-2, 1), R(-4, 1). Find the image of PQR under the following combinations of transformations. Write down the coordinates of the image of point P in each case.
  (a) $\mathbf{DF}^{-1}$(PQR)   (b) $\mathbf{EC}^{-1}$(PQR)
  (c) $\mathbf{D}^2\mathbf{F}$(PQR)   (d) **GA**(PQR)
  (e) $\mathbf{C}^{-1}\mathbf{G}^{-1}$(PQR).

9. Draw triangle LMN at L(-2, 4), M(-4, 1), N(-2, 1). Find the image of LMN under the following combinations of transformations. Write down the coordinates of the image of point L in each case.
  (a) **HE**(LMN)   (b) $\mathbf{EAG}^{-1}$(LMN)
  (c) **EDA**(LMN)   (d) $\mathbf{BG}^2\mathbf{F}$(LMN).

10. Draw triangle XYZ at X(1, 2), Y(1, 6), Z(3, 6).
  (a) Find the image of XYZ under each of the transformations **BC** and **CB**.
  (b) Describe fully the single transformation equivalent to **BC**.
  (c) Describe fully the transformation **M** such that **MCB** = **BC**.

174

8.6 TRANSFORMATIONS USING MATRICES

Example 1

Find the image of triangle ABC, with A(1, 1), B(3, 1), C(3, 2), under the transformation represented by the matrix $\mathbf{M} = \begin{pmatrix} 1 & 0 \\ 0 & -1 \end{pmatrix}$.

(a) Write the coordinates of A as a column vector and multiply this vector by **M**.

$$\begin{matrix} \mathbf{M} & A & A' \end{matrix}$$
$$\begin{pmatrix} 1 & 0 \\ 0 & -1 \end{pmatrix}\begin{pmatrix} 1 \\ 1 \end{pmatrix} = \begin{pmatrix} 1 \\ -1 \end{pmatrix}$$

A', the image of A, has coordinates (1, −1).

(b) Repeat for B and C.

$$\begin{matrix} \mathbf{M} & B & B' \end{matrix}$$
$$\begin{pmatrix} 1 & 0 \\ 0 & -1 \end{pmatrix}\begin{pmatrix} 3 \\ 1 \end{pmatrix} = \begin{pmatrix} 3 \\ -1 \end{pmatrix}$$
$$\begin{matrix} \mathbf{M} & C & C' \end{matrix}$$
$$\begin{pmatrix} 1 & 0 \\ 0 & -1 \end{pmatrix}\begin{pmatrix} 3 \\ 2 \end{pmatrix} = \begin{pmatrix} 3 \\ -2 \end{pmatrix}$$

(c) Plot A'(1, −1), B'(3, −1) and C'(3, −2).

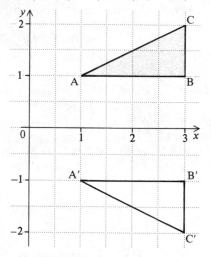

The transformation is a reflection in the x-axis.

Example 2

Find the image of L(1, 1), M(1, 3), N(2, 3) under the transformation represented by the matrix $\begin{pmatrix} 0 & -1 \\ 1 & 0 \end{pmatrix}$.

A quicker method is to write the three vectors for L, M and N in a single 2×3 matrix, and then perform the multiplication

$$\begin{matrix} & L\ M\ N & & L'\ M'\ N' \end{matrix}$$
$$\begin{pmatrix} 0 & -1 \\ 1 & 0 \end{pmatrix}\begin{pmatrix} 1 & 1 & 2 \\ 1 & 3 & 3 \end{pmatrix} = \begin{pmatrix} -1 & -3 & -3 \\ 1 & 1 & 2 \end{pmatrix}$$

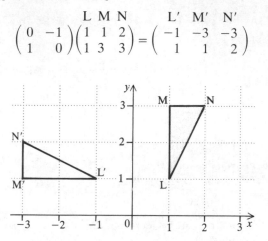

The transformation is a rotation, +90°, centre (0, 0).

Exercise 15

For questions 1 to 5 draw x- and y-axis with values from −8 to 8. Do all parts of each question on one graph.

1. Draw the triangle A(2, 2), B(6, 2), C(6, 4). Find its image under the transformations represented by the following matrices.

   (a) $\begin{pmatrix} 0 & -1 \\ 1 & 0 \end{pmatrix}$ (b) $\begin{pmatrix} -1 & 0 \\ 0 & 1 \end{pmatrix}$ (c) $\begin{pmatrix} 1 & 0 \\ 0 & -1 \end{pmatrix}$

   (d) $\begin{pmatrix} 0 & 1 \\ 1 & 0 \end{pmatrix}$ (e) $\begin{pmatrix} \frac{1}{2} & 0 \\ 0 & \frac{1}{2} \end{pmatrix}$.

2. Plot the object and image for the following:

	Object	Matrix
(a)	P(4, 2), Q(4, 4), R(0, 4)	$\begin{pmatrix} 2 & 0 \\ 0 & 2 \end{pmatrix}$
(b)	P(4, 2), Q(4, 4), R(0, 4)	$\begin{pmatrix} -\frac{1}{2} & 0 \\ 0 & -\frac{1}{2} \end{pmatrix}$
(c)	A(−6, 8), B(−2, 8), C(−2, 6)	$\begin{pmatrix} 0 & -1 \\ -1 & 0 \end{pmatrix}$
(d)	P(4, 2), Q(4, 4), R(0, 4)	$\begin{pmatrix} -2 & 0 \\ 0 & -2 \end{pmatrix}$

   Describe each as a *single* transformation.

3. Draw a trapezium at K(2, 2), L(2, 5), M(5, 8), N(8, 8). Find the images of KLMN under the transformations described by the following matrices:

$$A = \begin{pmatrix} 1 & 0 \\ 0 & -1 \end{pmatrix} \qquad E = \begin{pmatrix} 0 & -1 \\ -1 & 0 \end{pmatrix}$$

$$B = \begin{pmatrix} -1 & 0 \\ 0 & 1 \end{pmatrix} \qquad F = \begin{pmatrix} -1 & 0 \\ 0 & -1 \end{pmatrix}$$

$$C = \begin{pmatrix} 0 & 1 \\ 1 & 0 \end{pmatrix} \qquad G = \begin{pmatrix} 0 & -1 \\ 1 & 0 \end{pmatrix}$$

$$D = \begin{pmatrix} 0 & 1 \\ -1 & 0 \end{pmatrix} \qquad H = \begin{pmatrix} 1 & 0 \\ 0 & 1 \end{pmatrix}$$

Describe fully each of the eight transformations.

4. (a) Draw a quadrilateral at A(3, 4), B(4, 0), C(3, 1), D(0, 0). Find the image of ABCD under the transformation represented by the matrix $\begin{pmatrix} -2 & 0 \\ 0 & -2 \end{pmatrix}$.

   Find the ratio $\left(\dfrac{\text{area of image}}{\text{area of object}} \right)$.

5. (a) Draw axes so that both x and y can take values from −2 to +8.
   (b) Draw triangle ABC at A(2, 1), B(7, 1), C(2, 4).
   (c) Find the image of ABC under the transformation represented by the matrix $\begin{pmatrix} 1 & 1 \\ 1 & 1 \end{pmatrix}$ and plot the image on the graph.
   (d) The transformation is a rotation followed by an enlargement. Calculate the angle of the rotation and the scale factor of the enlargement.

6. (a) Draw axes to that x can take values from 0 to 15 and y can take values from −6 to +6.
   (b) Draw triangle PQR at P(2, 1), Q(7, 1), R(2, 4).
   (c) Find the image of PQR under the transformation represented by the matrix $\begin{pmatrix} 2 & 1 \\ -1 & 2 \end{pmatrix}$ and plot the image on the graph.
   (d) The transformation is a rotation followed by an enlargement. Calculate the angle of the rotation and the scale factor of the enlargement.

7. (a) On graph paper, draw the triangle T whose vertices are (2, 2), (6, 2) and (6, 4).
   (b) Draw the image U of T under the transformation whose matrix is $\begin{pmatrix} 0 & 1 \\ 1 & 0 \end{pmatrix}$.

   (c) Draw the image V of T under the transformation whose matrix is $\begin{pmatrix} 1 & 0 \\ 0 & -1 \end{pmatrix}$.
   (d) Describe the single transformation which would map U onto V.

8. (a) Find the images of the points (1, 0), (2, 1), (3, −1), (−2, 3) under the transformation with matrix $\begin{pmatrix} 1 & 3 \\ 2 & 6 \end{pmatrix}$.
   (b) Show that the images lie on a straight line, and find its equation.

9. The transformation with matrix $\begin{pmatrix} 2 & 3 \\ 6 & 9 \end{pmatrix}$ maps every point in the plane onto a line. Find the equation of the line.

10. Using a scale of 1 cm to one unit in each case draw x- and y-axes, taking values of x from −4 to +6 and values of y from 0 to 12.
   (a) Draw and label the quadrilateral OABC with O(0, 0), A(2, 0), B(4, 2), C(0, 2).
   (b) Find and draw the image of OABC under the transformation whose matrix is **R**, where $\mathbf{R} = \begin{pmatrix} 2 \cdot 4 & -1 \cdot 8 \\ 1 \cdot 8 & 2 \cdot 4 \end{pmatrix}$.
   (c) Calculate, in surd form, the lengths OB and O′B′.
   (d) Calculate the angle AOA′.
   (e) Given that the transformation **R** consists of a rotation about O followed by an enlargement, state the angle of the rotation and the scale factor of the enlargement.

11. The matrix $\mathbf{R} = \begin{pmatrix} \cos \theta & -\sin \theta \\ \sin \theta & \cos \theta \end{pmatrix}$ represents a positive rotation of $\theta°$ about the origin. Find the matrix which represents a rotation of:
   (a) 90°   (b) 180°   (c) 30°   (d) −90°
   (e) 60°   (f) 150°   (g) 45°   (h) 53·1°
   Confirm your results for parts (a), (e), (h) by applying the matrix to the quadrilateral O(0, 0), A(0, 2), B(4, 2), C(4, 0).

12. Using the matrix **R** given in question **11**, find the angle of rotation for the following:
   (a) $\begin{pmatrix} 0 & -1 \\ 1 & 0 \end{pmatrix}$ (b) $\begin{pmatrix} 0\cdot8 & -0\cdot6 \\ 0\cdot6 & 0\cdot8 \end{pmatrix}$
   (c) $\begin{pmatrix} 0\cdot5 & 0\cdot866 \\ -0\cdot866 & 0\cdot5 \end{pmatrix}$ (d) $\begin{pmatrix} 0\cdot6 & 0\cdot8 \\ -0\cdot8 & 0\cdot6 \end{pmatrix}$
   Confirm your results by applying each matrix to the quadrilateral O(0, 0), A(0, 2), B(4, 2), C(4, 0).

Exercise 16

1. Draw the rectangle (0, 0), (0, 1), (2, 1), (2, 0) and its image under the following transformations and describe the *single* transformation which each represents:

    (a) $\begin{pmatrix} 0 & 1 \\ 1 & 0 \end{pmatrix} \begin{pmatrix} x \\ y \end{pmatrix} + \begin{pmatrix} 1 \\ -1 \end{pmatrix}$

    (b) $\begin{pmatrix} 1 & 0 \\ 0 & -1 \end{pmatrix} \begin{pmatrix} x \\ y \end{pmatrix} + \begin{pmatrix} 0 \\ 2 \end{pmatrix}$

    (c) $\begin{pmatrix} 0 & 1 \\ -1 & 0 \end{pmatrix} \begin{pmatrix} x \\ y \end{pmatrix} + \begin{pmatrix} 4 \\ 0 \end{pmatrix}$

    (d) $\begin{pmatrix} 3 & 0 \\ 0 & 3 \end{pmatrix} \begin{pmatrix} x \\ y \end{pmatrix} + \begin{pmatrix} -4 \\ 2 \end{pmatrix}$

2. (a) Draw L(1, 1), M(3, 3), N(4, 1) and its image L′M′N′ under the matrix $\mathbf{A} = \begin{pmatrix} 1 & 0 \\ 0 & -1 \end{pmatrix}$.

    (b) Find and draw the image of L′M′N′ under matrix $\mathbf{B} = \begin{pmatrix} 0 & 1 \\ -1 & 0 \end{pmatrix}$ and label it L″M″N″.

    (c) Calculate the matrix product **BA**.

    (d) Find the image of LMN under the matrix **BA**, and compare with the result of performing **A** and then **B**.

3. (a) (i) Draw P(0, 0), Q(2, 2), R(4, 0) and its image P′Q′R′ under matrix $\mathbf{A} = \begin{pmatrix} 2 & 0 \\ 0 & 2 \end{pmatrix}$.

     (ii) Find and draw the image of P′Q′R′ under matrix $\mathbf{B} = \begin{pmatrix} 1 & 1 \\ 0 & 1 \end{pmatrix}$ and label it P″Q″R″.

     (iii) Calculate the matrix product **BA**.

     (iv) Find the image of PQR under the matrix **BA**, and compare with the result of performing **A** and then **B**.

4. (a) Draw L(1, 1), M(3, 3), N(4, 1) and its image L′M′N′ under matrix $\mathbf{K} = \begin{pmatrix} 2 & 0 \\ 0 & 2 \end{pmatrix}$.

    Find $\mathbf{K}^{-1}$, the inverse of **K**, and now find the image of L′M′N′ under $\mathbf{K}^{-1}$.

    (b) Repeat part (a) with $\mathbf{K} = \begin{pmatrix} 1 & 2 \\ 0 & 1 \end{pmatrix}$.

    (c) Repeat part (a) with $\mathbf{K} = \begin{pmatrix} 3 & 0 \\ 0 & 1 \end{pmatrix}$.

5. The image (x′, y′) of a point (x, y) under a transformation is given by

$$\begin{pmatrix} x' \\ y' \end{pmatrix} = \begin{pmatrix} 3 & 0 \\ 1 & -2 \end{pmatrix} \begin{pmatrix} x \\ y \end{pmatrix} + \begin{pmatrix} 2 \\ 5 \end{pmatrix}$$

    (a) Find the coordinates of the image of the point (4, 3),

    (b) The image of the point (m, n) is the point (11, 7). Write down two equations involving m and n and hence find the values of m and n.

    (c) The image of the point (h, k) is the point (5, 10). Find the values of h and k.

6. Draw A(0, 2), B(2, 2), C(0, 4) and its image under an enlargement, A′(2, 2), B′(6, 2), C′(2, 6).

    (a) What is the centre of enlargement?

    (b) Find the image of ABC under an enlargement, scale factor 2, centre (0, 0).

    (c) Find the translation which maps this image onto A′B′C′.

    (d) What is the matrix **X** and vector **v** which represents an enlargement scale factor 2, centre (−2, 2)?

7. Draw A(0, 1), B(1, 1), C(1, 3) and its image under a reflection A′(4, 1), B′(3, 1), C′(3, 3).

    (a) What is the equation of the mirror line?

    (b) Find the image of ABC under a reflection in the line x = 0.

    (c) Find the translation which maps this image onto A′B′C′.

    (d) What is the matrix **X** and vector **v** which represents a reflection in the line x = 2?

8. Use the same approach as in questions **6** and **7** to find the matrix **X** and vector **v** which represents the following transformations. (Start by drawing an object and its image under the transformation.)

    (a) Enlargement scale factor 2, centre (1, 3),

    (b) Enlargement scale factor 2, centre $(\frac{1}{2}, 1)$,

    (c) Reflection in y = x + 3,

    (d) Rotation 180°, centre $(1\frac{1}{2}, 2\frac{1}{2})$,

    (e) Reflection in y = 1,

    (f) Rotation −90°, centre (2, −2).

Describing a matrix using base vectors

It is possible to describe a transformation in matrix form by considering the effect on the *base vectors* $\begin{pmatrix} 1 \\ 0 \end{pmatrix}$ and $\begin{pmatrix} 0 \\ 1 \end{pmatrix}$.

We will let $\begin{pmatrix} 1 \\ 0 \end{pmatrix}$ be I and $\begin{pmatrix} 0 \\ 1 \end{pmatrix}$ be J.

The *columns* of a matrix give us the images of I and J after the transformation.

Example 3

Describe the transformation with matrix $\begin{pmatrix} 0 & 1 \\ -1 & 0 \end{pmatrix}$.

Column $\begin{pmatrix} 0 \\ -1 \end{pmatrix}$ represents I′ (the image of I).

Column $\begin{pmatrix} 1 \\ 0 \end{pmatrix}$ represents J′ (the image of J)

$$\begin{matrix} \text{I}' & \text{J}' \\ \begin{pmatrix} 0 & 1 \\ -1 & 0 \end{pmatrix} \end{matrix}$$

Draw I, J, I′ and J′ on a diagram.

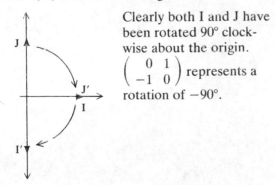

Clearly both I and J have been rotated 90° clockwise about the origin. $\begin{pmatrix} 0 & 1 \\ -1 & 0 \end{pmatrix}$ represents a rotation of −90°.

This method can be used to describe a reflection, rotation, enlargement, shear or stretch in which the origin remains fixed.

Exercise 17

In questions **1** to **12**, use base vectors to describe the transformation represented by each matrix.

1. $\begin{pmatrix} 0 & -1 \\ 1 & 0 \end{pmatrix}$ 2. $\begin{pmatrix} -1 & 0 \\ 0 & 1 \end{pmatrix}$ 3. $\begin{pmatrix} 0 & -1 \\ -1 & 0 \end{pmatrix}$

4. $\begin{pmatrix} 0 & 1 \\ 1 & 0 \end{pmatrix}$ 5. $\begin{pmatrix} 2 & 0 \\ 0 & 2 \end{pmatrix}$ 6. $\begin{pmatrix} \frac{1}{2} & 0 \\ 0 & \frac{1}{2} \end{pmatrix}$

7. $\begin{pmatrix} 3 & 0 \\ 0 & 1 \end{pmatrix}$ 8. $\begin{pmatrix} 1 & 1 \\ 0 & 1 \end{pmatrix}$ 9. $\begin{pmatrix} 1 & 0 \\ 2 & 1 \end{pmatrix}$

10. $\begin{pmatrix} 1 & 0 \\ 0 & 2 \end{pmatrix}$ 11. $\begin{pmatrix} -2 & 0 \\ 0 & -2 \end{pmatrix}$ 12. $\begin{pmatrix} -\frac{1}{2} & 0 \\ 0 & -\frac{1}{2} \end{pmatrix}$

In questions **13** to **22**, use base vectors to write down the matrix which represents each of the transformations.

13. Rotation +90° about (0, 0).
14. Reflection in $y = x$.
15. Reflection in y axis
16. Rotation 180° about (0, 0).
17. Enlargement, centre (0, 0), scale factor 3.
18. Reflection in $y = -x$.
19. Enlargement, centre (0, 0), scale factor −2.
20. Reflection in x axis.
21. Rotation −90° about (0, 0).
22. Enlargement, centre (0, 0), scale factor $\frac{1}{2}$.

Shear

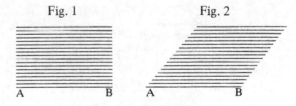

Fig. 1   Fig. 2

Fig. 1 shows a pack of cards stacked neatly into a pile.
Fig. 2 shows the same pack after a *shear* has been performed.

Note
(a) the card AB, at the bottom has not moved (we say the line AB is invariant).
(b) the distance moved by any card depends on its distance from the base card.
    The card at the top moves twice as far as the card in the middle.

Stretch

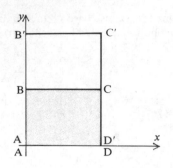

The rectangle ABCD has been *stretched* in the direction of the y-axis so that A′B′ is twice AB.

A stretch is fully described if we know
(a) the direction of the stretch,
(b) the ratio of corresponding lengths.

The matrix $\begin{pmatrix} k & 0 \\ 0 & 1 \end{pmatrix}$ represents a stretch parallel to the x-axis where the ratio of corresponding lengths is k.

Exercise 18

1. In the diagram, OABC has been mapped onto OA′B′C by a shear.

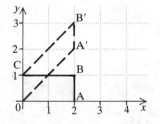

What is the invariant line of the shear?

2. Draw axes for values of x from -6 to $+9$ and for values of y from -2 to $+5$.

Find the coordinates of the image of each of the following shapes under the shear represented by the matrix $\begin{pmatrix} 1 & 2 \\ 0 & 1 \end{pmatrix}$.

Draw each object and image together on a diagram.
(a) $(0, 0)(0, 3)(2, 3)(2, 0)$
(b) $(0, 0)(-2, 0)(-2, -2)(0, -2)$
(c) $(0, 0)(2, 2)(3, 0)$
(d) $(1, 1)(1, 3)(3, 3)(3, 1)$.
What is the invariant line for this shear?

In questions **3** to **10**, plot the rectangle ABCD at A(0, 0), B(0, 2), C(3, 2), D(3, 0). Find and draw the image of ABCD under the transformation given and describe the transformation fully.

3. $\begin{pmatrix} 2 & 0 \\ 0 & 1 \end{pmatrix}$ 4. $\begin{pmatrix} 3 & 0 \\ 0 & 1 \end{pmatrix}$ 5. $\begin{pmatrix} 1 & 0 \\ 0 & 2 \end{pmatrix}$

6. $\begin{pmatrix} 1\frac{1}{2} & 0 \\ 0 & 1 \end{pmatrix}$ 7. $\begin{pmatrix} 1 & 1 \\ 0 & 1 \end{pmatrix}$ 8. $\begin{pmatrix} -2 & 0 \\ 0 & 1 \end{pmatrix}$

9. $\begin{pmatrix} 1 & 0 \\ 0 & 3 \end{pmatrix}$ 10. $\begin{pmatrix} \frac{1}{2} & 0 \\ 0 & 1 \end{pmatrix}$

11. (a) Find and draw the image of the square $(0, 0)$, $(1, 1)$, $(0, 2)$, $(-1, 1)$ under the transformation represented by the matrix $\begin{pmatrix} 4 & 3 \\ -3 & -2 \end{pmatrix}$.
    (b) Show that the transformation is a shear and find the equation of the invariant line.

12. (a) Find and draw the image of the square $(0, 0)$, $(1, 1)$, $(0, 2)$, $(-1, 1)$ under the shear represented by the matrix $\begin{pmatrix} 0\cdot5 & -0\cdot5 \\ 0\cdot5 & 1\cdot5 \end{pmatrix}$.
    (b) Find the equation of the invariant line.

13. Find and draw the image of the square $(0, 0)$, $(1, 0)$, $(1, 1)$, $(0, 1)$ under the transformation represented by the matrix $\begin{pmatrix} 3 & 0 \\ 0 & 2 \end{pmatrix}$.

This transformation is called a two-way stretch.

REVISION EXERCISE 8A

1. $\mathbf{A} = \begin{pmatrix} 3 & 2 \\ 1 & 4 \end{pmatrix}$, $\mathbf{B} = \begin{pmatrix} -1 & 3 \\ 0 & 2 \end{pmatrix}$,

Express as a single matrix
(a) $2\mathbf{A}$ (b) $\mathbf{A} - \mathbf{B}$ (c) $\frac{1}{2}\mathbf{A}$
(d) $\mathbf{AB}$ (e) $\mathbf{B}^2$

2. Evaluate

(a) $\begin{pmatrix} -3 & 0 \\ 1 & 2 \end{pmatrix} \begin{pmatrix} 3 & \frac{1}{3} \\ 1 & \frac{1}{2} \end{pmatrix}$ (b) $\begin{pmatrix} 3 \\ 1 \end{pmatrix}(4 \ 2)$

(c) $\begin{pmatrix} 3 & -2 \\ 4 & 1 \end{pmatrix} + 2\begin{pmatrix} 3 & 0 \\ -1 & -4 \end{pmatrix}$

3. $\mathbf{A} = \begin{pmatrix} 4 & 2 \\ 1 & 1 \end{pmatrix}$, $\mathbf{B} = (1 \ 5)$, $\mathbf{C} = \begin{pmatrix} -1 \\ 3 \end{pmatrix}$.

(a) Determine $\mathbf{BC}$ and $\mathbf{CB}$,

(b) If $\mathbf{AX} = \begin{pmatrix} 8 & 20 \\ 3 & 7 \end{pmatrix}$, where $\mathbf{X}$ is a (2×2)
matrix, determine $\mathbf{X}$.

4. Find the inverse of the matrix $\begin{pmatrix} 2 & -1 \\ 3 & 5 \end{pmatrix}$.

5. The determinant of the matrix $\begin{pmatrix} 3 & 2 \\ x & -1 \end{pmatrix}$ is -9.
Find the value of x and write down the inverse of the matrix.

6. $\mathbf{A} = \begin{pmatrix} 2 & 0 \\ 1 & 2 \end{pmatrix}$; h and k are numbers so that
$\mathbf{A}^2 = h\mathbf{A} + k\mathbf{I}$, where $\mathbf{I} = \begin{pmatrix} 1 & 0 \\ 0 & 1 \end{pmatrix}$.
Find the values of h and k.

7. $\mathbf{M} = \begin{pmatrix} a & 1 \\ 1 & -a \end{pmatrix}$.

(a) Find the values of a if $M^2 = 17\begin{pmatrix} 1 & 0 \\ 0 & 1 \end{pmatrix}$.

(b) Find the values of a if $|\mathbf{M}| = -10$.

8. Find the coordinates of the image of $(1, 4)$ under
(a) a clockwise rotation of $90°$ about $(0, 0)$,
(b) a reflection in the line $y = x$,
(c) a translation which maps $(5, 3)$ onto $(1, 1)$.

9. Draw x- and y-axes with values from -8 to $+8$.
Draw triangle $A(1, -1)$, $B(3, -1)$, $C(1, -4)$.
Find the image of ABC under the following enlargements
(a) scale factor 2, centre $(5, -1)$
(b) scale factor 2, centre $(0, 0)$
(c) scale factor $\frac{1}{2}$, centre $(1, 3)$
(d) scale factor $-\frac{1}{2}$, centre $(3, 1)$
(e) scale factor -2, centre $(0, 0)$.

10. Using the diagram below, describe the transformations for the following,
(a) $T_1 \rightarrow T_6$ (b) $T_4 \rightarrow T_5$ (c) $T_8 \rightarrow T_2$
(d) $T_4 \rightarrow T_1$ (e) $T_8 \rightarrow T_4$ (f) $T_6 \rightarrow T_8$

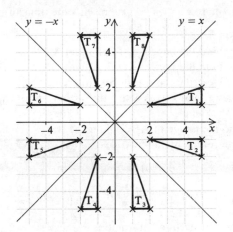

11. Describe the single transformation which maps
(a) $\triangle ABC$ onto $\triangle DEF$
(b) $\triangle ABC$ onto $\triangle PQR$
(c) $\triangle ABC$ onto $\triangle XYZ$.

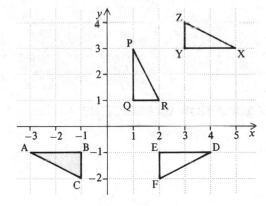

12. $\mathbf{M}$ is a reflection in the line $x + y = 0$.
$\mathbf{R}$ is an anticlockwise rotation of $90°$ about $(0, 0)$.
$\mathbf{T}$ is a translation which maps $(-1, -1)$ onto $(2, 0)$.
Find the image of the point $(3, 1)$ under
(a) $\mathbf{M}$ (b) $\mathbf{R}$ (c) $\mathbf{T}$
(d) $\mathbf{MR}$ (e) $\mathbf{RT}$ (f) $\mathbf{TMR}$

13. **A** is a rotation of 180° about (0, 0).
   B is a reflection in the line $x = 3$.
   C is a translation which maps (3, −1) onto
   (−2, −1).
   Find the image of the point (1, −2) under
   (a) **A**   (b) **A²**   (c) **BC**
   (d) **C⁻¹**   (e) **ABC**   (f) **C⁻¹B⁻¹A⁻¹**

14. The matrix $\begin{pmatrix} 0 & -1 \\ 1 & 0 \end{pmatrix}$ represents the transformation **X**.
   (a) Find the image of (5, 2) under **X**
   (b) Find the image of (−3, 4) under **X**
   (c) Describe the transformation **X**.

15. Draw x- and y-axes with values from −8 to +8.
   Draw triangle A(2, 2), B(6, 2), C(6, 4).
   Find the image of ABC under the transformations
   represented by the matrices,
   (a) $\begin{pmatrix} 0 & -1 \\ 1 & 0 \end{pmatrix}$ (b) $\begin{pmatrix} 1 & 0 \\ 0 & -1 \end{pmatrix}$ (c) $\begin{pmatrix} -1 & 0 \\ 0 & -1 \end{pmatrix}$
   (d) $\begin{pmatrix} 0 & 1 \\ -1 & 0 \end{pmatrix}$ (e) $\begin{pmatrix} 0 & -1 \\ -1 & 0 \end{pmatrix}$.
   Describe each transformation.

16. Using base vectors, describe the transformations
   represented by the following matrices.
   (a) $\begin{pmatrix} 0 & 1 \\ 1 & 0 \end{pmatrix}$ (b) $\begin{pmatrix} -1 & 0 \\ 0 & 1 \end{pmatrix}$ (c) $\begin{pmatrix} 3 & 0 \\ 0 & 3 \end{pmatrix}$

17. Using base vectors, write down the matrices which
   describe the following transformations.
   (a) Rotation 180°, centre (0, 0),
   (b) Reflection in the line $y = 0$,
   (c) Enlargement scale factor 4, centre (0, 0),
   (d) Reflection in the line $x = −y$,
   (e) Clockwise rotation 90°, centre (0, 0).

18. Transformation **N**, which is given by
   $$\begin{pmatrix} x' \\ y' \end{pmatrix} = \begin{pmatrix} 2 & 0 \\ 0 & 2 \end{pmatrix} \begin{pmatrix} x \\ y \end{pmatrix} + \begin{pmatrix} 5 \\ -2 \end{pmatrix},$$
   is composed of two single transformations.
   (a) Describe each of the transformations,
   (b) Find the image of the point (3, −1) under **N**,
   (c) Find the image of the point (−1, ½) under **N**,
   (d) Find the point which is mapped by **N** onto the
       point (7, 4).

19. **A** is the reflection in the line $y = x$.
   B is the reflection in the y-axis.
   Find the matrix which represents
   (a) **A**   (b) **B**   (c) **AB**   (d) **BA**.
   Describe the single transformations **AB** and **BA**.

EXAMINATION EXERCISE 8B

1.

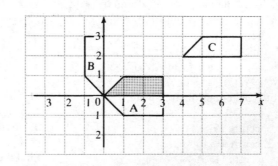

(a) Describe completely the single transformation that would move the shaded
    figure to
    (i) position A,   (ii) position B,   (iii) position C,
(b) What single transformation would move figure A to figure B directly?   [N]

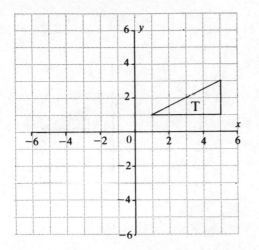

2. (a) A is the image of triangle T after it has been rotated through a half-turn about the origin (0, 0). Draw and label A.
   (b) B is the image of triangle T after it has been rotated through a half-turn about the point (3, 0). Draw and label B.
   (c) C is the image of triangle T after it has been reflected in the y-axis. Draw and label C.
   (d) Describe fully the transformation which maps
       (i) A onto B   (ii) A onto C

   [L]

3. It is given that

   $$M = \begin{pmatrix} 3 & 1 \\ 1 & 2 \end{pmatrix} \text{ and } N = \begin{pmatrix} 1 & 4 \\ -3 & 2 \end{pmatrix}.$$

   (a) Express $M + 3N$ as a single matrix.
   (b) Express M^2 as a single matrix.   [M]

4. Answer the whole of this question on graph paper.
   The vertices of a triangle ABC have coordinates A(4, 3), B(8, 6) and C(−3, 9).
   Triangle ABC is mapped onto triangle $A_1B_1C_1$ by the transformation represented by the matrix **M** where

   $$M = \begin{pmatrix} 0.8 & 0.6 \\ 0.6 & 0.8 \end{pmatrix}.$$

   (a) Find the coordinates of A_1, B_1 and C_1.
   (b) Taking 1 cm to represent 1 unit on each axis, draw and label the triangles ABC and $A_1B_1C_1$.
   (c) Describe fully the transformation represented by M.   [M]

5. On squared paper draw a triangle with vertices A(2, 2), B(2, 4) and C(5, 4).
   (a) Write as a column vector
       (i) the translation from the point B to the point C,
       (ii) the translation from the point B to the point A.
   (b) On the squared paper draw the images of the triangle ABC after the transformations determined by
       (i) $P = \begin{pmatrix} -1 & 0 \\ 0 & 1 \end{pmatrix}$ (label the image p)

       (ii) $Q = \begin{pmatrix} 0 & -1 \\ -1 & 0 \end{pmatrix}$ (label the image q)
   (c) Describe geometrically the single transformation determined by
       (i) **P**,
       (ii) **Q**,
       (iii) the triangle p mapped onto triangle q.   [S]

6. A transformation **S** is described by the matrix $\begin{pmatrix} 0 & -1 \\ -1 & 0 \end{pmatrix}$.

T is the translation with vector $\begin{pmatrix} 3 \\ 3 \end{pmatrix}$.

(a) In the diagram, show the image of the triangle ABC under the transformation **S**.

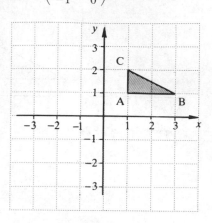

(b) Describe the transformation **S** in geometrical terms.
(c) Find the coordinates of P′ and Q′, the images of the points P, (3, 0) and Q, (2, 1) respectively, under the single transformation equivalent to **S** followed by **T**.
(d) Describe in geometrical terms the single transformation equivalent to **S** followed by **T**.   [M]

7. The matrix $\mathbf{A} = \begin{pmatrix} 2 & -3 \\ 4 & -4 \end{pmatrix}$ represents the transformation **T**: $\begin{pmatrix} x \\ y \end{pmatrix} \rightarrow \begin{pmatrix} 2 & -3 \\ 4 & -4 \end{pmatrix} \begin{pmatrix} x \\ y \end{pmatrix}$.

(a) Find the inverse matrix of **A**.
(b) Find the coordinates of the point which is mapped to (9, 16) under the transformation T.
(c) Find the matrix $\mathbf{A}^2$.
(d) Show that $\mathbf{A}^3$ is of the form $\begin{pmatrix} d & 0 \\ 0 & d \end{pmatrix}$ and give a geometrical description of the transformation whose matrix is $\mathbf{A}^3$.   [N]

8. The points O(0, 0), A(2, 0) and B(2, 1) are transformed by the matrix **M**, where

$$\mathbf{M} = \begin{pmatrix} 1 & -1 \\ 1 & 1 \end{pmatrix}, \text{ into O, A′ and B′ respectively.}$$

(a) Find the coordinates of A′ and B′.
(b) Using graph paper and taking a scale of 2 cm to 1 unit on each axis, draw and label triangle OAB and its image triangle OA′B′.
(c) The transformation whose matrix is **M** can be obtained by a combination of two separate transformations. By taking measurements from your graph, give a full description of each of these transformations.
(d) Find $\mathbf{M}^{-1}$ and hence, or otherwise, find the coordinates of the point whose image is (6, 2) under the transformation whose matrix is **M**.   [L]

9. The transformation **T** consists of a reflection in the *x*-axis followed by an enlargement with centre (0, 0) and scale factor 2. Find

(a) the image of $\begin{pmatrix} 1 \\ 0 \end{pmatrix}$ and $\begin{pmatrix} 0 \\ 1 \end{pmatrix}$ under **T**;
(b) the matrix associated with **T**;
(c) the image of $\begin{pmatrix} 2 \\ 3 \end{pmatrix}$ under **T**.   [N]

9 Sets, vectors and functions

Bertrand Russell (1872–1970) tried to reduce all mathematics to formal logic. He showed that the idea of a set of all sets which are not members of themselves leads to contradictions. He wrote to Gottlieb Frege just as he was putting the finishing touches to a book that represented his life's work, pointing out that Frege's work was invalidated. Russell's elder brother, the second Earl Russell, showed great foresight in 1903 by queueing overnight outside the vehicle licensing office in London to have his car registered as A1.

9.1 SETS

1. ∩ 'intersection'
   A ∩ B is shaded.

2. ∪ 'union'
   A ∪ B is shaded.

3. ⊂ 'is a subset of'
   A ⊂ B
   [B ⊄ A means 'B is *not* a subset of A']

4. ∈ 'is a member of'
   'belongs to'
   b ∈ X
   [e ∉ X means 'e is not a member of set X']

5. $\mathscr{E}$ 'universal set'
   $\mathscr{E} = \{a, b, c, d, e\}$

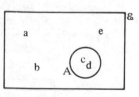

6. A′ 'complement of'
   'not in A'
   A′ is shaded
   $(A \cup A' = \mathscr{E})$

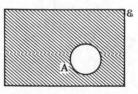

7. $n(A)$ 'the number of elements in set A'
   $n(A) = 3$

8. $A = \{x : x \text{ is an integer}, 2 \leqslant x \leqslant 9\}$

The $\boxed{\text{set of}}$ elements x $\boxed{\text{such that}}$ x is

an integer and $2 \leqslant x \leqslant 9$.
The set A is $\{2, 3, 4, 5, 6, 7, 8, 9\}$.

9. $\varnothing$ or $\{\}$ 'empty set'
(Note $\varnothing \subset A$ for any set A)

Exercise 1

1. In the Venn diagram,
  $\mathscr{E} = \{\text{people in an hotel}\}$
  $B = \{\text{people who like bacon}\}$
  $E = \{\text{people who like eggs}\}$

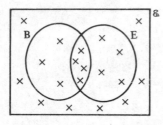

  (a) How many people like bacon?
  (b) How many people like eggs but not bacon?
  (c) How many people like bacon and eggs?
  (d) How many people are in the hotel?
  (e) How many people like neither bacon nor eggs?

2. In the Venn diagram,
  $\mathscr{E} = \{\text{boys in the fourth form}\}$
  $R = \{\text{ members of the rugby team}\}$
  $C = \{\text{ members of the cricket team}\}$

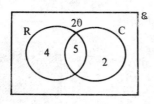

  (a) How many are in the rugby team?
  (b) How many are in both teams?
  (c) How many are in the rugby team but not in the cricket team?
  (d) How many are in neither team?
  (e) How many are there in the fourth form?

3. In the Venn diagram,
  $\mathscr{E} = \{\text{cars in a street}\}$
  $B = \{\text{blue cars}\}$
  $L = \{\text{cars with left hand drive}\}$
  $F = \{\text{cars with four doors}\}$

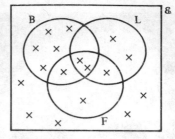

  (a) How many cars are blue?
  (b) How many blue cars have four doors?
  (c) How many cars with left hand drive have four doors?
  (d) How many blue cars have left hand drive?
  (e) How many cars are in the street?
  (f) How many blue cars with left hand drive do not have four doors?

4. In the Venn diagram,
  $\mathscr{E} = \{\text{houses in the street}\}$
  $C = \{\text{houses with central heating}\}$
  $T = \{\text{houses with a colour T.V.}\}$
  $G = \{\text{houses with a garden}\}$

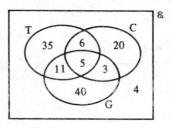

  (a) How many houses have gardens?
  (b) How many houses have a colour T.V. and central heating?
  (c) How many houses have a colour T.V. and central heating and a garden?
  (d) How many houses have a garden but not a T.V. or central heating?
  (e) How many houses have a T.V. and a garden but not central heating?
  (f) How many houses are there in the street?

5. In the Venn diagram

$\mathscr{E}$ = {children in a mixed school}
G = {girls in the school}
S = {children who can swim}
L = {children who are left-handed}

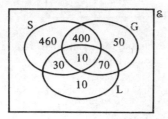

(a) How many left-handed children are there?
(b) How many girls cannot swim?
(c) How many boys can swim?
(d) How many girls are left-handed?
(e) How many boys are left-handed?
(f) How many left-handed girls can swim?
(g) How many boys are there in the school?

Example 1

$\mathscr{E}$ = {1, 2, 3   12}, A = {2, 3, 4, 5, 6} and
B = {2, 4, 6, 8, 10}.

(a) A∪B = {2, 3, 4, 5, 6, 8, 10}.
(b) A∩B = {2, 4, 6}.
(c) A′ = {1, 7, 8, 9, 10, 11, 12}.
(d) n(A∪B) = 7.
(e) B′∩A = {3, 5}

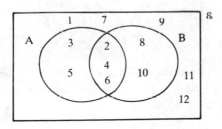

Exercise 2

In this exercise, be careful to use set notation only when the answer *is* a set.

1. If M = {1, 2, 3, 4, 5, 6, 7, 8}, N = {5, 7, 9, 11, 13}.
Find
(a) M∩N   (b) M∪N
(c) n(N)   (d) n(M∪N).
State whether true or false:
(e) 5∈M   (f) 7∈(M∪N)
(g) N⊂M   (h) {5, 6, 7}⊂M.

2. If A = {2, 3, 5, 7}, B = {1, 2, 3 . . . , 9}.
Find
(a) A∩B   (b) A∪B
(c) n(A∩B)   (d) {1, 4}∩A
State whether true or false:
(e) A∈B   (f) A⊂B
(g) 9⊂B   (h) 3∈(A∩B)

3. If X = {1, 2, 3, . . . 10}, Y = {2, 4, 6, . . . 20} and
Z = {x : x is an integer, $15 \leqslant x \leqslant 25$}.
Find
(a) X∩Y   (b) Y∩Z
(c) X∩Z   (d) n(X∪Y)
(e) n(Z)   (f) n(X∪Z)
State whether true or false:
(g) 5∈Y   (h) 20∈X
(i) n(X∩Y) = 5   (j) {15, 20, 25}⊂Z.

4. If D = {1, 3, 5}, E = {3, 4, 5}, F = {1, 5, 10}.
Find
(a) D∪E   (b) D∩F
(c) n(E∩F)   (d) (D∪E)∩F
(e) (D∩E)∪F   (f) n(D∪F)
State whether true or false:
(g) D⊂(E∪F)   (h) 3∈(E∩F)
(i) 4∉(D∩E).

5. Find
(a) n(E)   (b) n(F)   (c) E∩F
(d) E∪F   (e) n(E∪F)   (f) n(E∩F).

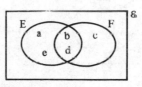

6. Find
(a) n(M∩N)   (b) n(N)   (c) M∪N
(d) M′∩N   (e) N′∩M   (f) (M∩N)′
(g) M∪N′   (h) N∪M′   (i) M′∪N′

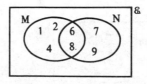

Example 2

On a Venn diagram, shade the regions
(a) $A \cap C$ (b) $(B \cap C) \cap A'$
where A, B, C are intersecting sets.

(a) $A \cap C$

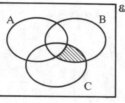

(b) $(B \cap C) \cap A'$
     [find $(B \cap C)$ first]

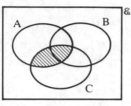

Exercise 3

1. Draw six diagrams similar to Figure 1 and shade
the following sets:
(a) $A \cap B$ (b) $A \cup B$ (c) A'
(d) $A' \cap B$ (e) $B' \cap A$ (f) $(B \cup A)'$.

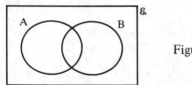

Figure 1

2. Draw four diagrams similar to Figure 2 and shade
the following sets:
(a) $A \cap B$ (b) $A \cup B$ (c) $B' \cap A$
(d) $(B \cup A)'$

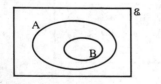

Figure 2

3. Draw four diagrams similar to Figure 3 and shade
the following sets:
(a) $A \cup B$ (b) $A \cap B$ (c) $A \cap B'$
(d) $(B \cup A)'$

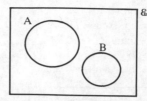

Figure 3

4. Draw eleven diagrams similar to Figure 4 and
shade the following sets:
(a) $A \cap B$ (b) $A \cup C$ (c) $A \cap (B \cap C)$
(d) $(A \cup B) \cap C$ (e) $B \cap (A \cup C)$
(f) $A \cap B'$ (g) $A \cap (B \cup C)'$
(h) $(B \cup C) \cap A$ (i) $C' \cap (A \cap B)$
(j) $(A \cup C) \cup B'$ (k) $(A \cup C) \cap (B \cap C)$

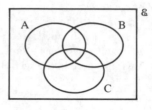

Figure 4

5. Draw nine diagrams similar to Figure 5 and shade
the following sets:
(a) $(A \cup B) \cap C$ (b) $(A \cap B) \cup C$
(c) $(A \cup B) \cup C$ (d) $A \cap (B \cup C)$
(e) $A' \cap C$ (f) $C' \cap (A \cup B)$
(g) $(A \cap B) \cap C$ (h) $(A \cap C) \cup (B \cap C)$
(i) $(A \cup B \cup C)'$

Figure 5

6. Copy each diagram and shade the region indicated.

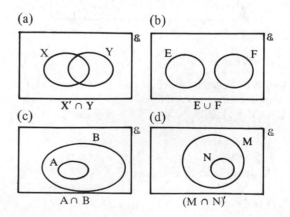

7. Describe the region shaded.

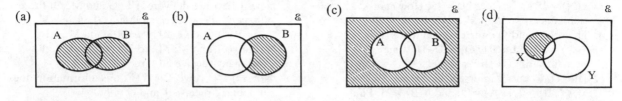

(a)   (b)   (c)   (d)

9.2   LOGICAL PROBLEMS

Example 1

In a form of 30 girls, 18 play netball and 14 play hockey, whilst 5 play neither.
Find the number who play both netball and hockey.

Let $\mathscr{E}$ = {girls in the form}
      N = {girls who play netball}
      H = {girls who play hockey}

and x = the number of girls who play both netball and hockey

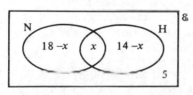

The number of girls in each portion of the universal set is shown in the Venn diagram.

Since $n(\mathscr{E}) = 30$
      $18 - x + x + 14 - x + 5 = 30$
                       $37 - x = 30$
                       $x = 7$

$\therefore$ seven girls play both netball and hockey.

Example 2

If   A = {sheep}
     B = {sheep dogs}
     C = {'intelligent' animals}
     D = {animals which make good pets}

(a) Express the following sentences in set language
    (i) No sheep are 'intelligent' animals.
    (ii) All sheep dogs make good pets.
    (iii) Some sheep make good pets.

(b) Interpret the following statements
    (i) $B \subset C$.
    (ii) $B \cup C = D$.

(a) (i) $A \cap C = \varnothing$.
     (ii) $B \subset D$.
     (iii) $A \cap D \neq \varnothing$.

(b) (i)   All sheep dogs are intelligent animals.
     (ii) Animals which make good pets are either sheep dogs or 'intelligent' animals (or both).

Exercise 4

1. In the Venn diagram $n(A) = 10$, $n(B) = 13$, $n(A \cap B) = x$ and $n(A \cup B) = 18$.

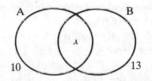

(a) Write in terms of x the number of elements in A but not in B,
(b) Write in terms of x the number of elements in B but not in A,
(c) Add together the number of elements in the three parts of the diagram to obtain the equation $10 - x + x + 13 - x = 18$,
(d) Hence find the number of elements in both A and B.

2. In the Venn diagram $n(A) = 21$, $n(B) = 17$, $n(A \cap B) = x$ and $n(A \cup B) = 29$.

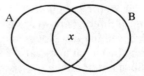

(a) Write down in terms of x the number of elements in each part of the diagram.
(b) Form an equation and hence find x.

3. The sets M and N intersect such that $n(M) = 31$, $n(N) = 18$ and $n(M \cup N) = 35$. How many elements are in both M and N?

4. The sets P and Q intersect such that $n(P) = 11$, $n(Q) = 29$ and $n(P \cup Q) = 37$. How many elements are in both P and Q?

5. The sets A and B intersect such that $n(A \cap B) = 7$, $n(A) = 20$ and $n(B) = 23$. Find $n(A \cup B)$.

6. Twenty boys in a form all play either football or basketball (or both). If thirteen play football and ten play basketball, how many play both sports?

7. Of the 53 staff at a school, 36 drink tea, 18 drink coffee and 10 drink neither tea nor coffee? How many drink both tea and coffee?

8. Of the 32 pupils in a class, 18 play golf, 16 play the piano and 7 play both. How many play neither?

9. Of the pupils in a class, 15 can spell 'parallel', 14 can spell 'Pythagoras', 5 can spell both words and 4 can spell neither. How many pupils are there in the class?

10. In a school, students must take at least one of these subjects: Maths, Physics or Chemistry. In a group of 50 students, 7 take all three subjects, 9 take Physics and Chemistry only, 8 take Maths and Physics only and 5 take Maths and Chemistry only. Of these 50 students, x take Maths only, x take Physics only and $x + 3$ take Chemistry only. Draw a Venn diagram, find x, and hence find the number taking Maths.

11. All of 60 different vitamin pills contain at least one of the vitamins A, B and C. Twelve have A only, 7 have B only, and 11 have C only. If 6 have all three vitamins and there are x having A and B only, B and C only and A and C only, how many pills contain vitamin A?

12. The 'O' level results of the 30 members of a Rugby squad were as follows:
All 30 players passed at least two subjects, 18 players passed at least three subjects, and 3 players passed four subjects or more. Calculate
(a) how many passed exactly two subjects,
(b) what fraction of the squad passed exactly three subjects.

13. In a group of 59 people, some are wearing hats, gloves or scarves (or a combination of these), 4 are wearing all three, 7 are wearing just a hat and gloves, 3 are wearing just gloves and a scarf and 9 are wearing just a hat and scarf. The number wearing only a hat or only gloves is x, and the number wearing only a scarf or none of the three items is $(x - 2)$. Find x and hence the number of people wearing a hat.

14. In a street of 150 houses, three different newspapers are delivered: T, G and M. Of these, 40 receive T, 35 receive G, and 60 receive M; 7 receive T and G, 10 receive G and M and 4 receive T and M; 34 receive no paper at all. How many receive all three?
Note: If '7 receive T and G', this information does not mean 7 receive T and G *only*.

15. If S = {Scottish men}, G = {good footballers}, express the following statements in words
(a) $G \subset S$
(b) $G \cap S = \varnothing$
(c) $G \cap S \neq \varnothing$.
(Ignore the truth or otherwise of the statements.)

16. Given that $\mathscr{E}$ = {pupils in a school}, B = {boys}, H = {hockey players}, F = {football players}, express the following in words:
(a) $F \subset B$ (b) $H \subset B'$
(c) $F \cap H \neq \varnothing$ (d) $B \cap H = \varnothing$.
Express in set notation
(e) No boys play football,
(f) All pupils play either football or hockey.

17. If $\mathscr{E}$ = {living creatures}, S = {spiders}, F = {animals that fly}, T = {animals which taste nice}, express in set notation:
(a) No spiders taste nice,
(b) All animals that fly taste nice,
(c) Some spiders can fly.
Express in words:
(d) $S \cup F \cup T = \mathscr{E}$ (e) $T \subset S$.

18. $\mathscr{E}$ = {tigers}, T = {tigers who believe in fairies}, X = {tigers who believe in Eskimos}, H = {tigers in hospital}. Express in words:
(a) $T \subset X$,   (b) $T \cup X = H$,
(c) $H \cap X = \varnothing$.
Express in set notation:
(d) All tigers in hospital believe in fairies,
(e) Some tigers believe in both fairies and Eskimos.

19. $\mathscr{E}$ = {school teachers}, P = {teachers called Peter}, B = {good bridge players}, W = {women teachers}. Express in words:
(a) $P \cap B = \varnothing$.
(b) $P \cup B \cup W = \mathscr{E}$,
(c) $P \cap W \neq \varnothing$.
Express in set notation:
(d) Women teachers cannot play bridge well,
(e) All good bridge players are women called Peter.

9.3   VECTORS

A vector quantity has both magnitude and direction. Problems involving forces, velocities and displacements are often made easier when vectors are used.

Addition of vectors

Vectors **a** and **b** represented by the line segments below can be added using the parallelogram rule or the 'nose-to-tail' method.

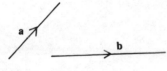

The 'tail' of vector **b** is joined to the 'nose' of vector **a**.

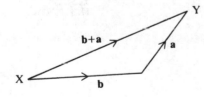

Alternatively the tail of **a** can be joined to the 'nose' of vector **b**.

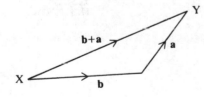

In both cases the vector $\overrightarrow{XY}$ has the same length and direction and therefore

$$\mathbf{a} + \mathbf{b} = \mathbf{b} + \mathbf{a}.$$

Multiplication by a scalar

A scalar quantity has magnitude but no direction (e.g. mass, volume, temperature). Ordinary numbers are scalars.

When vector **x** is multiplied by 2, the result is 2**x**.

When **x** is multiplied by -3 the result is $-3\mathbf{x}$.

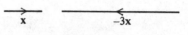

Note
(1) The negative sign reverses the direction of the vector.
(2) The results $\mathbf{a} - \mathbf{b}$ is $\mathbf{a} + -\mathbf{b}$.
    i.e. Subtracting **b** is equivalent to adding the negative of **b**.

Example 1

The diagram shows vectors **a** and **b**.

Find $\overrightarrow{OP}$ and $\overrightarrow{OQ}$ such that

$$\overrightarrow{OP} = 3\mathbf{a} + \mathbf{b}$$
$$\overrightarrow{OQ} = -2\mathbf{a} - 3\mathbf{b}$$

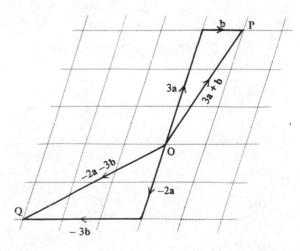

Exercise 5

In questions **1** to **26**, use the diagram below to describe the vectors given in terms of **c** and **d** where $\mathbf{c} = \overrightarrow{QN}$ and $\mathbf{d} = \overrightarrow{QR}$.

e.g. $\overrightarrow{QS} = 2\mathbf{d}$, $\overrightarrow{TD} = \mathbf{c} + \mathbf{d}$.

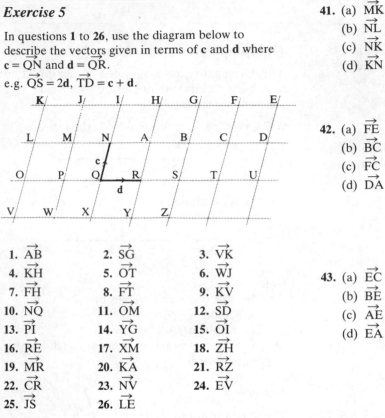

1. $\overrightarrow{AB}$	2. $\overrightarrow{SG}$	3. $\overrightarrow{VK}$
4. $\overrightarrow{KH}$	5. $\overrightarrow{OT}$	6. $\overrightarrow{WJ}$
7. $\overrightarrow{FH}$	8. $\overrightarrow{FT}$	9. $\overrightarrow{KV}$
10. $\overrightarrow{NQ}$	11. $\overrightarrow{OM}$	12. $\overrightarrow{SD}$
13. $\overrightarrow{PI}$	14. $\overrightarrow{YG}$	15. $\overrightarrow{OI}$
16. $\overrightarrow{RE}$	17. $\overrightarrow{XM}$	18. $\overrightarrow{ZH}$
19. $\overrightarrow{MR}$	20. $\overrightarrow{KA}$	21. $\overrightarrow{RZ}$
22. $\overrightarrow{CR}$	23. $\overrightarrow{NV}$	24. $\overrightarrow{EV}$
25. $\overrightarrow{JS}$	26. $\overrightarrow{LE}$	

In questions **27** to **38**, use the same diagram above to find vectors for the following in terms of the capital letters, starting from Q each time.

e.g. $3\mathbf{d} = \overrightarrow{QT}$, $\mathbf{c} + \mathbf{d} = \overrightarrow{QA}$.

27. $2\mathbf{c}$	28. $4\mathbf{d}$	29. $2\mathbf{c} + \mathbf{d}$
30. $2\mathbf{d} + \mathbf{c}$	31. $3\mathbf{d} + 2\mathbf{c}$	32. $2\mathbf{c} - \mathbf{d}$
33. $-\mathbf{c} + 2\mathbf{d}$	34. $\mathbf{c} - 2\mathbf{d}$	35. $2\mathbf{c} + 4\mathbf{d}$
36. $-\mathbf{c}$	37. $-\mathbf{c} - \mathbf{d}$	38. $2\mathbf{c} - 2\mathbf{d}$

In questions **39** to **43**, write each vector in terms of **a** and/or **b**.

39. (a) $\overrightarrow{BA}$
    (b) $\overrightarrow{AC}$
    (c) $\overrightarrow{DB}$
    (d) $\overrightarrow{AD}$

40. (a) $\overrightarrow{ZX}$
    (b) $\overrightarrow{YW}$
    (c) $\overrightarrow{XY}$
    (d) $\overrightarrow{XZ}$

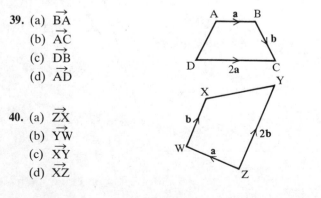

41. (a) $\overrightarrow{MK}$
    (b) $\overrightarrow{NL}$
    (c) $\overrightarrow{NK}$
    (d) $\overrightarrow{KN}$

42. (a) $\overrightarrow{FE}$
    (b) $\overrightarrow{BC}$
    (c) $\overrightarrow{FC}$
    (d) $\overrightarrow{DA}$

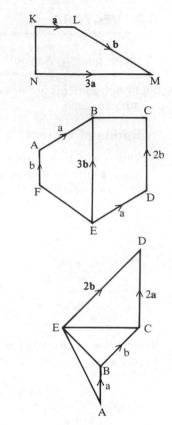

43. (a) $\overrightarrow{EC}$
    (b) $\overrightarrow{BE}$
    (c) $\overrightarrow{AE}$
    (d) $\overrightarrow{EA}$

In questions **44** to **46**, write each vector in terms of **a**, **b** and **c**.

44. (a) $\overrightarrow{FC}$
    (b) $\overrightarrow{GB}$
    (c) $\overrightarrow{AB}$
    (d) $\overrightarrow{HE}$
    (e) $\overrightarrow{CA}$

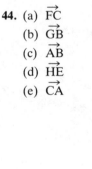

45. (a) $\overrightarrow{OF}$
    (b) $\overrightarrow{OC}$
    (c) $\overrightarrow{BC}$
    (d) $\overrightarrow{EB}$
    (e) $\overrightarrow{FB}$

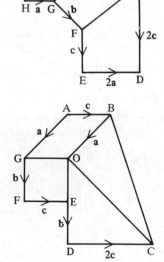

46. (a) $\overrightarrow{GD}$
(b) $\overrightarrow{GE}$
(c) $\overrightarrow{AD}$
(d) $\overrightarrow{AF}$
(e) $\overrightarrow{FE}$

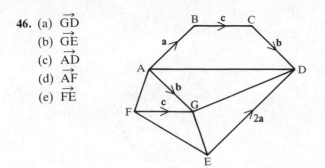

Example 2

Using Figure 1, express each of the following vectors in terms of **a** and/or **b**.

(a) $\overrightarrow{AP}$ (b) $\overrightarrow{AB}$ (c) $\overrightarrow{OQ}$ (d) $\overrightarrow{PO}$
(e) $\overrightarrow{PQ}$ (f) $\overrightarrow{PN}$ (g) $\overrightarrow{ON}$ (h) $\overrightarrow{AN}$
(i) $\overrightarrow{BP}$ (j) $\overrightarrow{QA}$

OA = AP
BQ = 3OB
N is the mid-point of PQ
$\overrightarrow{OA} = \mathbf{a}$, $\overrightarrow{OB} = \mathbf{b}$

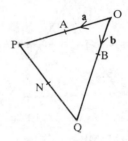

(a) $\overrightarrow{AP} = \mathbf{a}$
(b) $\overrightarrow{AB} = -\mathbf{a} + \mathbf{b}$
(c) $\overrightarrow{OQ} = 4\mathbf{b}$
(d) $\overrightarrow{PO} = -2\mathbf{a}$
(e) $\overrightarrow{PQ} = \overrightarrow{PO} + \overrightarrow{OQ}$
     $= -2\mathbf{a} + 4\mathbf{b}$
(f) $\overrightarrow{PN} = \frac{1}{2}\overrightarrow{PQ}$
     $= -\mathbf{a} + 2\mathbf{b}$
(g) $\overrightarrow{ON} = \overrightarrow{OP} + \overrightarrow{PN}$
     $= 2\mathbf{a} + (-\mathbf{a} + 2\mathbf{b})$
     $= \mathbf{a} + 2\mathbf{b}$
(h) $\overrightarrow{AN} = \overrightarrow{AP} + \overrightarrow{PN}$
     $= \mathbf{a} + (-\mathbf{a} + 2\mathbf{b})$
     $= 2\mathbf{b}$
(i) $\overrightarrow{BP} = \overrightarrow{BO} + \overrightarrow{OP}$
     $= -\mathbf{b} + 2\mathbf{a}$
(j) $\overrightarrow{QA} = \overrightarrow{QO} + \overrightarrow{OA}$
     $= -4\mathbf{b} + \mathbf{a}$

Exercise 6

In questions **1** to **6**, $\overrightarrow{OA} = \mathbf{a}$ and $\overrightarrow{OB} = \mathbf{b}$. Copy each diagram and use the information given to express the following vectors in terms of **a** and/or **b**.

(a) $\overrightarrow{AP}$ (b) $\overrightarrow{AB}$ (c) $\overrightarrow{OQ}$ (d) $\overrightarrow{PO}$
(e) $\overrightarrow{PQ}$ (f) $\overrightarrow{PN}$ (g) $\overrightarrow{ON}$ (h) $\overrightarrow{AN}$
(i) $\overrightarrow{BP}$ (j) $\overrightarrow{QA}$

1. A, B and N are mid-points of OP, OB and PQ respectively.

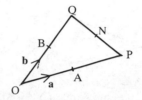

2. A and N are mid-points of OP and PQ; BQ = 2OB.

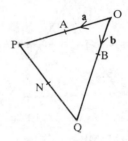

3. AP = 2OA, BQ = OB, PN = NQ.

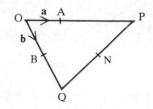

4. OA = 2AP, BQ = 3OB, PN = 2NQ.

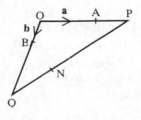

5. $AP = 5OA$, $OB = 2BQ$, $NP = 2QN$.

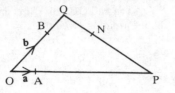

6. $OA = \frac{1}{5}OP$, $OQ = 3OB$, N is $\frac{1}{4}$ of the way along PQ.

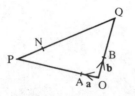

7. In $\triangle XYZ$, the mid-point of YZ is M. If $\vec{XY} = \mathbf{s}$ and $\vec{ZX} = \mathbf{t}$, find $\vec{XM}$ in terms of **s** and **t**.

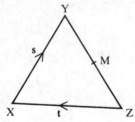

8. In $\triangle AOB$, $AM : MB = 2 : 1$. If $\vec{OA} = \mathbf{a}$ and $\vec{OB} = \mathbf{b}$, find $\vec{OM}$ in terms of **a** and **b**.

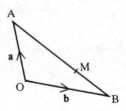

9. O is any point in the plane of the square ABCD. The vectors $\vec{OA}$, $\vec{OB}$ and $\vec{OC}$ are **a**, **b** and **c** respectively. Find the vector $\vec{OD}$ in terms of **a**, **b** and **c**.

10. ABCDEF is a regular hexagon with $\vec{AB}$ representing the vector **m** and $\vec{AF}$ representing the vector **n**. Find the vector representing $\vec{AD}$.

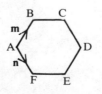

11. ABCDEF is a regular hexagon with centre O. $\vec{FA} = \mathbf{a}$ and $\vec{FB} = \mathbf{b}$.

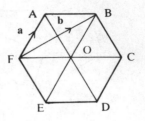

Express the following vectors in terms of **a** and/or **b**.

(a) $\vec{AB}$ (b) $\vec{FO}$ (c) $\vec{FC}$

(d) $\vec{BC}$ (e) $\vec{AO}$ (f) $\vec{FD}$

12. In the diagram, M is the mid-point of CD, $BP : PM = 2 : 1$, $\vec{AB} = \mathbf{x}$, $\vec{AC} = \mathbf{y}$ and $\vec{AD} = \mathbf{z}$.

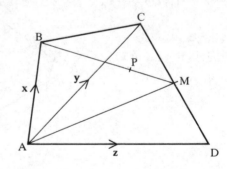

Express the following vectors in terms of **x**, **y** and **z**.

(a) $\vec{DC}$ (b) $\vec{DM}$ (c) $\vec{AM}$

(d) $\vec{BM}$ (e) $\vec{BP}$ (f) $\vec{AP}$.

9.4 COLUMN VECTORS

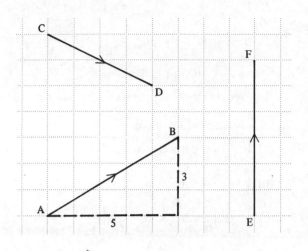

The vector $\vec{AB}$ may be written as a *column vector*. $AB = \begin{pmatrix} 5 \\ 3 \end{pmatrix}$.

The top number is the horizontal component of $\vec{AB}$ (i.e. 5) and the bottom number is the vertical component (i.e. 3).

Similarly $\vec{CD} = \begin{pmatrix} 4 \\ -2 \end{pmatrix}$

$\vec{EF} = \begin{pmatrix} 0 \\ 6 \end{pmatrix}$

Addition of vectors

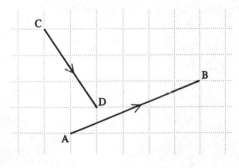

Figure 1

Suppose we wish to add vectors $\vec{AB}$ and $\vec{CD}$ in Figure 1.

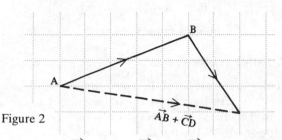

Figure 2 $\vec{AB} + \vec{CD}$

First move $\vec{CD}$ so that $\vec{AB}$ and $\vec{CD}$ join 'nose to tail' as in Figure 2. Remember that changing the *position* of a vector does not change the vector. A vector is described by its length and direction.

The broken line shows the result of adding $\vec{AB}$ and $\vec{CD}$.

In column vectors,

$$\vec{AB} + \vec{CD} = \begin{pmatrix} 5 \\ 2 \end{pmatrix} + \begin{pmatrix} 2 \\ -3 \end{pmatrix}$$

We see that the column vector for the broken line is $\begin{pmatrix} 7 \\ -1 \end{pmatrix}$. So we perform addition with vectors by adding together the corresponding components of the vectors.

Subtraction of vectors

Figure 3 shows $\vec{AB} - \vec{CD}$.

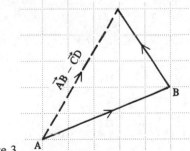

Figure 3   A

To subtract vector $\vec{CD}$ from $\vec{AB}$ we *add* the *negative* of $\vec{CD}$ to $\vec{AB}$.

So $\vec{AB} - \vec{CD} = \vec{AB} + (-\vec{CD})$

In column vectors,

$$\vec{AB} + (-\vec{CD}) = \begin{pmatrix} 5 \\ 2 \end{pmatrix} + \begin{pmatrix} -2 \\ 3 \end{pmatrix} = \begin{pmatrix} 3 \\ 5 \end{pmatrix}$$

Multiplication by a scalar

If $\mathbf{a} = \begin{pmatrix} 3 \\ -4 \end{pmatrix}$ then $2\mathbf{a} = 2\begin{pmatrix} 3 \\ -4 \end{pmatrix} = \begin{pmatrix} 6 \\ -8 \end{pmatrix}$.

Each component is multiplied by the number 2.

Parallel vectors

Vectors are parallel if they have the same direction. Both components of one vector must be in the same ratio to the corresponding components of the parallel vector.

e.g. $\begin{pmatrix} 3 \\ -5 \end{pmatrix}$ is parallel to $\begin{pmatrix} 6 \\ -10 \end{pmatrix}$.

because $\begin{pmatrix} 6 \\ -10 \end{pmatrix}$ may be written $2\begin{pmatrix} 3 \\ -5 \end{pmatrix}$.

In general the vector $k\begin{pmatrix} a \\ b \end{pmatrix}$ is parallel to $\begin{pmatrix} a \\ b \end{pmatrix}$.

Exercise 7

Questions **1** to **36** refer to the following vectors.

$\mathbf{a} = \begin{pmatrix} 3 \\ 4 \end{pmatrix}$ $\mathbf{b} = \begin{pmatrix} 1 \\ 4 \end{pmatrix}$ $\mathbf{c} = \begin{pmatrix} 4 \\ -3 \end{pmatrix}$ $\mathbf{d} = \begin{pmatrix} -1 \\ 1 \end{pmatrix}$

$\mathbf{e} = \begin{pmatrix} 5 \\ 12 \end{pmatrix}$ $\mathbf{f} = \begin{pmatrix} 3 \\ -2 \end{pmatrix}$ $\mathbf{g} = \begin{pmatrix} -4 \\ -2 \end{pmatrix}$ $\mathbf{h} = \begin{pmatrix} -12 \\ 5 \end{pmatrix}$

Draw and label the following vectors on graph paper (take 1 cm to 1 unit).

1. **c**   2. **f**   3. **2b**   4. **−a**
5. **−g**   6. **3a**   7. $\frac{1}{2}\mathbf{e}$ 8. **5d**
9. $-\frac{1}{2}\mathbf{h}$ 10. $\frac{3}{2}\mathbf{g}$ 11. $\frac{1}{5}\mathbf{h}$ 12. **−3b**

Find the following vectors in component form.

13. **b + h**   14. **f + g**
15. **e − b**   16. **a − d**
17. **g − h**   18. **2a + 3c**
19. **3f + 2d**   20. **4g − 2b**
21. $5\mathbf{a} + \frac{1}{2}\mathbf{g}$ 22. **a + b + c**
23. **3f − a + c**   24. **c + 2d + 3e**

In each of the following, find **x** in component form.

25. **x + b = e**   26. **x + d = a**
27. **c + x = f**   28 **x − g = h**
29. **2x + b = g**   30. **2x − 3d = g**
31. **2b = d − x**   32. **f − g = e − x**
33. **2x + b = x + e**   34. **3x − b = x + h**
35. **a + b + x = b + a**   36. **2x + e = 0**
                                       (zero vector)

37. (a) Draw and label each of the following vectors on graph paper.

$\mathbf{l} = \begin{pmatrix} -3 \\ -3 \end{pmatrix}$; $\mathbf{m} = \begin{pmatrix} 2 \\ 0 \end{pmatrix}$; $\mathbf{n} = \begin{pmatrix} 3 \\ 2 \end{pmatrix}$; $\mathbf{p} = \begin{pmatrix} 1 \\ -2 \end{pmatrix}$;

$\mathbf{q} = \begin{pmatrix} 3 \\ 0 \end{pmatrix}$; $\mathbf{r} = \begin{pmatrix} 6 \\ 4 \end{pmatrix}$; $\mathbf{s} = \begin{pmatrix} 2 \\ 2 \end{pmatrix}$; $\mathbf{t} = \begin{pmatrix} 2 \\ -4 \end{pmatrix}$;

$\mathbf{u} = \begin{pmatrix} -1 \\ -3 \end{pmatrix}$; $\mathbf{v} = \begin{pmatrix} 0 \\ 3 \end{pmatrix}$.

(b) Find four pairs of parallel vectors amongst the ten vectors.

38. State whether 'true' or 'false'.

(a) $\begin{pmatrix} 3 \\ -1 \end{pmatrix}$ is parallel to $\begin{pmatrix} 9 \\ -3 \end{pmatrix}$.

(b) $\begin{pmatrix} -2 \\ 0 \end{pmatrix}$ is parallel to $\begin{pmatrix} 4 \\ 0 \end{pmatrix}$.

(c) $\begin{pmatrix} -1 \\ 1 \end{pmatrix}$ is parallel to $\begin{pmatrix} 1 \\ -1 \end{pmatrix}$.

(d) $\begin{pmatrix} 5 \\ -15 \end{pmatrix} = 5\begin{pmatrix} 1 \\ -3 \end{pmatrix}$.

(e) $\begin{pmatrix} 4 \\ 0 \end{pmatrix}$ is parallel to $\begin{pmatrix} 0 \\ 6 \end{pmatrix}$.

(f) $\begin{pmatrix} 3 \\ -1 \end{pmatrix} + \begin{pmatrix} -4 \\ -2 \end{pmatrix} = \begin{pmatrix} -1 \\ 1 \end{pmatrix}$.

39. (a) Draw a diagram to illustrate the vector addition $\overrightarrow{AB} + \overrightarrow{CD}$.

(b) Draw a diagram to illustrate $\overrightarrow{AB} - \overrightarrow{CD}$.

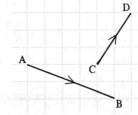

40. Draw separate diagrams to illustrate the following.

(a) $\overrightarrow{FE} + \overrightarrow{JI}$
(b) $\overrightarrow{HG} + \overrightarrow{FE}$
(c) $\overrightarrow{JI} - \overrightarrow{FE}$
(d) $\overrightarrow{HG} + \overrightarrow{JI}$.

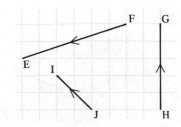

Exercise 8

1. If D has coordinates (7, 2) and E has coordinates (9, 0), find the column vector for $\overrightarrow{DE}$.

2. Find the column vector $\overrightarrow{XY}$ where X and Y have coordinates (−1, 4) and (5, 2) respectively.

3. In the diagram $\overrightarrow{AB}$ represents the vector $\begin{pmatrix} 5 \\ 2 \end{pmatrix}$ and $\overrightarrow{BC}$ represents the vector $\begin{pmatrix} 0 \\ 3 \end{pmatrix}$.

   (a) Copy the diagram and mark point D such that ABCD is a parallelogram.

   (b) Write $\overrightarrow{AD}$ and $\overrightarrow{CA}$ as column vectors.

   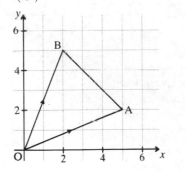

4. (a) On squared paper draw $\overrightarrow{AB} = \begin{pmatrix} 3 \\ -2 \end{pmatrix}$ and $\overrightarrow{BC} = \begin{pmatrix} 4 \\ 2 \end{pmatrix}$ and mark point D such that ABCD is a parallelogram.

   (b) Write $\overrightarrow{AD}$ and $\overrightarrow{CA}$ as column vectors.

5. Copy the diagram below in which $\overrightarrow{OA} = \begin{pmatrix} 5 \\ 2 \end{pmatrix}$, $\overrightarrow{OB} = \begin{pmatrix} 2 \\ 5 \end{pmatrix}$.

   M is the mid-point of AB. Express the following as column vectors.

   (a) $\overrightarrow{BA}$ (b) $\overrightarrow{BM}$

   (c) $\overrightarrow{OM}$ (use $\overrightarrow{OM} = \overrightarrow{OB} + \overrightarrow{BM}$).

   Hence write down the coordinates of M.

6. On a graph with origin at O, draw $\overrightarrow{OA} = \begin{pmatrix} 5 \\ -1 \end{pmatrix}$ and $\overrightarrow{OB} = \begin{pmatrix} 6 \\ -7 \end{pmatrix}$. Given that M is the mid-point of AB, express the following as column vectors.

   (a) $\overrightarrow{BA}$ (b) $\overrightarrow{BM}$ (c) $\overrightarrow{OM}$

   Hence write down the coordinates of M.

7. On a graph with origin at O, draw $\overrightarrow{OA} = \begin{pmatrix} -2 \\ 5 \end{pmatrix}$, $\overrightarrow{OB} = \begin{pmatrix} 4 \\ 2 \end{pmatrix}$ and $\overrightarrow{OC} = \begin{pmatrix} -2 \\ -4 \end{pmatrix}$.

   (a) Given that M divides AB such that AM : MB = 2 : 1, express the following as column vectors

   (i) $\overrightarrow{BA}$ (ii) $\overrightarrow{BM}$ (iii) $\overrightarrow{OM}$

   (b) Given that N divides AC such that AN : NC = 1 : 2, express the following as column vectors

   (i) $\overrightarrow{AC}$ (ii) $\overrightarrow{AN}$ (iii) $\overrightarrow{ON}$.

8. In square ABCD, side AB has column vector $\begin{pmatrix} 2 \\ 1 \end{pmatrix}$. Find two possible column vectors for $\overrightarrow{BC}$.

9. Rectangle KLMN has an area of 10 square units and $\overrightarrow{KL}$ has column vector $\begin{pmatrix} 5 \\ 0 \end{pmatrix}$. Find two possible column vectors for $\overrightarrow{LM}$.

10. In the diagram, ABCD is a trapezium in which $\overrightarrow{DC} = 2\overrightarrow{AB}$.

    If $\overrightarrow{AB} = \mathbf{p}$ and $\overrightarrow{AD} = \mathbf{q}$ express in terms of $\mathbf{p}$ and $\mathbf{q}$

    (a) $\overrightarrow{BD}$ (b) $\overrightarrow{AC}$

    (c) $\overrightarrow{BC}$

    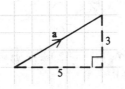

11. Find the image of the vector $\begin{pmatrix} 1 \\ 3 \end{pmatrix}$ after reflection in the following lines:

    (a) $y = 0$ (b) $x = 0$

    (c) $y = x$ (d) $y = -x$

Modulus of a vector

The modulus of a vector **a** is written $|\mathbf{a}|$ and represents the length (or magnitude) of the vector.

In the diagram above, $\mathbf{a} = \begin{pmatrix} 5 \\ 3 \end{pmatrix}$.

By Pythagoras' Theorem, $|\mathbf{a}| = \surd(5^2 + 3^2)$

$$|\mathbf{a}| = \surd 34 \text{ units.}$$

In general if $\mathbf{x} = \begin{pmatrix} m \\ n \end{pmatrix}$, $|\mathbf{x}| = \surd(m^2 + n^2)$.

Exercise 9

Questions **1** to **12** refer to the following vectors.

$$a = \begin{pmatrix} 3 \\ 4 \end{pmatrix} \quad b = \begin{pmatrix} 4 \\ 1 \end{pmatrix} \quad c = \begin{pmatrix} 5 \\ 12 \end{pmatrix} \quad d = \begin{pmatrix} -3 \\ 0 \end{pmatrix}$$

$$e = \begin{pmatrix} -4 \\ -3 \end{pmatrix} \quad f = \begin{pmatrix} -3 \\ 6 \end{pmatrix}$$

From the following, leaving the answer in square root form where necessary.

1. $|a|$
2. $|b|$
3. $|c|$
4. $|d|$
5. $|e|$
6. $|f|$
7. $|a + b|$
8. $|c - d|$
9. $|2e|$
10. $|f + 2b|$

11. (a) Find $|a + c|$.
    (b) Is $|a + c|$ equal to $|a| + |c|$?

12. (a) Find $|c + d|$.
    (b) Is $|c + d|$ equal to $|c| + |d|$?

13. If $\vec{AB} = \begin{pmatrix} 3 \\ -1 \end{pmatrix}$ and $\vec{BC} = \begin{pmatrix} 2 \\ 3 \end{pmatrix}$, find $|\vec{AC}|$.

14. If $\vec{PQ} = \begin{pmatrix} 5 \\ -2 \end{pmatrix}$ and $\vec{QR} = \begin{pmatrix} 0 \\ 1 \end{pmatrix}$, find $|\vec{PR}|$.

15. If $\vec{WX} = \begin{pmatrix} 1 \\ 3 \end{pmatrix}$, $\vec{XY} = \begin{pmatrix} -2 \\ 1 \end{pmatrix}$ and $\vec{YZ} = \begin{pmatrix} 2 \\ -1 \end{pmatrix}$, find $|\vec{WZ}|$.

16. Given that $\vec{OP} = \begin{pmatrix} 0 \\ 5 \end{pmatrix}$ and $\vec{OQ} = \begin{pmatrix} n \\ 3 \end{pmatrix}$, find
    (a) $|\vec{OP}|$
    (b) A value for n if $|\vec{OP}| = |\vec{OQ}|$.

17. Given that $\vec{OA} = \begin{pmatrix} 5 \\ 12 \end{pmatrix}$ and $\vec{OB} = \begin{pmatrix} 0 \\ m \end{pmatrix}$, find
    (a) $|\vec{OA}|$
    (b) A value for m if $|\vec{OA}| = |\vec{OB}|$.

18. Given that $\vec{LM} = \begin{pmatrix} -3 \\ 4 \end{pmatrix}$ and $\vec{MN} = \begin{pmatrix} -15 \\ p \end{pmatrix}$, find
    (a) $|\vec{LM}|$
    (b) A value for p if $|\vec{MN}| = 3|\vec{LM}| = 3$.

19. **a** and **b** are two vectors and $|a| = 3$.
    Find the value of $|a + b|$ when:
    (a) $b = 2a$
    (b) $b = -3a$
    (c) **b** is perpendicular to **a** and $|b| = 4$.

20. **r** and **s** are two vectors and $|r| = 5$.
    Find the value of $|r + s|$ when:
    (a) $s = 5r$
    (b) $s = -2r$
    (c) **r** is perpendicular to **s** and $|s| = 5$
    (d) **s** is perpendicular to $(r + s)$ and $|s| = 3$.

9.5   VECTOR GEOMETRY

Example 1

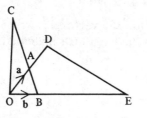

In the diagram, $\vec{OD} = 2\vec{OA}$, $\vec{OE} = 4\vec{OB}$, $\vec{OA} = a$ and $\vec{OB} = b$.

(a) Express $\vec{OD}$ and $\vec{OE}$ in terms of **a** and **b** respectively.

(b) Express $\vec{BA}$ in terms of **a** and **b**.

(c) Express $\vec{ED}$ in terms of **a** and **b**.

(d) Given that $\vec{BC} = 3\vec{BA}$, express $\vec{OC}$ in terms of **a** and **b**.

(e) Express $\vec{EC}$ in terms of **a** and **b**.

(f) Hence show that the points E, D and C lie on a straight line.

(a) $\vec{OD} = 2a$
    $\vec{OE} = 4b$

(b) $\vec{BA} = -b + a$

(c) $\vec{ED} = -4b + 2a$

(d) $\vec{OC} = \vec{OB} + \vec{BC}$
    $\vec{OC} = b + 3(-b + a)$
    $\vec{OC} = -2b + 3a$

(e) $\vec{EC} = \vec{EO} + \vec{OC}$
    $\vec{EC} = -4b + (-2b + 3a)$
    $\vec{EC} = -6b + 3a$

(f) Using the results for $\vec{ED}$ and $\vec{EC}$, we see that $\vec{EC} = \frac{3}{2}\vec{ED}$.

Since $\vec{EC}$ and $\vec{ED}$ are parallel vectors which both pass through the point E, the points E, D and C must lie on a straight line.

Exercise 10

1. $\vec{OD} = 2\vec{OA}$,
    $\vec{OE} = 3\vec{OB}$,
    $\vec{OA} = \mathbf{a}$
    $\vec{OB} = \mathbf{b}$.

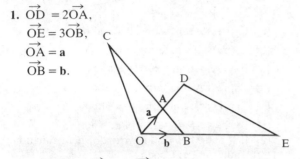

(a) Express $\vec{OD}$ and $\vec{OE}$ in terms of $\mathbf{a}$ and $\mathbf{b}$ respectively.

(b) Express $\vec{BA}$ in terms of $\mathbf{a}$ and $\mathbf{b}$.

(c) Express $\vec{ED}$ in terms of $\mathbf{a}$ and $\mathbf{b}$.

(d) Given that $\vec{BC} = 4\vec{BA}$, express OC in terms of $\mathbf{a}$ and $\mathbf{b}$.

(e) Express $\vec{EC}$ in terms of $\mathbf{a}$ and $\mathbf{b}$.

(f) Use the results for $\vec{ED}$ and $\vec{EC}$ to show that points E, D and C lie on a straight line.

2. $\vec{OY} = 2\vec{OB}$,
    $\vec{OX} = \frac{5}{2}\vec{OA}$,
    $\vec{OA} = \mathbf{a}$,
    $\vec{OB} = \mathbf{b}$.

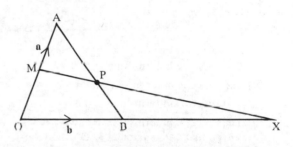

(a) Express $\vec{OY}$ and $\vec{OX}$ in terms of $\mathbf{b}$ and $\mathbf{a}$ respectively.

(b) Express $\vec{AB}$ in terms of $\mathbf{a}$ and $\mathbf{b}$.

(c) Express $\vec{XY}$ in terms of $\mathbf{a}$ and $\mathbf{b}$.

(d) Given that $\vec{AC} = 6\vec{AB}$, express $\vec{OC}$ in terms of $\mathbf{a}$ and $\mathbf{b}$.

(e) Express $\vec{XC}$ in terms of $\mathbf{a}$ and $\mathbf{b}$.

(f) Use the results for $\vec{XY}$ and $\vec{XC}$ to show that points X, Y and C lie on a straight line.

3. $\vec{OA} = \mathbf{a}$,
    $\vec{OB} = \mathbf{b}$;
    $\vec{AQ} = \frac{1}{2}\mathbf{a}$,
    $\vec{BR} = \mathbf{b}$,
    $\vec{AP} = 2\vec{BA}$.

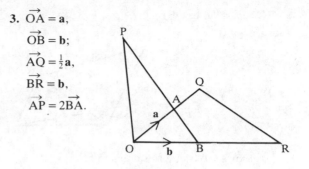

(a) Express $\vec{BA}$ and $\vec{BP}$ in terms of $\mathbf{a}$ and $\mathbf{b}$.

(b) Express $\vec{RQ}$ in terms of $\mathbf{a}$ and $\mathbf{b}$.

(c) Express $\vec{QA}$ and $\vec{QP}$ in terms of $\mathbf{a}$ and $\mathbf{b}$.

(d) Using the vectors for $\vec{RQ}$ and $\vec{QP}$, show that R, Q and P lie on a straight line.

4. In the diagram, $\vec{OA} = \mathbf{a}$ and $\vec{OB} = \mathbf{b}$, M is the mid-point of OA and P lies on AB such that $\vec{AP} = \frac{2}{3}\vec{AB}$.

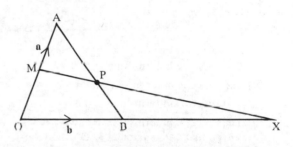

(a) Express $\vec{AB}$ and $\vec{AP}$ in terms of $\mathbf{a}$ and $\mathbf{b}$.

(b) Express $\vec{MA}$ and $\vec{MP}$ in terms of $\mathbf{a}$ and $\mathbf{b}$.

(c) If X lies on OB produced such that OB = BX, express $\vec{MX}$ in terms of $\mathbf{a}$ and $\mathbf{b}$.

(d) Show that MPX is a straight line.

5. $\vec{OP} = \mathbf{a}$,
    $\vec{OA} = 3\mathbf{a}$,
    $\vec{OB} = \mathbf{b}$ and
    M is the mid-point of AB.

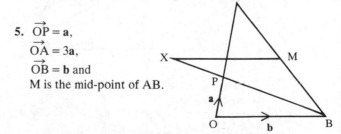

(a) Express $\vec{BP}$ and $\vec{AB}$ in terms of $\mathbf{a}$ and $\mathbf{b}$.

(b) Express $\vec{MB}$ in terms of $\mathbf{a}$ and $\mathbf{b}$.

(c) If X lies on BP produced so that $\vec{BX} = k \cdot \vec{BP}$, express $\vec{MX}$ in terms of $\mathbf{a}$, $\mathbf{b}$ and k.

(d) Find the value of k if MX is parallel to BO.

6. AC is parallel to OB,
$\overrightarrow{AX} = \frac{1}{4}\overrightarrow{AB}$;
$\overrightarrow{OA} = \mathbf{a}$,
$\overrightarrow{OB} = \mathbf{b}$,
$\overrightarrow{AC} = m\mathbf{b}$.

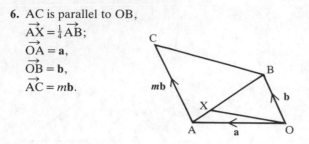

(a) Express $\overrightarrow{AB}$ in terms of **a** and **b**.
(b) Express $\overrightarrow{AX}$ in terms of **a** and **b**.
(c) Express $\overrightarrow{OX}$ in terms of **a**, and **b**.
(d) Express $\overrightarrow{BC}$ in terms of **a**, **b** and m.
(e) Given that OX is parallel to BC, find the value of m.

7. CY is parallel to OD
$\overrightarrow{CX} = \frac{1}{5}\overrightarrow{CD}$;
$\overrightarrow{OC} = \mathbf{c}$,
$\overrightarrow{OD} = \mathbf{d}$,
$\overrightarrow{CY} = n\mathbf{d}$.

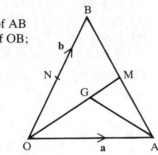

(a) Express $\overrightarrow{CD}$ in terms of **c** and **d**.
(b) Express $\overrightarrow{CX}$ in terms of **c** and **d**.
(c) Express $\overrightarrow{OX}$ in terms of **c** and **d**.
(d) Express $\overrightarrow{DY}$ in terms of **c**, **d** and n.
(e) Given that OX is parallel to DY, find the value of n.

8. M is the mid-point of AB
N is the mid-point of OB;
$\overrightarrow{OA} = \mathbf{a}$
$\overrightarrow{OB} = \mathbf{b}$.

(a) Express $\overrightarrow{AB}$, $\overrightarrow{AM}$ and $\overrightarrow{OM}$ in terms of **a** and **b**.
(b) Given that G lies on OM such that
OG : GM = 2 : 1, express $\overrightarrow{OG}$ in terms of **a** and **b**.
(c) Express $\overrightarrow{AG}$ in terms of **a** and **b**.
(d) Express $\overrightarrow{AN}$ in terms of **a** and **b**.
(e) Show that $\overrightarrow{AG} = m\overrightarrow{AN}$ and find the value of m.

9. M is the mid-point of AC and N is the mid-point of OB;
$\overrightarrow{OA} = \mathbf{a}$,
$\overrightarrow{OB} = \mathbf{b}$,
$\overrightarrow{OC} = \mathbf{c}$.

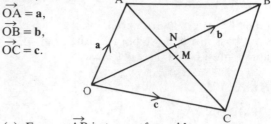

(a) Express $\overrightarrow{AB}$ in terms of **a** and **b**.
(b) Express $\overrightarrow{ON}$ in terms of **b**.
(c) Express $\overrightarrow{AC}$ in terms of **a** and **c**.
(d) Express $\overrightarrow{AM}$ in terms of **a** and **c**.
(e) Express $\overrightarrow{OM}$ in terms of **a** and **c**.
(f) Express $\overrightarrow{NM}$ in terms of **a**, **b** and **c**.
(g) If N and M coincide, write down an equation connecting **a**, **b** and **c**.

10. $\overrightarrow{OA} = \mathbf{a}$,
$\overrightarrow{OB} = \mathbf{b}$.

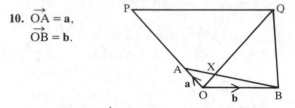

(a) Express $\overrightarrow{BA}$ in terms of **a** and **b**.
(b) Given that $\overrightarrow{BX} = m\overrightarrow{BA}$, show that
$\overrightarrow{OX} = m\mathbf{a} + (1 - m)\mathbf{b}$.
(c) Given that OP = 4**a** and $\overrightarrow{PQ} = 2\mathbf{b}$, express $\overrightarrow{OQ}$ in terms of **a** and **b**.
(d) Given that $\overrightarrow{OX} = n\overrightarrow{OQ}$ use the results for $\overrightarrow{OX}$ and $\overrightarrow{OQ}$ to find the values of m and n.

11. X is the mid-point of OD,
Y lies on CD such that
$\overrightarrow{CY} = \frac{1}{4}\overrightarrow{CD}$;
$\overrightarrow{OC} = \mathbf{c}$,
$\overrightarrow{OD} = \mathbf{d}$.

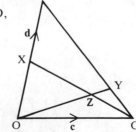

(a) Express $\overrightarrow{CD}$, $\overrightarrow{CY}$ and $\overrightarrow{OY}$ in terms of **c** and **d**.
(b) Express $\overrightarrow{CX}$ in terms of **c** and **d**.
(c) Given that $\overrightarrow{CZ} = h\overrightarrow{CX}$, express $\overrightarrow{OZ}$ in terms of **c**, **d** and h.
(d) If $\overrightarrow{OZ} = k\overrightarrow{OY}$, form an equation and hence find the values of h and k.

9.6 FUNCTIONS

The idea of a function is used in almost every branch of mathematics. The two common notations used are:

(a) $f(x) = x^2 + 4$,

(b) $f : x \mapsto x^2 + 4$.

We may interpret (b) as follows

'function f such that x is mapped onto $x^2 + 4$'.

Example 1

$f(x) = 3x - 1, \quad g(x) = 1 - x^2$

(a) $f(2) = 5$ (b) $f(-1) = -7$

(c) $g(0) = 1$ (d) $g(3) = -8$

(e) If $f(x) = 1$

     Then $3x - 1 = 1$

             $3x = 2$

              $x = \tfrac{2}{3}$

Flow diagrams

The function in Example 1 consisted of two simpler functions as illustrated by a flow diagram.

$x \longrightarrow$ | multiply by 3 | $\xrightarrow{3x}$ | subtract 1 | $\xrightarrow{3x-1}$

It is obviously important to 'multiply by 3' and 'subtract 1' in the correct order.

Example 2

Draw flow diagrams for the functions

(a) $f : x \mapsto (2x + 5)^2$,

(b) $g(x) = \dfrac{5 - 7x}{3}$

(a) $\xrightarrow{x}$ | multiply 2 | $\xrightarrow{2x}$ | add 5 | $\xrightarrow{(2x+5)}$ | square | $\xrightarrow{(2x+5)^2}$

(b) $\xrightarrow{x}$ | multiply by (−7) | $\xrightarrow{-7x}$ | add 5 | $\xrightarrow{5-7x}$ | divide by 3 | $\xrightarrow{\frac{5-7x}{3}}$

Exercise 11

1. Given the functions $h : x \mapsto x^2 + 1$ and $g : x \mapsto 10x + 1$. Find

    (a) $h(2), h(-3), h(0)$,

    (b) $g(2), g(10), g(-3)$,

In questions 2 to 15, draw a flow diagram for each function.

2. $f : x \mapsto 5x + 4$ 3. $f : x \mapsto 3(x - 4)$

4. $f : x \mapsto (2x + 7)^2$ 5. $f : x \mapsto \left(\dfrac{9 + 5x}{4}\right)$

6. $f : x \mapsto \dfrac{4 - 3x}{5}$ 7. $f : x \mapsto 2x^2 + 1$

8. $f : x \mapsto \dfrac{3x^2}{2} + 5$ 9. $f : x \mapsto \sqrt{(4x - 5)}$

10. $f : x \mapsto 4\sqrt{(x^2 + 10)}$ 11. $f : x \mapsto (7 - 3x)^2$

12. $f : x \mapsto 4(3x + 1)^2 + 5$ 13. $f : x \mapsto 5 - x^2$

14. $f : x \mapsto \dfrac{10\sqrt{(x^2 + 1)} + 6}{4}$

15. $f : x \mapsto \left(\dfrac{x^3}{4} + 1\right)^2 - 6$

For questions 16, 17 and 18, the functions f, g and h are defined as follows:

$f : x \mapsto 1 - 2x$

$g : x \mapsto \dfrac{x^3}{10}$

$h : x \mapsto \dfrac{12}{x}$

16. Find

    (a) $f(5), f(-5), f(\tfrac{1}{4})$

    (b) $g(2), g(-3), g(\tfrac{1}{2})$

    (c) $h(3), h(10), h(\tfrac{1}{3})$.

17. Find

    (a) x if $f(x) = 1$

    (b) x if $f(x) = -11$

    (c) x if $h(x) = 1$.

18. Find

    (a) y if $g(y) = 100$

    (b) z if $h(z) = 24$

    (c) w if $g(w) = 0 \cdot 8$.

For questions **19** and **20**, the functions k, l and m are defined as follows:

$$k : x \mapsto \frac{2x^2}{3}$$
$$l : y \mapsto \sqrt{[(y-1)(y-2)]}$$
$$m : x \mapsto 10 - x^2.$$

19. Find
   (a) $k(1)$, $k(6)$, $k(-3)$,
   (b) $l(2)$, $l(0)$, $l(4)$,
   (c) $m(4)$, $m(-2)$, $m(\frac{1}{2})$.

20. Find
   (a) x if $k(x) = 6$
   (b) x if $m(x) = 1$
   (c) y if $k(y) = 2\frac{2}{3}$
   (d) p if $m(p) = -26$.

21. $f(x)$ is defined as the product of the digits of x, e.g. $f(12) = 1 \times 2 = 2$.
   (a) Find   (i) $f(25)$ (ii) $f(713)$
   (b) If x is an integer with three digits, find
       (i) x such that $f(x) = 1$
       (ii) the largest x such that $f(x) = 4$
       (iii) the largest x such that $f(x) = 0$
       (iv) the smallest x such that $f(x) = 2$.

22. $g(x)$ is defined as the sum of the prime factors of x, e.g. $g(12) = 2 + 3 = 5$. Find
   (a) $g(10)$ (b) $g(21)$ (c) $g(36)$
   (d) $g(99)$ (e) $g(100)$ (f) $g(1000)$

23. $h(x)$ is defined as the number of letters in the English word describing x, e.g. $h(1) = 3$. Find
   (a) $h(2)$ (b) $h(11)$ (c) $h(18)$
   (d) the largest value of x for which $h(x) = 3$.

24. If $f : x \mapsto$ next prime number greater than x, find:
   (a) $f(7)$ (b) $f(14)$ (c) $f[f(3)]$

25. If $g : x \to 2^x + 1$, find:
   (a) $g(2)$ (b) $g(4)$ (c) $g(-1)$
   (d) the value of x if $g(x) = 9$

26. The function f is defined as $f : x \to ax + b$ where a and b are constants.
   If $f(1) = 8$ and $f(4) = 17$, find the values of a and b.

27. The function g is defined as $g(x) = ax^2 + b$ where a and b are constants.
   If $g(2) = 3$ and $g(-3) = 13$, find the values of a and b.

28. Functions h and k are defined as follows:
   $h : x \mapsto x^2 + 1$, $k : x \mapsto ax + b$, where a and b are constants.
   If $h(0) = k(0)$ and $k(2) = 15$, find the values of a and b.

Composite functions

The function $f : x \mapsto 3x + 2$ is itself a composite function, consisting of two simpler functions: 'multiply by 3' and 'add 2'.

If $f : x \mapsto 3x + 2$ and $g : x \mapsto x^2$ then fg is a composite function where g is performed first and then f is performed on the result of g.

The function fg may be found using a flow diagram.

Thus $fg : x \mapsto 3x^2 + 2$.

Inverse functions

If a function f maps a number n onto m, then the inverse function f^{-1} maps m onto n. The inverse of a given function is found using a flow diagram.

Example 3

Find the inverse of f where $f : x \to \dfrac{5x - 2}{3}$

(a) Draw a flow diagram for f

(b) Draw a new flow diagram with each operation replaced by its inverse. Start with x on the right.

Thus the inverse of f is given by

$$f^{-1} : x \mapsto \frac{3x + 2}{5}$$

Exercise 12

For questions **1** and **2**, the functions f, g and h are as follows: $f : x \mapsto 4x$

$$g : x \mapsto x + 5$$
$$h : x \mapsto x^2$$

1. Find the following in the form '$x \mapsto \ldots$'
  (a) fg (b) gf (c) hf
  (d) fh (e) gh (f) fgh
  (g) hfg

2. Find
  (a) x if $hg(x) = h(x)$
  (b) x if $fh(x) = gh(x)$

For questions **3**, **4** and **5**, the functions f, g and h are as follows: $f : x \mapsto 2x$

$$g x \mapsto x - 3$$
$$h x \mapsto x^2$$

3. Find the following in the form '$x \mapsto \ldots$'
  (a) fg (b) gf (c) gh
  (d) hf (e) ghf (f) hgf

4. Evaluate
  (a) $fg(4)$ (b) $gf(7)$ (c) $gh(-3)$
  (d) $fgf(2)$ (e) $ggg(10)$ (f) $hfh(-2)$

5. Find
  (a) x if $f(x) = g(x)$ (b) x if $hg(x) = gh(x)$
  (c) x if $gf(x) = 0$ (d) x if $fg(x) = 4$

For questions **6**, **7** and **8**, the functions, l, m and n are as follows: $l : x \mapsto 2x + 1$

$$m : x \mapsto 3x - 1$$
$$n : x \mapsto x^2$$

6. Find the following in the form '$x \mapsto \ldots$'
  (a) lm (b) ml (c) ln
  (d) nm (e) lnm (f) mln

7. Find
  (a) $lm(2)$ (b) $nl(1)$ (c) $mn(-2)$
  (d) $mm(2)$ (e) $nln(2)$ (f) $llm(0)$

8. Find
  (a) x if $l(x) = m(x)$
  (b) two values of x if $nl(x) = nm(x)$
  (c) x if $ln(x) = mn(x)$

In questions **9** to **26**, find the inverse of each function in the form '$x \mapsto \ldots$'

9. $f : x \mapsto 5x - 2$ **10.** $f : x \mapsto 5(x - 2)$

11. $f : x \mapsto 3(2x + 4)$ **12.** $g : x \mapsto \dfrac{2x + 1}{3}$

13. $f : x \mapsto \dfrac{3(x - 1)}{4}$ **14.** $g : x \mapsto 2(3x + 4) - 6$

15. $h : x \mapsto \frac{1}{2}(4 + 5x) + 10$

16. $k : x \mapsto -7x + 3$ **17.** $j : x \mapsto \dfrac{12 - 5x}{3}$

18. $l : x \mapsto \dfrac{4 - x}{3} + 2$ **19.** $m : x \mapsto \dfrac{\left[\dfrac{(2x - 1)}{4}\right] - 3}{5}$

20. $f : x \mapsto \dfrac{3(10 - 2x)}{7}$ **21.** $g : x \mapsto \left[\dfrac{\frac{x}{4} + 6}{5}\right] + 7$

22. A calculator has the following function buttons:
$x \mapsto x^2$; $x \mapsto \sqrt{x}$; $x \mapsto \frac{1}{x}$; $x \mapsto \log x$;
$x \mapsto \ln x$; $x \mapsto \sin x$; $x \mapsto \cos x$; $x \mapsto \tan x$; $x \mapsto x!$
Find which button was used for the following input/outputs:
  (a) $1000\,000 \to 1000$ (b) $1000 \to 3$
  (c) $3 \to 6$ (d) $0 \cdot 2 \to 0 \cdot 04$
  (e) $10 \to 0 \cdot 1$ (f) $45 \to 1$
  (g) $0 \cdot 5 \to 2$ (h) $64 \to 8$
  (i) $60 \to 0 \cdot 5$ (j) $1 \to 0$
  (k) $135 \to -1$ (l) $10 \to 3628\,800$
  (m) $0 \to 1$ (n) $30 \to 0 \cdot 5$
  (o) $90 \to 0$ (p) $0 \cdot 4 \to 2 \cdot 5$
  (q) $4 \to 24$ (r) $1000\,000 \to 6$

REVISION EXERCISE 9A

1. Given that $\mathscr{E} = \{1, 2, 3, 4, 5, 6, 7, 8\}$,
  $A = \{1, 3, 5\}$, $B = \{5, 6, 7\}$, list the members of the sets
  (a) $A \cap B$ (b) $A \cup B$
  (c) A' (d) $A' \cap B'$
  (e) $A \cup B'$

2. The set P and Q are such that $n(P \cup Q) = 50$, $n(P \cap Q) = 9$ and $n(P) = 27$. Find the value of $n(Q)$.

3. Draw three diagrams similar to Figure 1 below, and shade the following.
  (a) $Q \cap R'$
  (b) $(P \cup Q) \cap R$
  (c) $(P \cap Q) \cap R'$

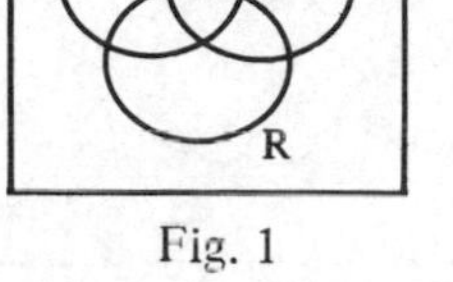

Fig. 1

4. Describe the shaded regions in Figures 2 and 3.

(a)

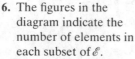

(b)

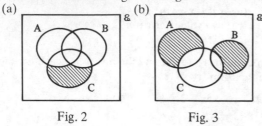

Fig. 2   Fig. 3

5. Given that $\mathscr{E} = \{$ people on a train$\}$, $M = \{$males$\}$,
$T = \{$people over 25 years old$\}$ and
$S = \{$snooker players$\}$
(a) express in set notation:
  (i) all the snooker players are over 25
  (ii) some snooker players are women.
(b) express in words $T \cap M' = \varnothing$.

6. The figures in the diagram indicate the number of elements in each subset of $\mathscr{E}$.
(a) Find $n(P \cap R)$
(b) Find $n(Q \cup R)'$
(c) Find $n(P' \cap Q')$.

7. In $\triangle OPR$, the mid-point of PR is M.
If $\overrightarrow{OP} = \mathbf{p}$ and $\overrightarrow{OR} = \mathbf{r}$,
find in terms of $\mathbf{p}$ and $\mathbf{r}$
(a) $\overrightarrow{PR}$ (b) $\overrightarrow{PM}$
(c) $\overrightarrow{OM}$

8. If $\mathbf{a} = \begin{pmatrix} 1 \\ 4 \end{pmatrix}$ and $\mathbf{b} = \begin{pmatrix} -3 \\ 4 \end{pmatrix}$, find

(a) $|\mathbf{b}|$ (b) $|\mathbf{a} + \mathbf{b}|$ (c) $|2\mathbf{a} - \mathbf{b}|$

9. If $4 \begin{pmatrix} 1 \\ 3 \end{pmatrix} + 2 \begin{pmatrix} 1 \\ m \end{pmatrix} = 3 \begin{pmatrix} n \\ -6 \end{pmatrix}$, find the values of m and n.

10. The points O, A and B have coordinates $(0, 0)$, $(5, 0)$ and $(-1, 4)$ respectively. Write as column vectors.
(a) $\overrightarrow{OB}$ (b) $\overrightarrow{OA} + \overrightarrow{OB}$ (c) $\overrightarrow{OA} - \overrightarrow{OB}$
(d) $\overrightarrow{OM}$ where M is the mid-point of AB.

11. In the parallelogram OABC, M is the mid-point of AB and N is the mid-point of BC.
If $\overrightarrow{OA} = \mathbf{a}$ and $\overrightarrow{OC} = \mathbf{c}$, express in terms of $\mathbf{a}$ and $\mathbf{c}$.
(a) $\overrightarrow{CA}$ (b) $\overrightarrow{ON}$ (c) $\overrightarrow{NM}$
Describe the relationship between CA and NM.

12. The vectors $\mathbf{a}, \mathbf{b}, \mathbf{c}$ are given by
$\mathbf{a} = \begin{pmatrix} 1 \\ 5 \end{pmatrix}$, $\mathbf{b} = \begin{pmatrix} -2 \\ 1 \end{pmatrix}$, $\mathbf{c} = \begin{pmatrix} -1 \\ 17 \end{pmatrix}$.
Find numbers m and n so that $m\mathbf{a} + n\mathbf{b} = \mathbf{c}$.

13. Given that $\overrightarrow{OP} = \begin{pmatrix} 3 \\ 2 \end{pmatrix}$, $\overrightarrow{OQ} = \begin{pmatrix} 0 \\ 4 \end{pmatrix}$ and that M is the mid-point of PQ, express as column vectors
(a) $\overrightarrow{PQ}$ (b) $\overrightarrow{PM}$ (c) $\overrightarrow{OM}$.

14. Given $f: x \mapsto 2x - 3$ and $g: x \mapsto x^2 - 1$, find
(a) $f(-1)$ (b) $g(-1)$
(c) $fg(-1)$ (d) $gf(3)$
Write the function ff in the form '$ff : x \mapsto \ldots$'

15. If $f: x \mapsto 3x + 4$ and $h: x \mapsto \dfrac{x-2}{5}$
express f^{-1} and h^{-1} in the form '$x \mapsto \ldots$'.
Find (a) $f^{-1}(13)$
      (b) the value of z if $f(z) = 20$.

16. Given that $f(x) = x - 5$, find
(a) the value of s such that $f(s) = -2$,
(b) the values of t such that $t \times f(t) = 0$.

EXAMINATION EXERCISE 9B

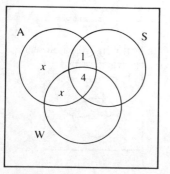

1. Nineteen people are employed in an office. The Venn diagram shows some details about the number who can do audio-typing (A), shorthand-typing (S) and use the word processor (W). They all have at least one of these skills.

(a) 11 office workers can do audio-typing. Find x.
(b) If 6 office workers cannot do either method of typing, how many can do shorthand-typing?
(c) Nobody does only shorthand-typing. Find:
  (i) How many can use the word-processor.
  (ii) How many can both use the word-processor and do shorthand-typing.
(d) Copy the Venn diagram and shade the region $W \cap A' \cap S'$.
  Give a brief description of this set.

[S]

2. In triangle OAL, the mid-point of OA is K and the mid-point of OL is B.

Given that OA = **a** and OB = **b**, express the vectors OK, OL, KB and AL in terms of **a** and **b**, and prove that KB is parallel to AL.

State the ratio of the area of triangle OAL to the area of triangle OKB. [N]

3. ABCDEF is a regular hexagon and O is its centre. The vectors **x** and **y** are such that $\overrightarrow{AB} = \mathbf{x}$ and $\overrightarrow{BC} = \mathbf{y}$.

Express in terms of **x** and **y** the vectors $\overrightarrow{AC}$, $\overrightarrow{AO}$, $\overrightarrow{CD}$ and $\overrightarrow{BF}$. [M]

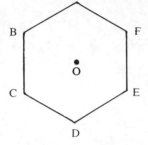

4. In a video game, a spot on the screen bounces off the four sides of a rectangular frame. The spot moves from A to B to C to D, as shown in the diagram below.

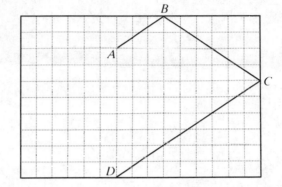

(a) What can you say about the angle at which the spot bounces off each side, compared with the angle at which it approached that side?

(b) The column vector describing the part AB of the movement is $\begin{pmatrix} 3 \\ 2 \end{pmatrix}$.

Write down the column vectors describing BC and CD.

(c) What do you notice about the parts of the path AB and CD? Justify your answer using the column vectors.

From D, the spot moves to a point E on the fourth side of the frame.

(d) Given that the spot continues in the same way, write down the vector DE.

(e) Given that the spot bounces off the fourth side at E and continues to move. describe its subsequent path. [W]

5. In the triangle OAB shown, OC = $\frac{2}{5}$OB, $\overrightarrow{OA}$ = **a**, $\overrightarrow{OB}$ = **b**.

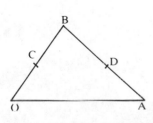

(a) Express the vectors $\overrightarrow{OC}$, $\overrightarrow{CB}$ and $\overrightarrow{BA}$ in terms of **a**, or **b**, or **a** and **b**.

The point D divides BA in the ratio 3:2.

(b) Find $\overrightarrow{BD}$ and $\overrightarrow{CD}$ in terms of **a**, or **b**, or **a** and **b**.

(c) Give the special name of the quadrilateral OCDA.

(d) Calculate the value of $\dfrac{\text{area of triangle OAB}}{\text{area of triangle CDB}}$.

(e) Given that the area of the quadrilateral OCDA is 48 cm², find the area of the triangle CBD. [L]

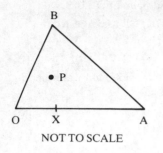

NOT TO SCALE

6. The point X lies on the side OA of the triangle OAB and $OX = \frac{1}{3}OA$. The vectors **a** and **b** are such that $\overrightarrow{OA} = \mathbf{a}$ and $\overrightarrow{OB} = \mathbf{b}$. The point P is such that $\overrightarrow{OP} = \frac{1}{5}\mathbf{a} + \frac{2}{5}\mathbf{b}$.

   (a) Express, in terms of **a** and **b**, the vectors
       (i) $\overrightarrow{BX}$,   (ii) $\overrightarrow{BP}$.
   (b) By considering the results of part (a), or otherwise, show that the line BX passes through P and write down the numerical value of $\dfrac{BP}{BX}$.
   (c) Given that the area of triangle OAB is 30 cm², calculate
       (i) the area of triangle ABX,
       (ii) the area of triangle ABP,
       (iii) the area of triangle OPA.   [M]

7. For positive values of x, $f(x) = x^2 - 8$ and $g(x) = x + 1$.
   (a) Find   (i) $gf(x)$;   (ii) $g^{-1}(x)$.
   (b) Form the equation $gf(x) = g^{-1}(x)$ and find the value of x.   [M]

8. A sketch of the graph of $f(x) = \sin x°$ for $0 \le x \le 360$ is shown below.

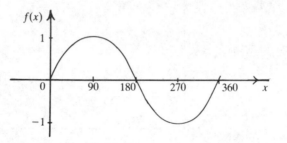

   (a) On the same axes, draw a line to show how you would find the set of solutions of the equation $f(x) = 0\cdot6$.
   (b) Sketch the graph of $g(x) = 1 - \sin x°$ for $0 \le x \le 360$.
   (c) Given that $h(x) = \dfrac{2}{x}$, express $hf(x)$ in terms of x and evaluate $hf(40)$.   [N]

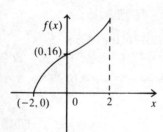

9. The function f, whose graph is shown, is given by $f : x \mapsto px^3 + q$, for $-2 \le x \le 2$.
   (a) State the value of q.
   (b) Calculate p.
   (c) Draw a flow chart, using your values of p and q, for calculating $f(x)$ from a given value of x.
   (d) Find the largest value of $f(x)$.
   (e) If $f = hg$, where g is the function given by $g : x \mapsto x^3$, express h in the form $h : x \mapsto \ldots.$ [L]

10 Statistics and probability

Blaise Pascal (1623–1662) suffered the most appalling ill-health throughout his short life. He is best known for his work with Fermat on probability. This followed correspondence with a gentleman gambler who was puzzled as to why he lost so much in betting on the basis of the appearance of a throw of dice. Pascal's work on probability became of enormous importance and showed for the first time that absolute certainty is not a necessity in mathematics and science. He also studied physics, but his last years were spent in religious meditation and illness.

10.1 DATA DISPLAY

Bar chart

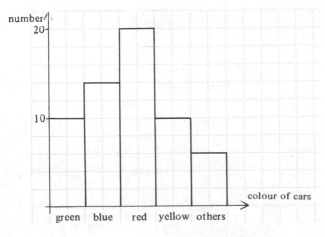

The length of each bar represents the quantity in question. The width of each bar has no significance. In the bar chart above, the number of the cars of each colour in a car park is shown. The bars can be joined together or separated.

Histogram

A histogram is similar to a bar chart but it is not the same. A histogram is a diagram which is used to represent a frequency distribution and consists of a set of rectangles whose *areas* represent the frequency of the various data. If all the rectangles have the same width, the frequencies will be represented by the heights of the rectangles.

Example 1

In a survey, people on a bus were asked how long they had waited before the bus arrived.

Time waiting (to the nearest minute)	3–7	8–12	13–17	18–22
Number of people	2	5	3	1

This information is shown on the histogram below.

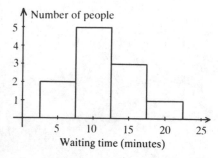

Notice: (a) There are no gaps between the rectangles.

(b) The bases of the rectangles are $2\frac{1}{2}$ to $7\frac{1}{2}$, $7\frac{1}{2}$ to $12\frac{1}{2}$, $12\frac{1}{2}$ to $17\frac{1}{2}$, $17\frac{1}{2}$ to $22\frac{1}{2}$.

Pie chart

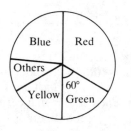

The information is displayed using sectors of a circle.

This pie chart shows the same information as the bar chart on the previous page.

The angles of the sectors are calculated as follows:

Total number of cars $= 10 + 14 + 20 + 10 + 6$
$= 60$

Angle representing green cars $= \dfrac{10}{60} \times 360°$
$= 60°$

Angle representing blue cars $= \dfrac{14}{60} \times 360°$, etc.

Exercise 1

1. The bar chart shows the number of children playing various games on a given day.

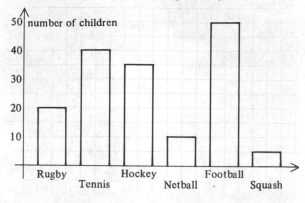

(a) Which game had the least number of players?

(b) What was the total number of children playing all the games?

(c) How many more footballers were there than tennis players?

2. The table shows the number of cars of different makes in a car park. Illustrate this data on a bar chart.

Make	Fiat	Renault	Leyland	Rolls Royce	Ford	Datsun
Number	14	23	37	5	42	18

3. The pie chart illustrates the values of various goods sold by a certain shop. If the total value of the sales was £24 000, find the sales value of

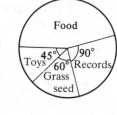

(a) toys

(b) grass seed

(c) records

(d) food.

4. The table shows the colours of a random selection of 'Smarties'. Calculate the angles on a pie chart corresponding to each colour.

colour	red	green	blue	yellow	pink
number	5	7	11	4	9

5. A quantity of scrambled eggs is made using the following recipe:

ingredient	eggs	milk	butter	cheese	salt/pepper
mass	450 g	20 g	39 g	90 g	1 g

Calculate the angles on a pie chart corresponding to each ingredient.

6. Calculate the angles on a pie chart corresponding to quantities A, B, C, D and E given in the tables.

   (a)
quantity	A	B	C	D	E
number	3	5	3	7	0

   (b)
quantity	A	B	C	D	E
mass	10 g	15 g	34 g	8 g	5 g

   (c)
quantity	A	B	C	D	E
length	7	11	9	14	11

7. A firm making artificial sand sold its products in four countries.

   5% were sold in Spain
   15% were sold in France
   15% were sold in Germany
   65% were sold in U.K.

   What would be the angles on a pie chart drawn to represent this information?

8. The weights of A, B, C are in the ratio $2 : 3 : 4$. Calculate the angles representing A, B and C on a pie-chart.

9. The cooking times for meals L, M and N are in the ratio $3 : 7 : x$. On a pie-chart, the angle corresponding to L is $60°$. Find x.

10. The results of an opinion poll of 2000 people are represented on a pie chart. The angle corresponding to 'don't know' is $18°$. How many people in the sample did not know?

11. The transfer fees of five players are as follows:
    Gibson   £4·50   Crisp   £9
    Campbell £6   Raynor   £80
    Hawley   50p
    Calculate the angles on a pie chart corresponding to Campbell and Hawley. (This example illustrates one of the limitations of a pie-chart: negative quantities cannot be displayed.)

12. The pie chart illustrates the sales of various makes of petrol.
    (a) What percentage of sales does 'Esso' have?
    (b) If 'Jet' accounts for $12\frac{1}{2}\%$ of total sales, calculate the angles x and y.

13. Thirty pupils in a class are weighed on the first day of term. The results are shown below.

Weight (kg)	36–40	41–45	46–50	51–55	56–60
Frequency	5	7	10	5	3

    Draw a histogram to illustrate the results.

14. The times taken by 19 runners in a race were recorded to the nearest minute as follows:

Time (min)	13–15	16–18	19–21	22–24
Frequency	3	5	7	4

    Draw a histogram to illustrate the results.

15. The diagram illustrates the production of apples in two countries.

U.K.
470
thousand
tonnes

FRANCE
950
thousand
tonnes

    In what way could the pictorial display be regarded as misleading?

16. The graph shows the performance of a company in the year in which a new manager was appointed. In what way is the graph misleading?

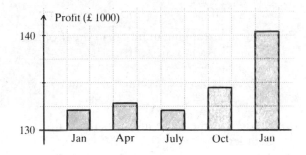

10.2   MEAN, MEDIAN AND MODE

(a) The *mean* of a series of numbers is obtained by adding the numbers and dividing the result by the number of numbers.

(b) The *median* of a series of numbers is obtained by arranging the numbers in ascending order and then choosing the number in the 'middle'. If there are *two* 'middle' numbers the median is the average (mean) of these two numbers.

(c) The *mode* of a series of numbers is simply the number which occurs most often.

Example 1

Find the mean, median and mode of the following numbers:
5, 4, 10, 3, 3, 4, 7, 4, 6, 5.

(a) Mean $= \dfrac{(5+4+10+3+3+4+7+4+6+5)}{10}$

$= \dfrac{51}{10} = 5 \cdot 1$

(b) Median: arranging numbers in order of size
3, 3, 4, 4, 4, 5, 5, 6, 7, 10
↑

The median is the 'average' of 4 and 5

∴   median = 4·5.

(c) Mode = 4 (there are more 4's than any other number).

Frequency tables

A frequency table shows a number x such as a mark or a score, against the frequency f or number of times that x occurs.

The next example shows how these symbols are used in calculating the mean, the median and the mode.

The symbol Σ (or sigma) means 'the sum of'.

Example 2

The marks obtained by 100 students in a test were as follows:

mark (x)	0	1	2	3	4
frequency (f)	4	19	25	29	23

Find
(a) the mean mark   (b) the median mark
(c) the modal mark

(a) Mean $= \dfrac{\Sigma xf}{\Sigma f}$

where Σxf means 'the sum of the products'
i.e. $\Sigma(\text{number} \times \text{frequency})$

and Σf means 'the sum of the frequencies'

Mean $= \dfrac{(0 \times 4)+(1 \times 19)+(2 \times 25)+(3 \times 29)+(4 \times 23)}{100}$

$= \dfrac{248}{100} = 2 \cdot 48$

(b) The median mark is the number between the 50th and 51st numbers. By inspection, both the 50th and 51st numbers are 3.

∴   Median = 3 marks.

(c) The modal mark = 3.

Exercise 2

1. Find the mean, median and mode of the following sets of numbers:
   (a) 3, 12, 4, 6, 8, 5, 4
   (b) 7, 21, 2, 17, 3, 13, 7, 4, 9, 7, 9
   (c) 12, 1, 10, 1, 9, 3, 4, 9, 7, 9
   (d) 8, 0, 3, 3, 1, 7, 4, 1, 4, 4.
2. Find the mean, median and mode of the following sets of numbers:
   (a) 3, 3, 5, 7, 8, 8, 8, 9, 11, 12, 12
   (b) 7, 3, 4, 10, 1, 2, 1, 3, 4, 11, 10, 4
   (c) −3, 4, 0, 4, −2, −5, 1, 7, 10, 5
   (d) $1, \frac{1}{2}, \frac{1}{2}, \frac{3}{4}, \frac{1}{4}, 2, \frac{1}{2}, \frac{1}{4}, \frac{3}{4}$.
3. The mean weight of five men is 76 kg. The weights of four of the men are 72 kg, 74 kg, 75 kg and 81 kg. What is the weight of the fifth man?
4. The mean length of 6 rods is 44·2 cm. The mean length of 5 of them is 46 cm. How long is the sixth rod?

5. (a) The mean of 3, 7, 8, 10 and x is 6. Find x.
   (b) The mean of 3, 3, 7, 8, 10, x and x is 7.
       Find x.
6. The mean height of 12 men is 1·70 m, and the
   mean height of 8 women is 1·60 m. Find
   (a) the total height of the 12 men,
   (b) the total height of the 8 women,
   (c) the mean height of the 20 men and women.
7. The total weight of 6 rugby players is 540 kg and
   the mean weight of 14 ballet dancers is 40 kg. Find
   the mean weight of the group of 20 rugby players
   and ballet dancers.
8. The mean weight of 8 boys is 55 kg and the mean
   weight of a group of girls is 52 kg. The mean
   weight of all the children is 53·2 kg. How many
   girls are there?
9. A group of 50 people were asked how many books
   they had read in the previous year; the results are
   shown in the frequency table below. Calculate the
   mean number of books read per person.

Number of books	0	1	2	3	4	5	6	7	8
Frequency	5	5	6	9	11	7	4	2	1

10. A number of people were asked how many coins
    they had in their pockets; the results are shown
    below. Calculate the mean number of coins per
    person.

Number of coins	0	1	2	3	4	5	6	7
Frequency	3	6	4	7	5	8	5	2

11. The following tables give the distribution of marks
    obtained by different classes in various tests. For
    each table, find the mean , median and mode.

   (a)
Mark	0	1	2	3	4	5	6
Frequency	3	5	8	9	5	7	3

   (b)
Mark	15	16	17	18	19	20
Frequency	1	3	7	1	5	3

   (c)
Mark	0	1	2	3	4	5	6
Frequency	10	11	8	15	25	20	11

12. One hundred golfers play a certain hole and their
    scores are summarised below.

Score	2	3	4	5	6	7	8
Number of players	2	7	24	31	18	11	7

   Find
   (a) the mean score
   (b) the median score.

13. The number of goals scored in a series of football
    matches was as follows:

Number of goals	1	2	3
Number of matches	8	8	x

   (a) If the mean number of goals is 2·04, find x.
   (b) If the modal number of goals is 3, find the
       smallest possible value of x.
   (c) If the median number of goals is 2, find the
       largest possible value of x.
14. In a survey of the number of occupants in a
    number of cars, the following data resulted.

Number of occupants	1	[2]	3	4
Number of cars	7	11	7	x

   (a) If the mean number of occupants is $2\frac{1}{3}$,
       find x.
   (b) If the mode is 2, find the largest possible value
       of x.
   (c) If the median is 2, find the largest possible
       value of x.
15. The numbers 3, 5, 7, 8 and N are arranged in
    ascending order. If the mean of the numbers is
    equal to the median, find N.
16. The mean of 5 numbers is 11. The numbers are in
    the ratio $1:2:3:4:5$. Find the smallest number.
17. The mean of a set of 7 numbers is 3·6 and the
    mean of a different set of 18 numbers is 5·1.
    Calculate the mean of the 25 numbers.
18. The number of illnesses suffered by the members
    of a herd of cows is as follows:

Number of illnesses	0	1	2	3
Number of cows	31	11	5	3

   (a) Find
       (i) the mean number of illnesses,
       (ii) the median number of illnesses.
   (b) In a different herd of 30 cows, the mean
       number of illnesses was 0·8. Find the mean
       number of illnesses for the 80 cows.
19. The median of five consecutive integers is N.
   (a) Find the mean of the five numbers.
   (b) Find the mean and the median of the squares
       of the integers.
   (c) Find the difference between these values.
20. The marks obtained by the members of a class are
    summarised in the table.

Mark	x	y	z
Frequency	a	b	c

   Calculate the mean mark in terms of a, b, c, x, y
   and z.

10.3 CUMULATIVE FREQUENCY

It is possible to record data in groups (class intervals) giving the frequency of occurrence in each group. A cumulative frequency curve (or ogive) shows the *median* at the 50th percentile of the cumulative frequency.
The value at the 25th percentile is known as the *lower quartile*, and that at the 75th percentile as the *upper quartile*.
A measure of the spread or dispersion of the data is given by the *inter-quartile range* where

$$\text{inter-quartile range} = \text{upper quartile} - \text{lower quartile}.$$

Example 1

The marks obtained by 80 students in an examination are shown below.

mark	frequency	cumulative frequency	marks represented by cumulative frequency
1–10	3	3	$\leqslant 10$
11–20	5	8	$\leqslant 20$
21–30	5	13	$\leqslant 30$
31–40	9	22	$\leqslant 40$
41–50	11	33	$\leqslant 50$
51–60	15	48	$\leqslant 60$
61–70	14	62	$\leqslant 70$
71–80	8	70	$\leqslant 80$
81–90	6	76	$\leqslant 90$
91–100	4	80	$\leqslant 100$

The table also shows the cumulative frequency.
Plot a cumulative frequency curve and hence estimate
(a) the median
(b) the inter-quartile range.

The points on the graph are plotted at the upper limit of each group of marks.

From the cumulative frequency curve
median = 55 marks
lower quartile = 37·5 marks
upper quartile = 68 marks

∴ inter-quartile range = 68 − 37·5
= 30·5 marks.

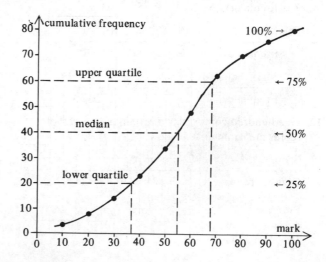

Exercise 3

1. Figure 1 shows the cumulative frequency curve for the marks of 60 students in an examination.

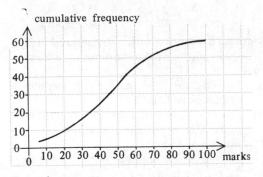

Figure 1

From the graph, estimate
(a) the median mark
(b) the mark at the lower quartile and at the upper quartile
(c) the inter-quartile range
(d) the pass mark if two-thirds of the students passed
(e) the number of students achieving less than 40 marks.

2. Figure 2 shows the cumulative frequency curve for the marks of 140 students in an examination.

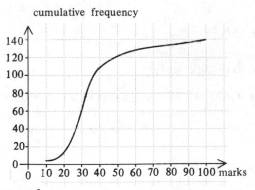

Figure 2

From the graph, estimate
(a) the median mark
(b) the mark at the lower quartile and at the upper quartile
(c) the inter-quartile range
(d) the pass mark if three-fifths of the students passed
(e) the number of students achieving more than 30 marks.

In questions **3** to **6**, draw a cumulative frequency curve, and find
(a) the median, (b) the interquartile range.

3.

mass (kg)	frequency
1–5	4
6–10	7
11–15	11
16–20	18
21–25	22
26–30	10
31–35	5
36–40	3

4.

length (cm)	frequency
41–50	6
51–60	8
61–70	14
71–80	21
81–90	26
91–100	14
101–110	7
111–120	4

5.

time (seconds)	frequency
36–45	3
46–55	7
56–65	10
66–75	18
76–85	12
86–95	6
96–105	4

6.

number of marks	frequency
1–10	0
11–20	2
21–30	4
31–40	10
41–50	17
51–60	11
61–70	3
71 80	3

7. In an experiment, 50 people were asked to guess the weight of a bunch of daffodils in grams. The guesses were as follows:

47	39	21	30	42	35	44	36	19	52
23	32	66	29	5	40	33	11	44	22
27	58	38	37	48	63	23	40	53	24
47	22	44	33	13	59	33	49	57	30
17	45	38	33	25	40	51	56	28	64

Construct a frequency table using intervals 0–9, 10–19, 20–29, etc. Hence draw a cumulative frequency curve and estimate
(a) the median weight
(b) the inter-quartile range
(c) the number of people who guessed a weight within 10 grams of the median.

8. In a competition, 30 children had to pick up as many paper clips as possible in one minute using a pair of tweezers. The results were as follows.

3	17	8	11	26	23	18	28	33	38
12	38	22	50	5	35	39	30	31	43
27	34	9	25	39	14	27	16	33	49

Construct a frequency table using intervals 1–10, 11–20, etc. and hence draw a cumulative frequency curve.
(a) From the curve, estimate the median number of clips picked up.
(b) From the frequency table, estimate the mean of the distribution using the mid-interval values 5·5, 15·5, etc.
(c) Calculate the exact value of the mean using the original data.
(d) Why is it possible only to estimate the mean in part (b)?

10.4   SIMPLE PROBABILITY

Probability theory is not the sole concern of people interested in betting, although it is true to say that a 'lucky' poker player is likely to be a player with a sound understanding of probability. All major airlines regularly overbook aircraft because they can predict with accuracy the probability that a certain number of passengers will fail to arrive for the flight.

Suppose a 'trial' can have n equally likely results and suppose that a 'success' can occur in s ways (from the n). Then the probability of a 'success' $= \dfrac{s}{n}$.

Example 1

A single card is drawn from a pack of 52 playing cards. Find the probability of the following results:

(a) the card is an ace
(b) the card is the ace of hearts
(c) the card is a spade
(d) the card is a picture card.

We will use the notation 'p (an ace)' to represent 'the probability of selecting an ace'.

(a) $p \text{ (an ace)} = \dfrac{4}{52} = \dfrac{1}{13}$.

(b) $p \text{ (ace of hearts)} = \dfrac{1}{52}$.

(c) $p \text{ (a spade)} = \dfrac{13}{52} = \dfrac{1}{4}$.

(d) $p \text{ (a picture card)} = \dfrac{12}{52} = \dfrac{3}{13}$.

In each case, we have counted the number of ways in which a 'success' can occur and divided by the number of possible results of a 'trial'.

Example 2

A black die and a white die are thrown at the same time. Display all the possible outcomes. Find the probability of obtaining:
(a) a total of 5,
(b) a total of 11,
(c) a 'two' on the black die and a 'six' on the white die.

It is convenient to display all the possible outcomes on a grid.

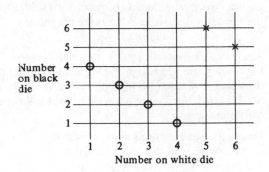

There are 36 possible outcomes, shown where the lines cross.

(a) There are four ways of obtaining a total of 5 on the two dice. They are shown circled on the diagram.

$$\therefore \quad \text{Probability of obtaining a total of } 5 = \frac{4}{36}$$

(b) There are two ways of obtaining a total of 11. They are shown with a cross on the diagram.

$$\therefore \quad p \text{ (total of 11)} = \frac{2}{36} = \frac{1}{18}$$

(c) There is only one way of obtaining a 'two' on the black die and a 'six' on the white die.

$$\therefore \quad p \text{ (2 on black and 6 on white)} = \frac{1}{36}.$$

Exercise 4

1. One card is drawn at random from a pack of 52 playing cards. Find the probability of drawing
   (a) a 'King',
   (b) a red card,
   (c) the seven of clubs,
   (d) either the King, Queen or Jack of diamonds.

2. A fair die is thrown once. Find the probability of obtaining
   (a) a six,
   (b) an even number,
   (c) a number greater than 3,
   (d) a three or a five.

3. A 10p and a 5p coin are tossed at the same time. List all the possible outcomes. Find the probability of obtaining
   (a) two heads,   (b) a head and a tail.

4. A bag contains 6 red balls and 4 green balls.
   (a) Find the probability of selecting at random:
       (i) a red ball   (ii) a green ball.
   (b) One red ball is removed from the bag. Find the new probability of selecting at random
       (i) a red ball   (ii) a green ball.

5. A 'hand' of 13 cards contains the cards shown.

A card is selected at random from the 13. Find the probability of selecting:
   (a) any card of the heart suit,
   (b) any card of the club suit,
   (c) a 'six' of any suit,
   (d) any 'picture' card [not including an ace],
   (e) the 'four' of clubs,
   (f) an 'eight' of any suit,
   (g) any 'six' or 'four'.

6. One letter is selected at random from the word 'UNNECESSARY'. Find the probability of selecting
   (a) an R   (b) an E
   (c) an O   (d) a C

7. The King, Queen and Jack of clubs are removed from a pack of 52 playing cards. One card is selected at random from the remaining cards. Find the probability that the card is:
   (a) a heart   (b) a King
   (c) a club   (d) the 10 of hearts.

8. Three coins are tossed at the same time. List all the possible outcomes. Find the probability of obtaining
   (a) three heads,
   (b) two heads and one tail,
   (c) no heads,
   (d) at least one head.

9. A bag contains 10 red balls, 5 blue balls and 7 green balls. Find the probability of selecting at random:
   (a) a red ball,
   (b) a green ball,
   (c) a blue *or* a red ball,
   (d) a red *or* a green ball.

10. Cards with the numbers 2 to 101 are placed in a hat. Find the probability of selecting:
    (a) an even number,
    (b) a number less than 14,
    (c) a square number,
    (d) a prime number less than 20.

11. A red die and a blue die are thrown at the same time. List all the possible outcomes in a systematic way . Find the probability of obtaining:
    (a) a total of 10,
    (b) a total of 12,
    (c) a total less than 6,
    (d) the same number of both dice,
    (e) a total more than 9.
    What is the most likely total?

12. A die is thrown; when the result has been recorded, the die is thrown a second time. Display all the possible outcomes of the two throws. Find the probability of obtaining:
    (a) a total of 4 from the two throws,
    (b) a total of 8 from the two throws,
    (c) a total between 5 and 9 inclusive from the two throws,
    (d) a number on the second throw which is double the number on the first throw,
    (e) a number on the second throw which is four times the number on the first throw.

13. Find the probability of the following:
    (a) throwing a number less than 8 on a single die,
    (b) obtaining the same number of heads and tails when five coins are tossed,
    (c) selecting a square number from the set A = {4, 9, 16, 25, 36, 49},
    (d) selecting a prime number from the set A.

14. Four coins are tossed at the same time. List all the possible outcomes in a systematic way. Find the probability of obtaining:
    (a) two heads and two tails,
    (b) four tails,
    (c) at least one tail,
    (d) three heads and one tail.

15. Louise buys five raffle tickets out of 1000 sold. She does not win first prize. What is the probability that she wins second prize?

16. Tickets numbered 1 to 1000 were sold in a raffle for which there was one prize. Mr Kahn bought all the tickets containing at least one '3' because '3' was his lucky number. What was the probability of Mr Kahn winning?

17. One ball is selected at random from a bag containing 12 balls of which x are white.
    (a) What is the probability of selecting a white ball?
    When a further 6 white balls are added the probability of selecting a white ball is doubled.
    (b) Find x.

18. Two dice and two coins are thrown at the same time. Find the probability of obtaining:
    (a) two heads and a total of 12 on the dice,
    (b) a head, a tail and a total of 9 on the dice,
    (c) two tails and a total of 3 on the dice.
    What is the most likely outcome?

19. A red, a blue and a white die are all thrown at the same time. Display all the possible outcomes in a suitable way. Find the probability of obtaining:
    (a) a total of 18 on the three dice,
    (b) a total of 4 on the three dice,
    (c) a total of 10 on the three dice,
    (d) a total of 15 on the three dice,
    (e) a total of 7 on the three dice,
    (f) the same number on each die.

10.5 EXCLUSIVE AND INDEPENDENT EVENTS

Two events are *exclusive* if they cannot occur at the same time:
e.g. Selecting an 'ace' or selecting a 'ten' from a pack of cards.

The 'OR' rule:

For exclusive events A and B

$p(\text{A or B}) = p(\text{A}) + p(\text{B})$

Two events are *independent* if the occurrence of one event is unaffected by the occurrence of the other.
e.g. Obtaining a 'head' on one coin, and a tail on another coin when the coins are tossed at the same time.

The 'AND' rule:

$p(\text{A } and \text{ B}) = p(\text{A}) \cdot p(\text{B})$

where $p(\text{A})$ = probability of A occurring etc.
This is the multiplication law.

Example 1

One ball is selected at random from a bag containing 5 red balls, 2 yellow balls and 4 white balls. Find the probability of selecting a red ball or a white ball.

The two events are exclusive.

$p(\text{red ball } or \text{ white ball}) = p(\text{red}) + p(\text{white})$
$$= \tfrac{5}{11} + \tfrac{4}{11}$$
$$= \tfrac{9}{11}.$$

Example 2

A fair coin is tossed and a fair die is rolled. Find the probability of obtaining a 'head' and a 'six'.

The two events are independent

$p(\text{head } and \text{ six}) = p(\text{head}) \times p(\text{six})$
$$= \tfrac{1}{2} \times \tfrac{1}{6}$$
$$= \tfrac{1}{12}.$$

Exercise 5

1. A card is drawn from a pack of playing cards and a die is thrown. Events A and B are as follows:
   A : 'a Jack is drawn from the pack'
   B : 'a three is thrown on the die'.
   (a) Write down the values of $p(\text{A})$, $p(\text{B})$.
   (b) Write down the value of $p(\text{A and B})$.

2. A coin is tossed and a die is thrown. Write down the probability of obtaining
   (a) a 'head' on the coin,
   (b) an odd number on the die,
   (c) a 'head' on the coin and an odd number on the die.

3. A card is drawn from a pack of playing cards. Events X and Y are as follows:
   X : 'a King is drawn'
   Y : 'a Queen is drawn'.
   Write down the values of
   (a) $p(\text{X})$ (b) $p(\text{Y})$ (c) $p(\text{X or Y})$.

4. In an experiment, a card is drawn from a pack of playing cards and a die is thrown.
   Find the probability of obtaining:
   (a) A card which is an ace and a six on the die,
   (b) the king of clubs and an even number on the die,
   (c) a heart and a 'one' on the die.

5. A card is taken at random from a pack of playing cards and replaced. After shuffling, a second card is selected. Find the probability of obtaining:
   (a) two cards which are clubs,
   (b) two Kings,
   (c) two picture cards,
   (d) a heart and a club in any order,
   (e) the ace of diamonds and the ace of spades in any order.

6. A ball is selected at random from a bag containing 3 red balls, 4 black balls and 5 green balls. The first ball is replaced and a second is selected. Find the probability of obtaining:
   (a) two red balls,
   (b) two green balls.

7. The letters of the word 'INDEPENDENT' are written on individual cards and the cards are put into a box. A card is selected and then replaced and then a second card is selected. Find the probability of obtaining:
   (a) the letter 'P' twice,
   (b) the letter 'E' twice.

8. Three coins are tossed and two dice are thrown at the same time. Find the probability of obtaining:
   (a) three heads and a total of 12 on the dice,
   (b) three tails and a total of 9 on the dice.

9. When a golfer plays any hole, he will take 3, 4, 5, 6, or 7 strokes with probabilities of $\frac{1}{10}, \frac{1}{5}, \frac{2}{5}, \frac{1}{5}$ and $\frac{1}{10}$ respectively. He never takes more than 7 strokes. Find the probability of the following events:
   (a) scoring 4 on each of the first three holes,
   (b) scoring 3, 4 and 5 (in that order) on the first three holes,
   (c) scoring a total of 28 for the first four holes,
   (d) scoring a total of 10 for the first three holes,
   (e) scoring a total of 20 for the first three holes.

10. If a hedgehog crosses a certain road before 7.00 a.m., the probability of being run over is $\frac{1}{10}$. After 7.00 a.m., the corresponding probability is $\frac{3}{4}$. The probability of the hedgehog waking up early enough to cross before 7.00 a.m., is $\frac{4}{5}$.

What is the probability of the following events:
   (a) the hedgehog waking up too late to reach the road before 7.00 a.m.,
   (b) the hedgehog waking up early and crossing the road in safety,
   (c) the hedgehog waking up late and crossing the road in safety,
   (d) the hedgehog waking up early and being run over,
   (e) the hedgehog crossing the road in safety.

11. A coin is biased so that it shows 'Heads' with a probability of $\frac{2}{3}$. The same coin is tossed three times. Find the probability of obtaining.
   (a) two tails on the first two tosses,
   (b) a head, a tail and a head (in that order),
   (c) two heads and one tail (in any order).

10.6   TREE DIAGRAMS

Example 1

A bag contains 5 red balls and 3 green balls. A ball is drawn at random and then replaced. Another ball is drawn.

What is the probability that both balls are green?

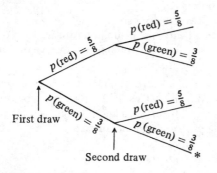

The branch marked* involves the selection of a green ball twice.

The probability of this event is obtained by simply multiplying the fractions on the two branches.

$$\therefore \quad p\,(\text{two green balls}) = \tfrac{3}{8} \times \tfrac{3}{8} = \tfrac{9}{64}$$

Example 2

A bag contains 5 red balls and 3 green balls. A ball is selected at random and not replaced. A second ball is then selected. Find the probability of selecting
(a) two green balls
(b) one red ball and one green ball.

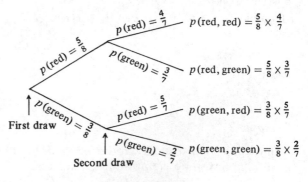

(a) $p\,(\text{two green balls}) = \tfrac{3}{8} \times \tfrac{2}{7}$

$\qquad\qquad\qquad\qquad = \tfrac{3}{28}$.

(b) $p\,(\text{one red, one green}) = (\tfrac{5}{8} \times \tfrac{3}{7}) + (\tfrac{3}{8} \times \tfrac{5}{7})$

$\qquad\qquad\qquad\qquad\qquad = \tfrac{15}{28}$.

Exercise 6

1. A bag contains 10 discs; 7 are black and 3 white. A disc is selected, and then replaced. A second disc is selected. Copy and complete the tree diagram showing all the probabilities and outcomes.

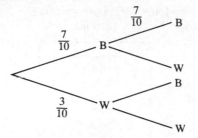

Find the probability of the following:
(a) both discs are black,
(b) both discs are white.

2. A bag contains 5 red balls and 3 green balls. A ball is drawn and then replaced before a ball is drawn again. Draw a tree diagram to show all the possible outcomes. Find the probability that
(a) two green balls are drawn,
(b) the first ball is red and the second is green.

3. A bag contains 7 green discs and 3 blue discs. A disc is drawn and *not* replaced.
A second disc is drawn. Copy and complete the tree diagram.

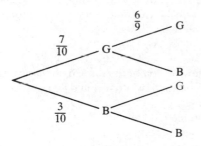

Find the probability that
(a) both discs are green,
(b) both discs are blue.

4. A bag contains 5 red balls, 3 blue balls and 2 yellow balls. A ball is drawn and not replaced. A second ball is drawn. Find the probability of drawing:
(a) two red balls,
(b) one blue ball and one yellow ball,
(c) two yellow balls,
(d) two balls of the same colour.

5. A bag contains 4 red balls, 2 green balls and 3 blue balls. A ball is drawn and not replaced. A second ball is drawn. Find the probability of drawing:
(a) two blue balls,
(b) two red balls,
(c) one red ball and one blue ball,
(d) one green ball and one red ball.

6. A six-sided die is thrown three times. Draw a tree diagram, showing at each branch the two events: 'six' and 'not six'. What is the probability of throwing a total of:
(a) three sixes,
(b) no sixes,
(c) one six,
(d) at least one six (use part (b)).

7. A card is drawn at random from a pack of 52 playing cards. The card is replaced and a second card is drawn. This card is replaced and a third card is drawn. What is the probability of drawing:
(a) three hearts?
(b) at least two hearts?
(c) exactly one heart?

8. A bag contains 6 red marbles and 4 blue marbles. A marble is drawn at random and not replaced. Two further draws are made, again without replacement. Find the probability of drawing:
(a) three red marbles,
(b) three blue marbles,
(c) no red marbles,
(d) at least one red marble.

9. When a cutting is taken from a geranium the probability that it grows is $\frac{3}{4}$. Three cuttings are taken. What is the probability that
(a) all three grow,
(b) none of them grow?

10. A die has its six faces marked 0, 1, 1, 1, 6, 6. Two of these dice are thrown together and the total score is recorded. Draw a tree diagram.
(a) How many different totals are possible?
(b) What is the probability of obtaining a total of 7?

11. A coin is biased so that the probability of a 'head' is $\frac{3}{4}$. Find the probability that, when tossed three times, it shows:
(a) three tails,
(b) two heads and one tail,
(c) one head and two tails,
(d) no tails.
Write down the sum of the probabilities in (a), (b), (c) and (d).

12. A teacher decides to award exam grades A, B or C by a new fairer method. Out of 20 children, three are to receive A'S, five B's and the rest C's. She writes the letters A, B and C on 20 pieces of paper and invites the pupils to draw their exam result, going through the class in alphabetical order. Find the probability that:
   (a) the first three pupils all get grade 'A'.
   (b) the first three pupils all get grade 'B',
   (c) the first three pupils all get different grades,
   (d) the first four pupils all get grade B.
   (Do not cancel down the fractions.)

13. The probability that an amateur golfer actually hits the ball is (regrettably for all concerned) only $\frac{1}{10}$. If four separate attempts are made, find the probability that the ball will be hit:
   (a) four times,
   (b) at least twice,
   (c) not at all.

14. A box contains x milk chocolates and y plain chocolates. Two chocolates are selected at random. Find, in terms of x and y, the probability of choosing:
   (a) a milk chocolate on the first choice,
   (b) two milk chocolates,
   (c) one of each sort,
   (d) two plain chocolates.

15. A pack of z cards contains x 'winning cards'. Two cards are selected at random. Find, in terms of x and z, the probability of choosing:
   (a) a 'winning' card on the first choice,
   (b) two 'winning cards' in the two selections,
   (c) exactly one 'winning' card in the pair.

16. Bag A contains 3 red balls and 3 blue balls. Bag B contains 1 red ball and 3 blue balls.

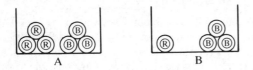

   A ball is taken at random from bag A and placed in bag B. A ball is then chosen from bag B. What is the probability that the ball taken from B is red?

17. On a Monday or a Thursday, Mr Gibson paints a 'masterpiece' with a probability of $\frac{1}{5}$. On any other day, the probability of producing a 'masterpiece' is $\frac{1}{100}$. In common with other great painters, Gibson never knows what day it is. Find the probability that on one day chosen at random, he will in fact paint a masterpiece.

18. Two dice, each with four faces marked 1, 2, 3 and 4, are thrown together.
   (a) What is the most likely total score on the faces pointing downwards?
   (b) What is the probability of obtaining this score on three successive throws of the two dice?

19. In the Venn diagram, $\mathscr{E}$ = {pupils in a class of 15}, G = {girls}, S = {swimmers}, F = {pupils who believe in Father Christmas}.

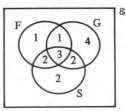

   A pupil is chosen at random. Find the probability that the pupil:
   (a) can swim,
   (b) is a girl swimmer,
   (c) is a boy swimmer who believes in Father Christmas.
   Two pupils are chosen at random. Find the probability that:
   (d) both are boys,
   (e) neither can swim,
   (f) both are girl swimmers who believe in Father Christmas.

20. A bag contains 3 red, 4 white and 5 green balls. Three balls are selected without replacement. Find the probability that the three balls chosen are:
   (a) all red,
   (b) all green,
   (c) one of each colour.
   If the selection of the three balls was carried out 1100 times, how often would you expect to choose:
   (d) three red balls?
   (e) one of each colour?

21. There are 1000 components in a box of which 10 are known to be defective. Two components are selected at random. What is the probability that:
   (a) both are defective,
   (b) neither are defective,
   (c) just one is defective?
   (Do *not* simplify your answers)

22. There are 10 boys and 15 girls in a class. Two children are chosen at random. What is the probability that:
   (a) both are boys,
   (b) both are girls,
   (d) one is a boy and one is a girl?

23. There are 500 ball bearings in a box of which 100 are known to be undersize. Three ball bearings are selected at random. What is the probability that:
(a) all three are undersize,
(b) none are undersize?
Give your answers as decimals correct to three significant figures.

24. There are 9 boys and 15 girls in a class. Three children are chosen at random. What is the probability that:
(a) all three are boys,
(b) all three are girls,
(c) one is a boy and two are girls?
Give your answers as fractions.

REVISION EXERCISE 10A

1. A pie chart is drawn with sectors to represent the following percentages:

   20%, 45%, 30%, 5%.

What is the angle of the sector which represents 45%?

2. The pie chart shows the numbers of votes for candidates A, B and C in an election.
What percentage of the votes were cast in favour of candidate C?

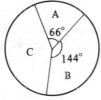

3. A pie chart is drawn showing the expenditure of a football club as follows:

Wages	£41,000
Travel	£9,000
Rates	£6,000
Miscellaneous	£4,000

What is the angle of the sector showing the expenditure on travel?

4. The mean of four numbers is 21.
(a) Calculate the sum of the four numbers,
Six other numbers have a mean of 18.
(b) Calculate the mean of the ten numbers.

5. Find
(a) the mean,
(b) the median,
(c) the mode,
of the numbers 3, 1, 5, 4, 3, 8, 2, 3, 4, 1.

6.

Marks	3	4	5	6	7	8
Number of pupils	2	3	6	4	3	2

The table shows the number of pupils in a class who scored marks 3 to 8 in a test. Find
(a) the mean mark,
(b) the modal mark,
(c) the median mark.

7. The mean height of 10 boys is 1·60 m and the mean height of 15 girls is 1·52 m. Find the mean height of the 25 boys and girls.

8.

Mark	3	4	5
Number of pupils	3	x	4

The table shows the number of pupils who scored marks 3, 4 or 5 in a test. Given that the mean mark is 4·1, find x.

9. When two dice are thrown simultaneously, what is the probability of obtaining the same number on both dice?

10. A bag contains 20 discs of equal size of which 12 are red, x are blue and the rest are white.
(a) If the probability of selecting a blue disc is $\frac{1}{4}$, find x.
(b) A disc is drawn and then replaced. A second disc is drawn. Find the probability that neither disc is red.

11. Three dice are thrown. What is the probability that none of them shows a 1 or a 6?

12. A coin is tossed four times. What is the probability of obtaining at least three 'heads'?

13. A bag contains 8 balls of which 2 are red and 6 are white. A ball is selected and not replaced. A second ball is selected. Find the probability of obtaining:
(a) two red balls,
(b) two white balls,
(d) one ball of each colour.

14. A bag contains x green discs and 5 blue discs. A disc is selected and replaced. A second disc is drawn. Find, in terms of x, the probability of selecting:
(a) a green disc on the first draw,
(b) a green disc on the first and second draws.
Find the corresponding probabilities when the first disc is *not* replaced.

15. In a group of 20 people, 5 cannot swim. If two people are selected at random, what is the probability that neither of them can swim?

16. (a) What is the probability of winning the toss in five consecutive hockey matches?
    (b) What is the probability of winning the toss in all the matches in the FA cup from the first round to the final (i.e. 8 matches)?

17. Mr and Mrs Stringer have three children. What is the probability that:
    (a) all the children are boys,
    (b) there are more girls than boys?
    (Assume that a boy is as likely as a girl.)

18. The probability that it will be wet today is $\frac{1}{6}$. If it is dry today, the probability that it will be wet tomorrow is $\frac{1}{8}$. What is the probability that both today and tomorrow will be dry?

19. Two dice are thrown. What is the probability that the *product* of the numbers on top is
    (a) 12,   (b) 4,   (c) 11?

20. The probability of snow on Christmas day is $\frac{1}{20}$. What is the probability that snow will fall on the next three Christmas days?

EXAMINATION EXERCISE 10B

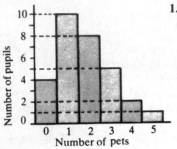

1. Each of the 30 pupils in a class were asked how many pets they had. The results are illustrated in the diagram.

   (a) Find the number of pupils who had two or more pets.
   (b) Find the total number of pets that the 30 pupils had.
   (c) Calculate the mean number of pets.   [M]

2. Sarah takes home £120 per week in her pay-packet. She divides her money up as follows:

   Rent   £25   Car Expenses   £20   Clothes   £15
   Food   £35   Entertainment   £5   Save what is left.

   (a) How much does she save each week?
   (b) Draw a pie-chart to show how she distributes her £120.
   In December she receives a £50 bonus with her pay. She treats herself to a few luxuries and spends all she has left on Christmas presents.
   The bar chart below shows her spending except presents.

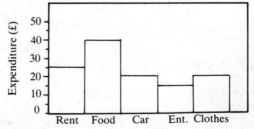

   (c) How much did she spend on presents?
   In January she is to receive a 10% increase in her pay. At the same time she increases all her spending by 10%.
   (d) What angle will savings be on a pie-chart showing how she uses her money?
                                                                  [S]

3. A student asked 30 people arriving at a football ground how long, to the nearest minute, it had taken them to reach the ground. The times they gave (in minutes) are listed below.

35 41 22 15 31 19 12 12 23 30
30 38 36 24 14 20 20 16 15 22
34 28 25 13 19   9 27 17 21 25

(a) (i) Copy and complete the following frequency table.

Time taken in minutes (to nearest minute)	8–12	13–17	18–22	23–27	28–32	33–37	38–42
Number of people	3	6	7	5	4		

   (ii) Draw a histogram to represent the information in the frequency table.

(b) Of the 30 people questioned,
   6 paid £2 each to see the football match,
   8 paid £3 each,
   4 paid £4 each,
   10 paid £5 each and
   2 paid £6 each.
   (i) Calculate the total amount paid by these 30 people.
   (ii) Calculate the mean amount paid by these 30 people. [M]

4. (*The whole of this question should be answered on graph paper.*)
Potatoes are supplied to a greengrocer's shop in 50 kg bags. Each of the potatoes in one of these bags was weighed to the nearest gram, and the following table drawn up.

Mass (g)	50–99	100–149	150–199	200–249	250–299	300–349
Number of potatoes	5	53	87	73	33	3

(a) (i) Complete the sequence 5, 58, ... of cumulative frequencies for these classes.
   (ii) Using scales of 2 cm to 50 g and 2 cm to 50 potatoes draw a cumulative frequency graph for the data.
(b) From your graph estimate
   (i) the median mass,
   (ii) the number of potatoes weighing at least 225 g each.
(c) For a party, Harry requires 50 baking potatoes, each weighing at least 225 g. The greengrocer sells the potatoes by the kilogram without special selection.
   (i) Use your answer to (b)(ii) to estimate how many kilograms Harry will have to buy.
   (ii) Taking the mean mass of the 50 baking potatoes to be 260 g, estimate how many kilograms of potatoes he will have left over. [N]

5. In an examination there were 5000 candidates and the marks they obtained are summarised in the table below.

Mark obtained	20 or less	30 or less	40 or less	50 or less	60 or less	70 or less	80 or less	100 or less
Number of candidates	250	500	1000	2000	3200	4000	4500	5000

(a) Draw a cumulative frequency diagram to represent these results. (Use a scale of 2 cm to represent 20 marks on the *x*-axis and 2 cm to represent 500 candidates on the cumulative frequency axis.)

(b) Using the cumulative frequency diagram , or otherwise, find
  (i) the number of candidates who scored 82 marks or less,
  (ii) the median mark,
  (iii) the interquartile range,
  (iv) the percentage of candidates who scored more than 70 marks,
  (v) the minimum mark required for a pass if 80% of the candidates passed the examination.   [M]

6.

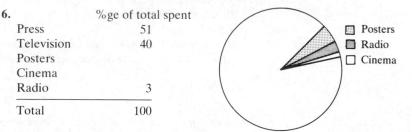

	%ge of total spent
Press	51
Television	40
Posters	
Cinema	
Radio	3
Total	100

In Britain money was spent on advertisements in 1983 in the press, TV, posters, etc. The incomplete table and pie-chart show the way this was divided between the media.

(a) Calculate the angle of the sector representing television, and complete the pie-chart.

(b) The angle of the sector representing posters is 18°. Calculate the percentage spent on posters, and hence complete the table.   [L]

7. (a) Mary is one of the children in a class of 30. 7 of the children are left-handed and 10 of them wear glasses.
    (i) Write down the probability of each of the following:
    Mary is left handed;
    Mary is right handed;
    Mary wears glasses;
    Mary does not wear glasses.
    (ii) Which one of the above events is the most likely?
    (iii) Which one of the above events is the least likely?

  (b) The local rugby club was allocated 10 stand tickets and 15 enclosure tickets for the Wales v England match. The tickets are placed in a bag and then drawn at random from the bag for distribution.
  Find the probability that
    (i) the first ticket drawn is an enclosure ticket,
    (ii) the second ticket drawn is a stand ticket given that the first ticket was an enclosure ticket,
    (iii) the two first tickets drawn are stand tickets.   [W]

8. A bag contains five sweets. They all look the same. However, two have hard centres (H) and three have soft centres (S). Balbir chooses a sweet and eats it. Then Winston chooses a sweet and eats it.

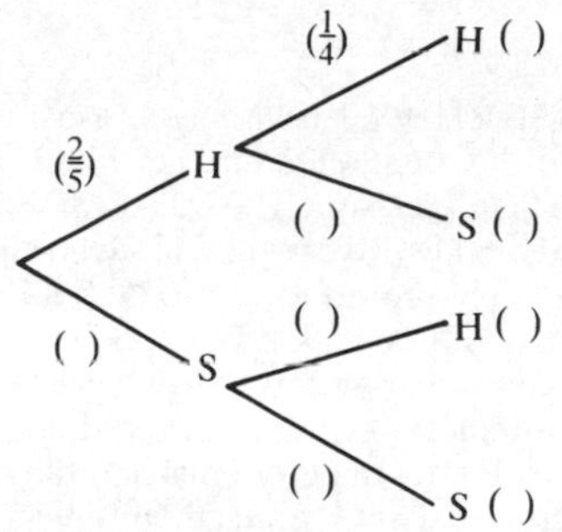

(a) Copy and complete the tree diagram showing clearly all probabilities and outcomes.

(b) Find the probabilities of the outcome when
   (i) both choose soft sweets;
   (ii) one has a hard sweet and one has a soft sweet.

[M]

9. In a board game a counter is moved on a rectangular grid according to the score on a die, as in the table below.

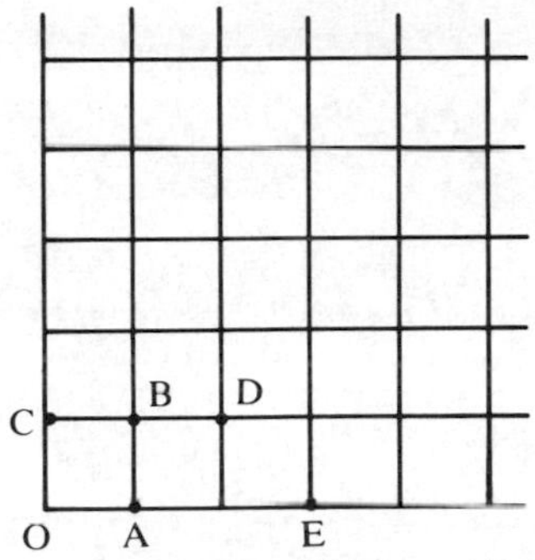

Score	Move
1, 3 or 5	One place to right
2 or 4	One place upwards
6	One place to right then one place upwards

The counter starts at O, the bottom left hand corner of the board. The diagram shows part of the board.

(a) Write down the probability that after one throw the counter is at
   (i) A   (ii) B   (iii) C
(b) Find the probability that after two throws the counter is at D.
(c) Find the probability that after three throws the counter is at E.   [N]

10. In order to decide the order of play in a tennis competition, the names of the six competitors are drawn at random from a hat. The competitors are 2 boys — Alan and Imran — and 4 girls — Carol, Debbie, Aysha and Fiona.
(a) Find the probability that the first name drawn is Alan.
(b) Find the probability that the first name drawn is that of a boy.
(c) Find the probability that the first two names drawn are both girls.   [M]

11. In a game to select a winner from three friends Arshad, Belinda and Connie, Arshad and Belinda both roll a normal die. If Arshad scores a number greater than 2 *and* Belinda throws an odd number, then Arshad is the winner. Otherwise Arshad is eliminated and Connie then rolls the die. If the die shows an odd number Connie is the winner, otherwise Belinda is the winner.
(a) Calculate the probability that
   (i) Arshad will be the winner,
   (ii) Connie will roll the die,
   (iii) Connie will be the winner.
(b) Is this a fair game? Give a reason for your answer.   [L]

11 Investigations, Practical problems, Puzzles

William Shockley Every time you use a calculator you are making use of integrated circuits which were developed from the first transistor. The transistor was invented by William Shockley, working with two scientists, in 1947. The three men shared the 1956 Nobel Prize for physics. The story of the invention is a good example of how mathematics can be used to solve practical problems.

The first electronic computers did not make use of transistors or integrated circuits and they were so big that they occupied whole rooms themselves. A modern computer which can carry out just the same functions can be carried around in a brief case.

INVESTIGATIONS

There are a large number of possible starting points for investigations here so it may be possible to allow students to choose investigations which appeal to them. On other occasions the same investigation may be set to a whole class.

Here are a few guidelines for pupils:

(a) If the set problem is too complicated try an easier case;

(b) Draw your own diagrams;

(c) Make tables of your results and be systematic;

(d) Look for patterns;

(e) Is there a rule or formula to describe the results?

(f) Can you *predict* further results?

(g) Can you *prove* any rules which you may find?

1. Opposite corners

Here the numbers are arranged in 10 columns.

1	2	3	4	5	6	7	8	9	10
11	12	13	14	15	16	17	18	19	20
21	22	23	24	25	26	27	28	29	30
31	32	33	34	35	36	37	38	39	40
41	42	43	44	45	46	47	48	49	50
51	52	53	54	55	56	57	58	59	60
61	62	63	64	65	66	67	68	69	70
71	72	73	74	75	76	77	78	79	80
81	82	83	84	85	86	87	88	89	90
91	92	93	94	95	96	97	98	99	100

In the 2×2 square

$7 \times 18 = 126$
$8 \times 17 = 136$

the difference between them is 10.

7	8
17	18

In the 3×3 square

$12 \times 34 = 408$

$14 \times 32 = 448$

the difference between them is 40.

12	13	14
22	23	24
32	33	34

Investigate to see if you can find any rules or patterns connecting the size of square chosen and the difference.

If you find a rule, use it to *predict* the difference for larger squares.

Test your rule by looking at squares like 8×8 or 9×9.

Can you *generalise* the rule?
[What is the difference for a square of size $n \times n$?]

Can you *prove* the rule?
Hint:
In a 3×3 square . . .

x	?
?	?

What happens if the numbers are arranged in six columns or seven columns?

1	2	3	4	5	6
7	8	9	10	11	12
13	14	15	16	17	18
19					

1	2	3	4	5	6	7
8	9	10	11	12	13	14
15	16	17	18	19	20	21
22						

2. Weighing scales

In the diagram we are measuring the weight of the package *x* using two weights.

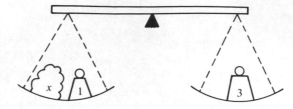

If the scales are balanced, *x* must be 2 kg.

Show how you can measure all the weights from 1 kg to 10 kg using three weights: 1 kg, 3 kg, 6 kg.

It is possible to measure all the weights from 1 kg to 13 kg using a different set of three weights. What are the three weights?

It is possible to measure all the weights from 1 kg to 40 kg using four weights. What are the weights?

3. Buying stamps

You have one 1p, one 2p, one 5p and one 10p coin.

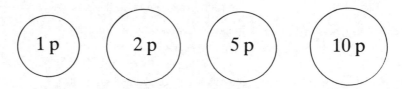

You can buy stamps of any value you like, but you must give the exact money.

How many different value stamps can you buy?

Suppose you now have one 1p, one 2p, one 5p, one 10p, one 20p, one 50p and one £1 coin.

How many different value stamps can you buy now?

4. Frogs

This is a game invented by a French mathematician called Lucas.

Object: To swap the positions of the discs so that they end up the other way round (with a space in the middle).

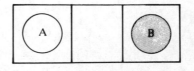

Rules 1. A disc can slide one square in either direction onto an empty square.
2. A disc can hop over one adjacent disc of the other colour provided it can land on an empty square.

Example (a) Slide Ⓐ one square to the right.

(b) Ⓑ hops over Ⓐ to the left.

(c) Slide Ⓐ one square to the right.

We took 3 moves.

1. Look at the diagram. What is the smallest number of moves needed for two discs of each colour?

2. Now try three discs of each colour. Can you complete the task in 15 moves?

3. Try four discs of each colour.
    Now look at your results and try to find a formula which gives the least number of moves needed for any number of discs x. It may help if you count the number of 'hops' and 'slides' separately.

4. Try the game with a different number of discs on each side. Say two reds and three blues. Play the game with different combinations and again try to find a formula giving the number of moves for x discs of one colour and y discs of another colour.

5. Triples

In this investigation a *triple* consists of three whole numbers in a definite order. For example, (4, 2, 1) is a triple and (1, 4, 2) is a different triple.

The three numbers in a triple do not have to be different. For example, (2, 2, 3) is a triple but (2, 0, 1) is not a triple because 0 is not allowed.

The *sum* of a triple is found by adding the three numbers together. So the sum of (4, 2, 1) is 7.

Investigate how many different triples there are with a given sum.
See what happens to the number of different triples as the sum is changed.

If you find any pattern, try to explain why it occurs.

How many different triples are there whose sum is 22?

6. Mystic rose

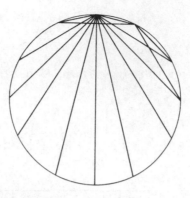

Straight lines are drawn between each of the 12 points on the circle. Every point is joined to every other point. How many straight lines are there?

Suppose we draw a mystic rose with 24 points on the circle. How many straight lines are there?

How many straight lines would there be with n points on the circle?

7. Knockout competition

Eight teams reach the 'knockout' stage of the World Cup.

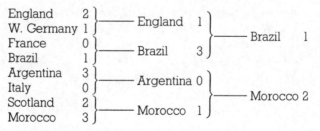

How would you organise a knockout competition if there were 12 teams? Or 15?

How many matches are played up to and including the final if there are
(a) 8 teams,
(b) 12 teams,
(c) 15 teams,
(d) 23 teams,
(e) n teams?

In a major tournament like Wimbledon, the better players are seeded from 1 to 16. Can you organise a tournament for 32 players so that, if they win all their games,
(a) seeds 1 and 2 can meet in the final,
(b) seeds 1, 2, 3 and 4 can meet in the semi-finals,
(c) seeds 1, 2, 3, 4, 5, 6, 7, 8 can meet in the quarter-finals?

8. Discs

(a) You have five black discs and five white discs which are arranged in a line as shown.

We want to get all the black discs to the right-hand end and all the white discs to the left-hand end.

The only move allowed is to interchange two neighbouring discs.

  becomes

How many moves does it take?
How many moves would it take if we had fifty black discs and fifty white discs arranged atternately?

(b) Suppose the discs are arranged in pairs

 ... etc.

How many moves would it take if we had fifty black discs and fifty white discs arranged like this?

[Hint: In both cases work with a smaller number of discs until you can see a pattern].

(c) Now suppose you have three colours black, white and green arranged alternately.

(B) (W) (G) (B) (W) (G) (B) ... etc.

You want to get all the black discs to the right, the green discs to the left and the white discs in the middle.
How many moves would it take if you have 30 discs of each colour?

9. Chess board

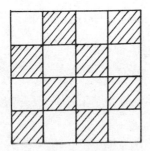

Start with a small board, just 4 × 4.

How many squares are there? [It is not just 16!]

How many squares are there on an 8 × 8 chess board?

How many squares are there on an $n \times n$ chess board?

10. Area and perimeter

This is about finding different shapes in which the area is numerically equal to the perimeter.

This rectangle has an area of 10 square units and a perimeter of 14 units, so we will have to try another one.

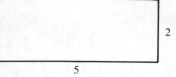

There are some suggestions below but you can investigate shapes of your own choice if you prefer.

(a) Find rectangles with equal area and perimeter. After a while you can try adding on bits like this.

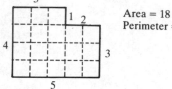

Area = 18
Perimeter = 18

(b) Suppose one dimension of the rectangle is fixed.

In this rectangle the length is 5 units.

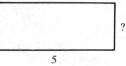

(c) Try right-angled triangles and equilateral triangles.

(d) Try circles, semi-circles and so on.

(e) How about three-dimensional shapes? Now we are looking for cuboids, spheres, cylinders in which the volume is numerically equal to the surface area.

(f) Can you find any connection between the square with equal area and perimeter and the circle with equal area and perimeter. How about the equilateral triangle with equal area and perimeter?

11. Happy numbers (and more)

(a) Take the number 23.
Square the digits and add.

$$2 \quad 3$$
$$2^2 + 3^2 = 1 \quad 3$$
$$\qquad 1^2 + 3^2 = 1 \quad 0$$
$$\qquad\qquad 1^2 + 0^2 = 1$$

The sequence ends at 1 and we call 23 a 'happy' number.

Investigate for other numbers. Here are a few suggestions: 70, 85, 49, 44, 14, 15, 94.

(b) Now change the rule. Instead of squaring the digits we will cube them.

$$2 \quad 1$$
$$2^3 + 1^3 = 0 \quad 9$$
$$\qquad 0^3 + 9^3 = 7 \quad 2 \quad 9$$
$$\qquad\qquad 7^3 + 2^3 + 9^3 = 1 \quad 0 \quad 8 \quad 0$$
$$\qquad\qquad\qquad 1^3 + 0^3 + 8^3 + 0^3 = 5 \quad 1 \quad 3$$
$$\qquad\qquad\qquad\qquad 5^3 + 1^3 + 3^3 = 153$$

And now we are stuck because 153 leads to 153 again.

Investigate for numbers of your own choice. Do any numbers lead to 1?

12. Prime numbers

Write all the numbers from 1 to 104 in eight columns and draw a ring around the prime numbers 2, 3, 5 and 7.

1	②	③	4	⑤	6	⑦	8
9	10	11	12	13	14	15	16
17	18	19	20	21	22	23	24
25							

If we cross out all the multiples of 2, 3, 5 and 7, we will be left with all the prime numbers below 104. Can you see why this works?

Draw *four* lines to eliminate the multiples of 2.
Draw *six* lines to eliminate the multiples of 3.
Draw *two* lines to eliminate the multiples of 7.
Cross out all the numbers ending in 5.

Put a ring around all the prime numbers less than 104.
[Check there are 27 numbers].

Many prime numbers can be written as the sum of two squares. For example $5 = 2^2 + 1^2$, $13 = 3^2 + 2^2$. Find all the prime numbers in your table which can be written as the sum of two squares. Draw a red ring around them in the table. What do you notice?
Check any 'gaps' you may have found.

Extend the table up to 200 and see if the pattern continues. In this case you will need to eliminate the multiples of 11 and 13 as well.

13. Squares

For this investigation you need either dotted paper or squared paper.

The shaded square
has an area
of 1 unit.

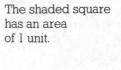
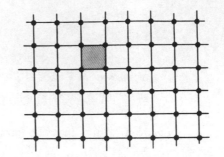

Can you draw a square, with its corners on the dots, with an area of 2 units?
Can you draw a square with an area of 3 units?
Can you draw a square with an area of 4 units?

Investigate for squares up to 100 units.

For which numbers x can you draw a square of area x units?

14. Painting cubes

The large cube below consists of 27 unit cubes.

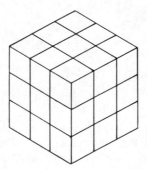

All six faces of the large cube are painted green.

How many unit cubes have 3 green faces?
How many unit cubes have 2 green faces?
How many unit cubes have 1 green face?
How many unit cubes have 0 green faces?

Suppose the large cube is $20 \times 20 \times 20$.
Answer the four questions above.

Answer the four questions for the cube which is $n \times n \times n$.

15. Final score

The final score in a football match was 3-2. How many different scores were possible at half-time?

Investigate for other final scores where the difference between the teams is always one goal. [1-0, 5-4 etc]. Is there a pattern or rule which would tell you the number of possible half-time scores in a game which finished 58-57?

Suppose the game ends in a draw. Find a rule which would tell you the number of possible half-time scores if the final score was 63-63.

Investigate for other final scores [3-0, 5-1, 4-2 etc].

16. Cutting paper

The rectangle ABCD is cut in half to give two smaller rectangles.

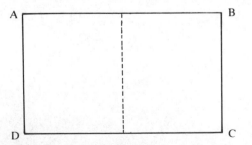

Each of the smaller rectangles is mathematically similar to the large rectangle. Find a rectangle which has this property.

What happens when the small rectangles are cut in half? Do they have the same property?

Why is this a useful shape for paper used in business?

17. Matchstick shapes

(a) Here we have a sequence of matchstick shapes

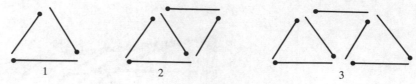

Can you work out the number of matches in the 10th member of the sequence? Or the 20th member of the sequence?
How about the nth member of the sequence?

(b) Now try to answer the same questions for the patterns below. Or you may prefer to design patterns of your own.

(i)

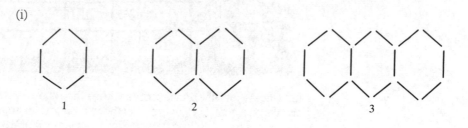

(ii)

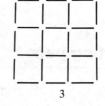

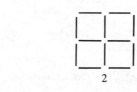

(iii)

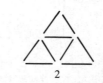

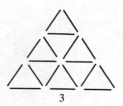

18. Maximum box

(a) You have a square sheet of card 24 cm by 24 cm.
You can make a box (without a lid) by cutting squares from the corners and
folding up the sides.

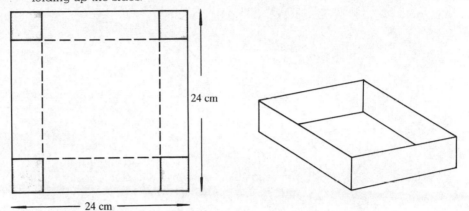

What size corners should you cut out so that the volume of the box is as large
as possible?
Try different sizes for the corners and record the results in a table:

Length of the side of the corner square (cm)	Dimensions of the open box (cm)	Volume of the box (cm^3)
1	$22 \times 22 \times 1$	484
2		
–		
–		

Now consider boxes made from different sized cards:
15 cm $\times$ 15 cm and 20 cm by 20 cm.
What size corners should you cut out this time so that the volume of the box
is as large as possible?

Is there a connection between the size of the corners cut out and the size of
the square card?

(b) Investigate the situation when the card is not square. Take rectangular cards
where the length is twice the width (20 $\times$ 10, 12 $\times$ 6, 18 $\times$ 9 etc.)
Again, for the maximum volume is there a connection between the size of
the corners cut out and the size of the original card?

19. Digit sum

Take the number 134.
Add the digits $1 + 3 + 4 = 8$.
The digit sum of 134 is 8.

Take the number 238.
$2 + 3 + 8 = 13$ [We continue if the sum is more than 9].
    $1 + 3 = 4$
The digit sum of 238 is 4.

Consider the multiples of 3:

Number	3	6	9	12	15	18	21	24	27	30	33	36
Digit sum	3	6	9	3	6	9	3	6	9	3	6	9

The digit sum is always 3, 6, or 9.
These numbers can be shown on a circle.

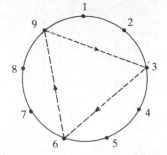

Investigate the pattern of the digit sums for multiples of:

(a) 2   (b) 5   (c) 6   (d) 7   (e) 8
(f) 9   (g) 11   (h) 12   (i) 13.

Is there any connection between numbers where the pattern of the digit sums is the same?

Can you (without doing all the usual working) predict what the pattern would be for multiples of 43? Or 62?

20. An expanding diagram

Look at the series of diagrams below.

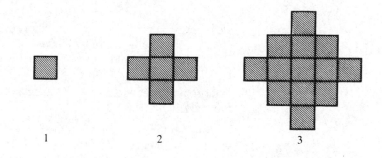

1   2   3

Each time new squares are added all around the outside of the previous diagram.

Draw the next few diagrams in the series and count the number of squares in each one.

How many squares are there in diagram number 15 or in diagram number 50?

What happens if we work in three dimensions? Instead of adding squares we add cubes all around the outside. How many cubes are there in the fifth member of the series or the fifteenth?

21. Fibonacci sequence

Fibonacci was the nickname of the Italian mathematician Leonardo de Pisa (A.D. 1170–1250). The sequence which bears his name has fascinated mathematicians for hundreds of years. You can if you like join the Fibonacci Association which was formed in 1963.

Here is the start of the sequence
1, 1, 2, 3, 5, 8, 13, 21, 34, 55, 89, 144, . . .

There are no prizes for working out the next term!

The sequence has many interesting properties to investigate.
Here are a few suggestions.

(a) Add three terms.
$$1 + 1 + 2, 1 + 2 + 3, \text{ etc.}$$
Add four terms.

(b) Add squares of terms
$$1^2 + 1^2, 1^2 + 2^2, 2^2 + 3^2, \ldots$$

(c) Ratios
$$\frac{1}{1} = 1, \frac{2}{1} = 2, \frac{3}{2} = 1 \cdot 5, \ldots$$

(d) In fours $\boxed{2\ 3\ 5\ 8}$
$$2 \times 8 = 16, 3 \times 5 = 15$$

(e) In threes $\boxed{3\ 5\ 8}$
$$3 \times 8 = 24, 5^2 = 25$$

(f) In sixes $\boxed{1\ 1\ 2\ 3\ 5\ 8}$
square and add the first five numbers
$$1^2 + 1^2 + 2^2 + 3^2 + 5^2 = 40$$
$$5 \times 8 = 40.$$
Now try seven numbers from the sequence, or eight . . .

(g) Take a group of 10 consecutive terms. Compare the sum of the 10 terms with the seventh member of the group.

22. Alphabetical order

A teacher has four names on a piece of paper which are in no particular order (say Smith, Jones, Biggs, Eaton). He wants the names in alphabetical order.

One way of doing this is to interchange each pair of names which are clearly out of order.
So he could start like this;   S   J   B   E
the order becomes: J   S   B   E
He would then interchange S and B.

Using this method, what is the largest number of interchanges he could possibly have to make?

What if he had thirty names, or fifty?

23. Mr Gibson's job

Mr Gibson's job is counting tiles of the black or white variety. When he is bored Mr Gibson counts the tiles by placing them in a pattern consisting of alternate black and white tiles. This one is five tiles across and altogether there are 13 tiles in the pattern.

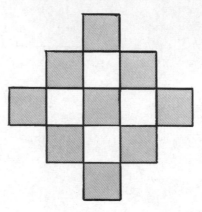

He makes the pattern so that there are always black tiles all around the outside. Draw the pattern which is nine tiles across. You should find that there are 41 tiles in the pattern.

How many tiles are there in the pattern which is 101 tiles across?

24. Diagonals

In a 4 × 7 rectangle the diagonal passes through 10 squares.

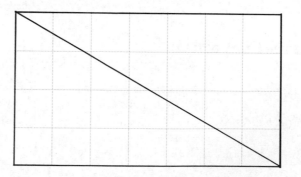

Draw rectangles of your own choice and count the number of squares through which the diagonal passes.

A rectangle is 640 × 250. How many squares will the diagonal pass through?

25. Biggest number

A calculator has the following buttons:

Also the only digits buttons which work are the '1', '2' and '3'.

(a) You can press any button, but only once.
What is the biggest number you can get?

(b) Now the '1', '2', '3' and '4' buttons are working.
What is the biggest number you can get?

(c) Investigate what happens as you increase the number of digits which you can use.

26. What shape tin?

We need a cylindrical tin which will contain a volume of 600 cm³ of drink.

What shape should we make the tin so that we use the minimum amount of metal?
In other words, for a volume of 600 cm³, what is the smallest possible surface area?

Hint: Make a table.

r	h	A
2	?	?
3	?	?
.		
.		
.		

What shape tin should we design to contain a volume of 1000 cm³?

27. Find the connection

Work through the flow diagram several times, using a calculator.

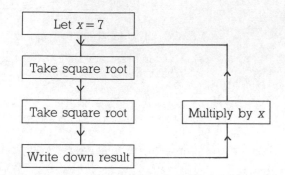

·What do you notice?
Try different numbers for x (suggestions: 11, 5, 8, 27)
What do you notice?

What happens if you take the square root three times?

Suppose in the flow diagram you change
'Multiply by x' to 'Divide by x'. What happens now?

Suppose in the flow diagram you change
'Multiply by x' to 'Multiply by x^2'. What happens now?

28. Spotted shapes

For this investigation you need dotted paper. If you have not got any, you can make your own using a felt tip pen and squared paper.

The rectangle in Diagram 1 has
10 dots on the perimeter ($p = 10$)
and 2 dots inside the shape ($i = 2$).
The area of the shape is
6 square units ($A = 6$)

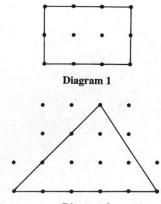

Diagram 1

The triangle in Diagram 2 has
9 dots on the perimeter ($p = 9$)
and 4 dots inside the shape ($i = 4$).
The area of the triangle
is $7\frac{1}{2}$ square units ($A = 7\frac{1}{2}$)

Diagram 2

Draw more shapes of your own design
and record the values for p, i and A
in a table. Make some of your shapes
more difficult like the one in Diagram 3.

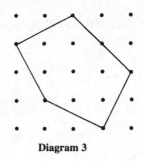

Diagram 3

Can you find a formula connecting p, i and A?
[Hint: $\frac{1}{2}i$, $\frac{1}{2}p$?]
Try out your formula with some more shapes to see if it always works.

29. Stopping distances

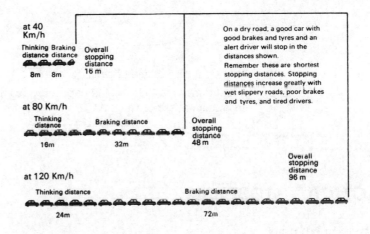

at 40
Km/h

Thinking Braking   Overall
distance distance   stopping
                    distance
8m   8m   16 m

On a dry road, a good car with
good brakes and tyres and an
alert driver will stop in the
distances shown.
Remember these are shortest
stopping distances. Stopping
distances increase greatly with
wet slippery roads, poor brakes
and tyres, and tired drivers.

at 80 Km/h

Thinking   Braking distance   Overall
distance   stopping
                                         distance
16m   32m   48 m

                                              Overall
                                              stopping
                                              distance
                                              96 m

at 120 Km/h

Thinking distance   Braking distance
24m   72m

This diagram from the Highway Code gives the overall stopping distances for
cars travelling at various speeds.

What is meant by 'thinking distance'?
Work out the thinking distance for a car travelling at a speed of 90 km/h.
What is the formula which connects the speed of the car and the thinking
distance?

(More difficult)
Try to find a formula which connects the speed of the car and the *overall*
stopping distance. It may help if you draw a graph of speed (across the page)
against *braking* distance (up the page).
What curve are you reminded of?

Check that your formula gives the correct answer for the overall stopping
distance at a speed of:
(a)   40 km/h   (b)   120 km/h.

30. Maximum cylinder

A rectangular piece of paper has a fixed perimeter of 40 cm. It could for example be 7 cm × 13 cm.

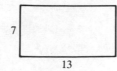

This paper can make a hollow cylinder of height 7 cm or of height 13 cm.

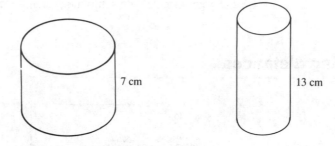

Work out the volume of each cylinder.

What dimensions should the paper have so that it can make a cylinder of the maximum possible volume?

PRACTICAL PROBLEMS

1. Timetabling

(a) Every year a new timetable has to be written for the school. We will look at the problem of writing the timetable for one department (mathematics). The department allocates the teaching periods as follows:

	U6	2 sets (at the same times); 8 periods in 4 doubles.
	L6	2 sets (at the same times); 8 periods in 4 doubles.
Year	5	6 sets (at the same times); 5 single periods.
Year	4	6 sets (at the same times); 5 single periods.
Year	3	6 sets (at the same times); 5 single periods.
Year	2	6 sets (at the same times); 5 single periods.
Year	1	5 mixed ability forms; 5 single periods not necessarily at the same times.

Here are the teachers and the maximum number of maths periods which they can teach.

A	33	
B	33	
C	33	
D	20	
E	20	
F	15	(Must be Years 5, 4, 3)
G	10	(Must be Years 2, 1)
H	10	(Must be Years 2, 1)
I	5	(Must be Year 3)

Furthermore, to ensure some continuity of teaching, teachers B and C must teach the U6 and teachers A, B, C, D, E, F must teach year 5.

Here is a timetable form which has been started

M 5						U6 B, C	U6 B, C		
Tu	5	U6 B, C	U6 B, C						
W					5				
Th						5	U6 B, C	U6 B, C	
F U6 B, C	U6 B, C		5						

Your task is to write a complete timetable for the mathematics department subject to the restrictions already stated.

(b) If that was too easy, here are some changes.

U6 and L6 have 4 sets each (still 8 periods)
Two new teachers: J 20 periods maximum
K 15 periods maximum but cannot teach on Mondays.

Because of games lessons: A cannot teach Wednesday afternoon
B cannot teach Tuesday afternoon
C cannot teach Friday afternoon

Also: A, B, C and E must teach U6
A, B, C, D, E, F must teach year 5

For the pupils, games afternoons are as follows:
Monday year 2; Tuesday year 3; Wednesday 5 L6, U6; Thursday year 4; Friday year 1.

2. Hiring a car

You are going to hire a car for one week (7 days).
Which of the firms below should you choose?

Gibson car hire	Snowdon rent-a-car	Hav-a-car
£170 per week unlimited mileage	£10 per day 6·5p per mile	£60 per week 500 miles without charge 22p per mile over 500 miles.

Work out as detailed an answer as possible.

3. Running a business

Mr Singh runs a small business making two sorts of steam cleaner: the basic model B and the deluxe model D.
Here are the details of the manufacturing costs:

	model B	model D
Assembly time (in man-hours)	20 hours	30 hours
Component costs	£35	£25
Selling price	£195	£245

He employs 10 people and pays them each £160 for a 40-hour week. He can spend up to £525 per week on components.

(a) In one week the firm makes and sells six cleaners of each model. Does he make a profit?
[Remember he has to pay his employees for a full week.]

(b) What number of each model should he make so that he makes as much profit as possible? Assume he can sell all the machines which he makes.

4. Income tax

Here are the details about income tax in 1978 (Labour) and
1983 (Conservative).

	1978		1983	
Allowances:				
Married man	1,535		2,445	
Single person	985		1,565	
Rates of tax on	1–750	25%	1–12,800	30%
taxable income	751–8,000	33%	12,801–15,100	40%
	8,001–9,000	40%	15,101–19,100	45%
	9,001–10,000	45%	19,101–25,300	50%
	10,001–11,000	50%	25,301–31,500	55%
	11,001–12,500	55%	more than 31,500	60%
	12,501–14,000	60%		
	14,001–16,000	65%		
	16,001–18,500	70%		
	18,501–24,000	75%		
	more than 24,000	83%		

In 1978 the average income of an adult was £2,479.
In 1983 the average income of an adult was £4,532.

What difference did a change of Government make for a person earning the
average income?
How about someone earning ten times the average income? How about twenty
times?
What happened to someone earning only half the average income?

Decide for yourself what would be a fair way of comparing the tax paid in the
two years.

This is of course a simplified model. We could extend the problem and consider
other allowances such as the allowance for people buying their homes with a
mortgage. This makes the problem far more complicated but if you are
interested have a try. You can obtain much useful data from the reference
section in your Public library. Look for the 'Annual Abstract of Statistics'.
(H.M.S.O.)

5. How many of each?

A shop owner has room in her shop for up to 20 televisions. She can buy either type A for £150 each or type B for £300 each.

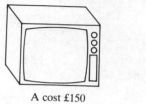

A cost £150

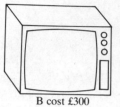

B cost £300

She has a total of £4500 she can spend and she must have at least 6 of each type in stock. She makes a profit of £80 on each television of type A and a profit of £100 on each of type B.

How many of each type should she buy so that she makes the maximum profit?

PUZZLES AND EXPERIMENTS

1. Cross numbers

(a) Copy out the cross number pattern.
(b) Fit all the given numbers into the correct spaces. Tick off the numbers from the lists as you write them in the square.

1.

2 digits	3 digits	4 digits	5 digits	6 digits
11	121	2104	14700	216841
17	147	2356	24567	588369
18	170	2456	25921	846789
19	174	3714	26759	861277
23	204	4711	30388	876452
31	247	5548	50968	
37	287	5678	51789	
58	324	6231	78967	
61	431	6789	98438	
62	450	7630		
62	612	9012		7 digits
70	678	9921		6645678
74	772			
81	774			
85	789			
94	870			
99				

2.

2 digits		3 digits	4 digits	5 digits	6 digits
12	47	129	2096	12641	324029
14	48	143	3966	23449	559641
16	54	298	5019	33111	956782
18	56	325	5665	33210	
20	63	331	6462	34509	
21	67	341	7809	40551	
23	81	443	8019	41503	
26	90	831	8652	44333	*7 digits*
27	91	923		69786	1788932
32	93			88058	5749306
38	98			88961	
39	99			90963	
46				94461	
				99654	

2. Estimating game

This is a game for two players. On squared paper draw an answer grid with the numbers shown below.

Answer grid

891	7047	546	2262	8526	429
2548	231	1479	357	850	7930
663	1078	2058	1014	1666	3822
1300	1950	819	187	1050	3393
4350	286	3159	442	2106	550
1701	4050	1377	4900	1827	957

The players now take turns to choose two numbers from the question grid below and multiply them on a calculator.

Question grid

11	26	81
17	39	87
21	50	98

The game continues until all the numbers in the answer grid have been crossed out. The object is to get four answers in a line (horizontally, vertically or diagonally). The winner is the player with most lines of four.
A line of *five* counts as *two* lines of four.
A line of *six* counts as *three* lines of four.

3. The chess board problem

(a) On the 4 × 4 square below we have placed four objects subject to the restriction that nowhere are there two objects on the same row, column or diagonal.

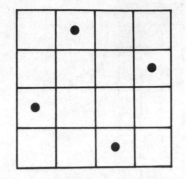

Subject to the same restrictions:
  (i) find a solution for a 5 × 5 square, using five objects,
  (ii) find a solution for a 6 × 6 square, using six objects,
  (iii) find a solution for a 7 × 7 square, using seven objects,
  (iv) find a solution for a 8 × 8 square, using eight objects.

It is called the chess board problem because the objects could be 'Queens' which can move any number of squares in any direction.

(b) Suppose we remove the restriction that no two Queens can be on the same row, column or diagonal. Is it possible to attack every square on an 8 × 8 chess board with less than eight Queens?
Try the same problem with other pieces like knights or bishops.

4. Creating numbers

Using only the numbers 1, 2, 3 and 4 once each and the operations $+, -, \times, \div, !$ create every number from 1 to 100.

You can use the numbers as powers and you must use all of the numbers 1, 2, 3 and 4.

[4! is pronounced 'four factorial' and means $4 \times 3 \times 2 \times 1$ (i.e. 24)
similarly $3! = 3 \times 2 \times 1 = 6$
                $5! = 5 \times 4 \times 3 \times 2 \times 1 = 120$]

Examples: $1 = (4 - 3) \div (2 - 1)$
                $20 = 4^2 + 3 + 1$
                $68 = 34 \times 2 \times 1$
                $100 = (4! + 1)(3! - 2!)$

5. Pentominoes

A pentomino is a set of five squares joined along their edges. Here are three of the twelve different pentomino designs.

(a) Find the other nine pentomino designs to make up the complete set of twelve. Reflections or rotations of other pentominoes are not allowed.

(b) On squared paper draw an 8×8 square. It is possible to fill up the 8×8 square with the twelve different pentominoes together with a 2×2 square. Here we have made a possible start.

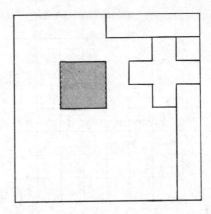

There are in fact many different ways in which this can be done.

(c) Now draw a 10×6 rectangle.
Try to fill up the rectangle with as many different pentominoes as you can. This problem is more difficult than the previous one but it is possible to fill up the rectangle with the twelve different pentominoes.

6. Calculator words

On a calculator work out $9508^2 + 192^2 + 10^2 + 6$.

If you turn the calculator upside down and use a little imagination, you can see the word 'HEDGEHOG'.

Which letters of the alphabet can be used to write calculator words in this way? Write some words and then work out mathematical clues which can be used to find them. Try to make the clues complicated.

Here is another one: $\dfrac{27 \times 2000 - 2}{0 \cdot 63 \div 0 \cdot 09}$.

7. Find your reaction time

Here is an experiment in which we are going to measure Derek's 'reaction time'. Jim holds a ruler vertically between two fingers. Derek waits with his fingers over the zero mark on the ruler but not touching the ruler. Jim releases the ruler without warning and when he does Derek catches it as quickly as he can.

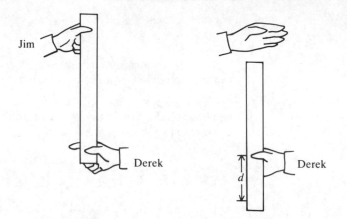

Do this experiment and record the distance d through which the ruler has fallen. Repeat the experiment many times to obtain a mean value for d. We can now calculate the reaction time from the formula $t = \sqrt{\dfrac{d}{490}}$, where d is measured in centimetres.

8. Probability pi

For this experiment you need a thin stick such as a cocktail stick or a needle and a large piece of paper.
On the paper draw a series of parallel lines distance d apart.

Now drop the stick at random anywhere on the paper on its end so it 'bounces'. Count the number of times you drop it and the number of times it falls across one of your lines

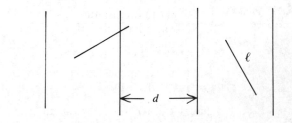

If you drop the stick n times and it crosses a line s times, you can find a value for π from the equation $\dfrac{s}{n} = \dfrac{2\ell}{\pi d}$ where ℓ is the length of the stick you drop.

12 Mental arithmetic and Revision tests

MENTAL ARITHMETIC

In each test the questions are read out by the teacher while all the pupils' books are open at pages 264 and 265.

Each question is repeated once and then the answer is written down.

The tests appear in the book so that pupils can check afterwards to see where they made mistakes.

Test 1

1. I bought an article costing 63p and paid with a one pound coin. My change consisted of four coins. What were they?
2. I bought two books costing £2·50 and £1·90. How much did I spend altogether?
3. What is the cost of six items at thirty-five pence each?
4. Tickets for a concert cost £6·50 each. What is the cost of four tickets?
5. It takes me 24 minutes to walk to school. I cycle three times as fast as I walk. How long do I take to cycle to school?
6. Work out as a single number, four squared plus three squared.
7. Write down an approximate value for forty-nine times eleven.
8. Write one metre as a fraction of one kilometre.
9. What number is exactly half-way between 2·5 and 2·8?
10. The First World War started in 1914. How long ago was that?
11. Lottery tickets cost £2·50 each. How much is raised from the sale of six thousand tickets?
12. Train fares are increased by ten per cent. If the old fare was £3·50, what is the new fare?
13. If a man earns £5·50 per hour, how much does he earn in five hours?
14. A beer crate holds twelve bottles. How many crates are needed for 90 bottles?
15. When playing darts you score double ten, double twenty and treble eight. What is your total score?
16. How many inches are there in three yards?
17. A petrol pump delivered $2\frac{1}{2}$ litres in 5 seconds. How many litres will it deliver in one minute?
18. A square has sides of length 5 cm. How long is a diagonal to the nearest centimetre?

For questions **19** to **25** look at pages 264 and 265.

19. (Fig. C). Mr Wren posts two 2nd class letters, each weighing 625 g. What is the total cost?
20. (Fig. B). How many minutes does the 'Paul Daniels Magic Show' last?
21. (Fig. D). Mrs Eaton buys three 6 × 4 pine fence panels. What is the total cost?
22. (Fig. G). Mr Wheeler takes his car, which is 3·60 m long, on the ferry from Dover to Calais on tariff C. How much does it cost for the car?

23. (Fig. F). A train arrives at Moor Park at 20 23. When did it leave Baker Street?
24. (Fig. J). Estimate the size of angle BDC.
25. (Fig. E). What is the capacity of the fuel tank of a Fiat Regata?

Test 2

1. How much more than 119 is 272?
2. What is the cube root of 64?
3. Find the average of 4, 8 and 9.
4. A rectangular lawn is 7 yards wide and 15 yards long. What area does it cover?
5. How many centimetres are there in 20 km?
6. A ship was due at noon on Tuesday, but arrived at 15 00 on Thursday. How many hours late was it?
7. A litre of wine fills 9 glasses. How many litre bottles are needed to fill 50 glasses?
8. Work out 15% of £40.
9. A cake weighs two pounds. How many ounces is that?
10. How many seconds are there in $2\frac{1}{2}$ minutes?
11. How many days are there altogether in 19 weeks?
12. If the eighth of May is a Monday, what day of the week is the seventeenth?
13. I buy three pounds of oranges for £1·02. How much do they cost per pound?
14. A rectangular pane of glass is 3 feet long and 2 feet wide. Glass costs £1·50 per square foot. How much will the pane cost?
15. A car journey of 150 miles took $2\frac{1}{2}$ hours. What was the average speed?
16. Add 218 to 84.
17. Pencils cost 5 pence each. How many can I buy with £2·50?
18. Write down the next prime number after 31.

For questions 19 to 25 look at pages 264 and 265.

19. (Fig. A). What is my monthly payment if I borrow £2500 over 60 months?
20. (Fig. F). How many minutes does it take the 20 10 train from Marylebone to reach Chesham?
21. (Fig. H). What share of the market did Gas have in 1983?
22. (Fig. I). How much will I pay for 7 nights in Elma starting on the 6th of June?
23. (Fig. J). A ship sails from A to B. On approximately what bearing is it sailing?
24. (Fig. D). Mavis buys one Larch 6 × 6 fence panel and one 8 foot post. What is the total cost?
25. (Fig. A). Steven repays a loan of £5500 over 10 years. How much does he pay each month?

Test 3

1. What is the angle between the hands of a clock at two o'clock?
2. What is a half of a half of 0·2?
3. In a test Paul got 16 out of 20. What percentage is that?
4. Work out $2 \times 20 \times 200$.
5. Two friends share a bill for £33·80. How much does each person pay?
6. Work out $\frac{1}{2}$ plus $\frac{1}{5}$ and give the answer as a decimal.
7. How long will it take a car to travel 320 miles at an average speed of 60 m.p.h.?
8. What is $\frac{1}{8}$ as a percentage?
9. What is the height of a triangle with base 12 cm and area 36 cm^2?
10. Work out 0·1 cubed.
11. Between which two consecutive whole numbers does the square root of 58 lie?
12. What is eight per cent of £25?
13. A car has a 1795 c.c. engine. What is that approximately in litres?
14. The mean of four numbers is 12·3. What is their sum?
15. Find the cost of smoking 40 cigarettes a day for five days if a packet of 20 costs £1·25.
16. How many minutes are there in $2\frac{3}{4}$ hours?
17. A pie chart has a red sector representing 20% of the whole chart. What is the angle of the sector?
18. How many five pence coins are needed to make £12?

For questions **19** to **25** look at pages 264 and 265.

19. (Fig. E). What is the top speed of a Toyota Camry?
20. (Fig. B). A man video tapes 'Perfect Strangers', 'Jim'll Fix it' and 'Hi-de-hi'. How long will it last on his tape?
21. (Fig. A). Mrs White is paying £62·67 each month. How much has she borrowed?
22. (Fig. F). I arrive at Finchley Road station at half-past seven and take the next train. When will I arrive at Amersham?
23. (Fig. I). How much will I pay for 14 nights in Mimosa starting on the 24th of April?
24. (Fig. H). What share of the market did Coal and Electricity have together in 1979?
25. (Fig. G). The Jameson family consists of 2 adults and 3 children. Their car is 3·6 m long. How much will it cost to travel from Folkestone to Boulogne on tariff B?

Test 4

1. Two angles of a triangle are 42° and 56°. What is the third angle?
2. Telephone charges are increased by 20%. What is the new charge for a call which previously cost 60p?
3. What number is exactly half way between 0·1 and 0·4?
4. A boat sails at a speed of 18 knots for five hours. How far does it go?
5. How many 23p stamps can be bought for £2?
6. The mean age of three girls is 12 years. If two of the girls are aged 9 and 16 years, how old is the third girl?
7. Multiply $3\frac{1}{4}$ by 100.
8. What is a quarter of a third?
9. A prize of five million pounds is shared between 200 people. How much does each person receive?
10. The attendance at an athletics meeting was forty-eight thousand, seven hundred and eleven. Write this number correct to two significant figures.
11. Work out 0·1 multiplied by 63.
12. Find the cost of 6 litres of wine at £1·45 per litre.
13. Three people agree to share a bill equally. The cost comes to £7·20. How much does each person pay?
14. A pump removes water at a rate of 6 gallons per minute. How many hours will it take to remove 1800 gallons?
15. Work out three-eighths of £100.
16. A metal rod of length 27·1 cm is cut exactly in half. How long is each piece?
17. A square has sides of length 7 cm. How long is a diagonal to the nearest centimetre?
18. The cost of five tins of salmon is £7·50. How much will six tins cost?

For questions **19** to **25** look at pages 264 and 265.

19. (Fig. D). Mr Jones needs twenty 6 × 4 fence panels. How much more will it cost if he buys Larch rather than Pine?
20. (Fig. E). How much change from £10 000 would I get if I bought a Toyota Camry?
21. (Fig. B). Steve has 2 hours left on a video tape. How much will be left after he tapes the film?
22. (Fig. C). Mrs James sends three 1st class letters weighing 20 g, 35 g and 80 g. Find the total cost.
23. (Fig. J). A ship sails from A to C. On approximately what bearing is it sailing?
24. (Fig. I). Two adults stay at Buganvilia for 14 nights starting on the 16th of May. What is the total cost?
25. (Fig. H). If the 1978 figures were displayed on a pie chart, approximately what angle would represent 'Gas'?

Test 5

1. What number is a thousand times as big as 0·2?
2. Pencils cost five pence each. How much will two dozen pencils cost?
3. How many fours are there in a thousand?
4. Work out the area, in square metres, of a rectangular field of width twenty metres and length twenty-five metres.
5. A packet of peanuts costs 65 pence. I buy two packets and pay with a ten pound note. Find the change.
6. What is a half of a half of 0·1?
7. I bought three kilograms of flour and I use four hundred and fifty grams of it. How many grams of flour do I have left?
8. A bingo prize of two hundred thousand pounds is shared equally between five people. How much does each person receive?
9. What is the angle between the hands of a clock at 5 o'clock?
10. Five boys and three girls share £240. How much do the boys get altogether?
11. How many 17p stamps can I buy for £2?
12. A milk crate has space for 24 bottles. How many crates are needed for 200 bottles?
13. A ruler costs 37 pence. What is the total cost of three rulers?
14. A salesman receives commission of $1\frac{1}{2}\%$ on sales. How much commission does he receive when he sells a computer for £1000?
15. How many edges does a cube have?
16. Between which two consecutive whole numbers does the square root of 80 lie?
17. A coat is marked at a sale price of £60 after a reduction of 25%. What was the original price?
18. Theatre tickets cost £3·45 each. How much will four tickets cost?

For questions **19** to **25** look at pages 264 and 265.

19. (Fig. E). How long does it take a Mazda 626 to cover $\frac{1}{4}$ mile?
20. (Fig. B). For how many hours and minutes is 'Grandstand' on the air?
21. (Fig. A). I want to borrow £1000 and I can repay over either 60 months or 90 months. What is the difference in the monthly payments?
22. (Fig. D). Mr Reynolds bought ten 6 foot posts and caps. How much will it cost him?
23. (Fig. C). Brian posts two 1st class letters weighing 420 g and 650 g. Find the total cost.
24. (Fig. G). John pays for himself and his 4·2 m car to travel from Dover to Calais on Tariff C. How much does it cost him?
25. (Fig. F). How many minutes does it take the 19 22 train from Harrow-on-the-Hill to reach Aylesbury?

Test 6

1. What four coins make seventy-six pence?
2. How many minutes are there between 7.44 p.m. and 9.10 p.m.?
3. How many 15 pence rulers can be bought for one pound?
4. A spark plug for a car costs £1·25. How much does a set of six plugs cost?
5. A pile of 15 boxes is 3 metres high. What is the depth of each box?
6. What is the smallest number which must be added to 77 to make it exactly divisible by 9?
7. How many 2p coins are worth the same as forty 5p coins?
8. Find the difference between $11\frac{1}{4}$ and 19.
9. The profit on one drink is 4 pence. How many must be sold to make a profit of £2?
10. By how much is three kilometres longer than three metres?
11. A gallon of a valuable liquid costs £200. How much does one pint cost?
12. Work out $2 \times 3 \times 4 \times 5$.
13. How many prime numbers are there between 30 and 40?
14. I pay £3 for four rolls and one cake. How much is each roll if the cake costs 60p?
15. V.A.T. is charged at 15%. How much V.A.T. do I pay on a tyre costing £80?
16. One angle in an isosceles triangle is 102°. What are the other angles?
17. How many minutes are there in a day?
18. A circle has an area of 300 cm². Approximately what is the radius?

For questions **19** to **25** look at pages 264 and 265.

19. (Fig. E). Which car has the best fuel consumption at a steady 70 m.p.h.?
20. (Fig. B). Tina has $1\frac{1}{2}$ hours left on a video tape. How much will be left after taping 'Bergerac'?
21. (Fig. A). Hannah borrows £1000 and starts repaying over 90 months. How much has she paid after 5 months?
22. (Fig. F). Yahangir arrives at Northwood at twenty past nine. At what time did his train leave Northwick Park?
23. (Fig. J). An aircraft flies from C to D. On approximately what bearing is it flying?
24. (Fig. I). How much do I pay for a child to stay in Elma for 14 nights starting on the 9th of May?
25. (Fig. H). If the 1984 figures were displayed in a pie chart, approximately what angle would represent 'Oil'?

Test 7

1. Find two ways of making 64 pence using five coins.
2. How long will it take to drive 100 miles at an average speed of 60 m.p.h.?
3. Three tickets cost £6·30 altogether. How much will five tickets cost?
4. A watch ticks once every second. How many times will it tick in two hours?
5. Work out as a single number, five squared plus six squared.
6. Philip weighs ten stone six pounds. How many pounds is that?
7. How many centimetres are there in 25 km?
8. A rectangular pane of glass measures 4 feet by 3 feet. How much will the pane cost if the price of glass is £1·50 per square foot?
9. A dress is marked at a sale price of £40 after a reduction of 20%. What was the original price?
10. An estate agent charges commission of $1\frac{1}{2}$ per cent on sales. How much does he receive when a house is sold for £50 000?
11. My grandfather celebrated his ninety-second birthday in 1980. In what year was he born?
12. A small glass holds 50 millilitres of spirit. How many glasses can be filled from a one litre bottle?
13. One drilling machine weighs 35·8 kg. What is the weight of two machines?
14. Steve pays £2·30 for six drinks and one sandwich. How much is each drink if the sandwich costs 50p?
15. How many prime numbers are there between 36 and 46?
16. A rectangle measures 5 cm by 3 cm. How long is a diagonal to the nearest centimetre?
17. A ship's pump delivered 2·5 litres in 3 seconds. How many litres will it deliver in one minute?
18. What is the smallest number which must be added to 57 to make it exactly divisible by 7?

For questions **19** to **25** look at pages 264 and 265.

19. (Fig. C). I have two letters each weighing 380 g. How much more will it cost to send them 1st class rather than 2nd class?
20. (Fig. E). The fuel tank of a standard Toyota Camry is increased by 20% for a rally. What is the new size?
21. (Fig. G). Mr and Mrs Campbell take their 5 m car on the ferry from Folkestone to Boulogne. How much will it cost for them and the car on tariff B?
22. (Fig. J). A submarine sails from C to A. On approximately what bearing is it sailing?
23. (Fig. I). I want to go on holiday for 14 nights starting on the 8th of September. How much more expensive is Mimosa rather than Elma?
24. (Fig. B). A video is set up to record 'Bob's Full House' and 'Cagney and Lacey'. How much tape will be used?
25. (Fig. A). I can repay a loan for £4000 over either 10 years or 90 months. What is the difference in the monthly payments?

REVISION TESTS

Test 1

1. How many mm are there in 1 m 1 cm?

    A 1001
    B 1110
    C 1010
    D 1100

2. The circumference of a circle is 16π cm. The radius, in cm, of the circle is

    A 2
    B 4
    C $\frac{4}{\pi}$
    D 8

3. In the triangle below the value of $\cos x$ is

    A 0·8
    B 1·333
    C 0·75
    D 0·6

4. The line $y = 2x - 1$ cuts the x-axis at P. The coordinates of P are

    A $(0, -1)$
    B $(\frac{1}{2}, 0)$
    C $(-\frac{1}{2}, 0)$
    D $(-1, 0)$

5. The formula $b + \dfrac{x}{a} = c$ is rearranged to make x the subject. What is x?

    A $a(c - b)$
    B $ac - b$
    C $\dfrac{c - b}{a}$
    D $ac + ab$

6. The mean weight of a group of 11 men is 70 kg. What is the mean weight of the remaining group when a man of weight 90 kg leaves?

    A 80 kg
    B 72 kg
    C 68 kg
    D 62 kg

7. How many lines of symmetry has this shape?

    A 0
    B 1
    C 2
    D 4

8. In standard form the value of $2000 \times 80\,000$ is

    A 16×10^6
    B $1·6 \times 10^9$
    C $1·6 \times 10^7$
    D $1·6 \times 10^8$

9. The solutions of the equation $(x - 3)(2x + 1) = 0$ are

    A $-3, \frac{1}{2}$
    B $3, -2$
    C $3, -\frac{1}{2}$
    D $-3, -2$

10. In the triangle the size of angle x is

    A 35°
    B 70°
    C 110°
    D 40°

11. A man paid tax on £9000 at 30%. He paid the tax in 12 equal payments. Each payment was

    A £2·25
    B £22·50
    C £225
    D £250

12. The approximate value of $\dfrac{3·96 \times (0·5)^2}{97·1}$ is

    A 0·01
    B 0·02
    C 0·04
    D 0·1

13. Given that $\dfrac{3}{n} = 5$, then $n =$

    A 2
    B -2
    C $1\frac{2}{3}$
    D 0·6

14. Cube A has side 2 cm. Cube B has side 4 cm. $\left(\dfrac{\text{Volume of B}}{\text{Volume of A}}\right) =$

    A 2
    B 4
    C 8
    D 16

15. How many tiles of side 50 cm will be needed to cover the floor shown?

    A 16
    B 32
    C 64
    D 84

16. The equation $ax^2 + x - 6 = 0$ has a solution $x = -2$ What is a?

    A 1
    B -2
    C $\sqrt{2}$
    D 2

17. Which of the following is/are correct?
    1. $\sqrt{0·16} = 0·4$
    2. $0·2 \div 0·1 = 0·2$
    3. $\frac{4}{7} > \frac{3}{5}$

    A **1** only
    B **2** only
    C **3** only
    D **1** and **2**

18. How many prime numbers are there between 30 and 40?

    A 0
    B 1
    C 2
    D 3

19. A man is paid £180 per week after a pay rise of 20%. What was he paid before?

A £144
B £150
C £160
D £164

20. A car travels for 20 minutes at 45 m.p.h. and then for 40 minutes at 60 m.p.h. The average speed for the whole journey is

A $52\frac{1}{2}$ m.p.h.
B 50 m.p.h.
C 54 m.p.h.
D 55 m.p.h.

21. The point $(3, -1)$ is reflected in the line $y = 2$. The new coordinates are

A $(3, 5)$
B $(1, -1)$
C $(3, 4)$
D $(0, -1)$

22. Two discs are randomly taken from a bag containing 3 red discs and 2 blue discs. What is the probability of taking 2 red discs?

A $\frac{9}{25}$
B $\frac{1}{10}$
C $\frac{3}{10}$
D $\frac{2}{5}$

23. The shaded area, in cm², is

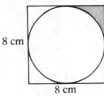

8 cm

8 cm

A $16 - 2\pi$
B $16 - 4\pi$
C $\frac{4}{\pi}$
D $64 - 8\pi$

24. Given the equation $5^x = 120$, the best approximate solution is $x =$

A 2
B 3
C 4
D 25

25. What is the sine of 45°?

1

45°

A 1
B $\frac{1}{2}$
C $\frac{1}{\sqrt{2}}$
D $\sqrt{2}$

Test 2

1. What is the value of the expression $(x - 2)(x + 4)$ when $x = -1$?

A 9
B -9
C 5
D -5

2. The perimeter of a square is 36 cm. What is its area?

A 36 cm²
B 324 cm²
C 81 cm²
D 9 cm²

3. AB is a diameter of the circle. Find the angle BCO.

A 70°
B 20°
C 60°
D 50°

4. The gradient of the line $2x + y = 3$ is

A 3
B -2
C $\frac{1}{2}$
D $-\frac{1}{2}$

5. A firm employs 1200 people, of whom 240 are men. The percentage of employees who are men is

A 40%
B 10%
C 15%
D 20%

6. A car is travelling at a constant speed of 30 m.p.h. How far will the car travel in 10 minutes?

A $\frac{1}{3}$ mile
B 3 miles
C 5 miles
D 6 miles

7. What are the coordinates of the point $(1, -1)$ after reflection in the line $y = x$?

A $(-1, 1)$
B $(1, 1)$
C $(-1, -1)$
D $(1, -1)$

8. $\frac{1}{3} + \frac{2}{5} =$

A $\frac{2}{8}$
B $\frac{3}{8}$
C $\frac{3}{15}$
D $\frac{11}{15}$

9. In the triangle the size of the largest angle is

A 30°
B 90°
C 120°
D 80°

10. 800 decreased by 5% is

A 795
B 640
C 760
D 400

11. Which of the statements is (are) true?
  1. $\tan 60° = 2$
  2. $\sin 60° = \cos 30°$
  3. $\sin 30° > \cos 30°$

   A **1** only
   B **2** only
   C **3** only
   D **2** and **3**

12. Given $a = \frac{3}{5}$, $b = \frac{1}{3}$, $c = \frac{1}{2}$ then

   A $a < b < c$
   B $a < c < b$
   C $a > b > c$
   D $a > c > b$

13. The *larger* angle between South-West and East is

   A $225°$
   B $240°$
   C $135°$
   D $315°$

14. Each exterior angle of a regular polygon with n sides is $10°$; $n =$

   A 9
   B 18
   C 30
   D 36

15. What is the value of $1 - 0.05$ as a fraction?

   A $\frac{1}{20}$
   B $\frac{9}{10}$
   C $\frac{19}{20}$
   D $\frac{5}{100}$

16. Find the length x.

   A 5
   B 6
   C 8
   D $\sqrt{50}$

17. Given that $m = 2$ and $n = -3$, what is mn^2?

   A -18
   B 18
   C -36
   D 36

18. The graph of $y = (x - 3)(x - 2)$ cuts the y-axis at P. The coordinates of P are

   A $(0, 6)$
   B $(6, 0)$
   C $(2, 0)$
   D $(3, 0)$

19. £240 is shared in the ratio $2 : 3 : 7$. The largest share is

   A £130
   B £140
   C £150
   D £160

20. Adjacent angles in a parallelogram are $x°$ and $3x°$. Th smallest angles in the parallelogram are each

   A $30°$
   B $45°$
   C $60°$
   D $120°$

21. When the sides of a square are increased by 10% the area is increased by

   A 10%
   B 20%
   C 21%
   D 15%

22. The volume, in cm³, of the cylinder is

   A 9π
   B 12π
   C 600π
   D 900π

23. A car travels for 10 minutes at 30 m.p.h. and then for 20 minutes at 45 m.p.h. The average speed for the whole journey is

   A 40 m.p.h.
   B $37\frac{1}{2}$ m.p.h.
   C 20 m.p.h.
   D 35 m.p.h.

24. Four people each toss a coin. What is the probability that the fourth person will toss a 'tail'?

   A $\frac{1}{2}$
   B $\frac{1}{4}$
   C $\frac{1}{8}$
   D $\frac{1}{16}$

25. A rectangle 8 cm by 6 cm is inscribed inside a circle. What is the area, in cm², of the circle?

   A 10π
   B 25π
   C 49π
   D 100π

Test 3

1. The price of a T.V. changed from £240 to £300. What is the percentage increase?

   A 15%
   B 20%
   C 60%
   D 25%

2. Find the length x.

   A 6
   B 5
   C $\sqrt{44}$
   D $\sqrt{18}$

3. The bearing of A from B is $120°$. What is the bearing of B from A?

   A $060°$
   B $120°$
   C $240°$
   D $300°$

4. Numbers m, x and y satisfy the equation $y = mx^2$. When $m = \frac{1}{2}$ and $x = 4$ the value of y is

   A 4
   B 8
   C 1
   D 2

5. A school has 400 pupils, of whom 250 are boys. The ratio of boys to girls is

    A 5 : 3
    B 3 : 2
    C 3 : 5
    D 8 : 5

6. A train is travelling at a speed of 30 km per hour. How long will it take to travel 500 m?

    A 2 minutes
    B $\frac{3}{50}$ hour
    C 1 minute
    D $\frac{1}{2}$ hour

7. The approximate value of $\dfrac{9 \cdot 65 \times 0 \cdot 203}{0 \cdot 0198}$ is

    A 99
    B 9·9
    C 0·99
    D 180

8. Which point does *not* lie on the curve $y = \dfrac{12}{x}$?

    A (6, 2)
    B ($\frac{1}{2}$, 24)
    C (−3, −4)
    D (3, −4)

9. $t = \dfrac{c^3}{y}$, $y =$

    A $\dfrac{t}{c^3}$
    B $c^3 t$
    C $c^3 - t$
    D $\dfrac{c^3}{t}$

10. The largest number of 1 cm cubes which will fit inside a cubical box of side 1 m is

    A 10^3
    B 10^6
    C 10^8
    D 10^{12}

11. The shaded area in the Venn diagram represents

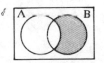

    A $A' \cup B$
    B $A \cap B'$
    C $A' \cap B$
    D $(A \cap B)'$

12. Which of the following has the largest value?

    A $\sqrt{100}$
    B $\sqrt{\dfrac{1}{0 \cdot 1}}$
    C $\sqrt{1000}$
    D $\dfrac{1}{0 \cdot 01}$

13. Two dice numbered 1 to 6 are thrown together and their scores are added. The probability that the sum will be 12 is

    A $\frac{1}{6}$
    B $\frac{1}{12}$
    C $\frac{1}{18}$
    D $\frac{1}{36}$

14. The length, in cm, of the minor arc is

    A 2π
    B 3π
    C 6π
    D $13\frac{1}{2}\pi$

15. Metal of weight 84 kg is made into 40 000 pins. What is the weight, in kg, of one pin?

    A 0·0021
    B 0·0036
    C 0·021
    D 0·21

16. What is the value of x which satisfies both equations?
$3x + y = 1$
$x - 2y = 5$

    A −1
    B 1
    C −2
    D 2

17. What is the new fare when the old fare of £250 is increased by 8%?

    A £258
    B £260
    C £270
    D £281·25

18. What is the area of this triangle?

    A $12x^2$
    B $15x^2$
    C $16x^2$
    D $30x^2$

19. What values of x satisfy the inequality $2 - 3x > 1$?

    A $x < -\frac{1}{3}$
    B $x > -\frac{1}{3}$
    C $x > \frac{1}{3}$
    D $x < \frac{1}{3}$

20. A right-angled triangle has sides in the ratio 5 : 12 : 13. The tangent of the smallest angle is

    A $\frac{12}{5}$
    B $\frac{12}{13}$
    C $\frac{5}{13}$
    D $\frac{5}{12}$

21. The area of $\triangle ABE$ is 4 cm². The area of $\triangle ACD$ is

    A 10 cm²
    B 6 cm²
    C 25 cm²
    D 16 cm²

22. Given $2^x = 3$ and $2^y = 5$, the value of 2^{x+y} is

    A 15
    B 8
    C 4
    D 125

23. The probability of an event occurring is 0·35. The probability of the event *not* occurring is

    A $\dfrac{1}{0 \cdot 35}$
    B 0·65
    C 0·35
    D 0

24. What fraction of the area of the rectangle is the area of the triangle?

A $\frac{1}{4}$
B $\frac{1}{8}$
C $\frac{1}{16}$
D $\frac{1}{32}$

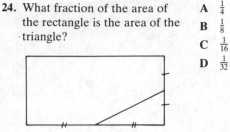

25. On a map a distance of 36 km is represented by a line of 1·8 cm. What is the scale of the map?

A 1 : 2000
B 1 : 20 000
C 1 : 200 000
D 1 : 2 000 000

Test 4

1. What is the value of x satisfying the simultaneous equations
$3x + 2y = 13$
$x - 2y = -1$?

A 7
B 3
C $3\frac{1}{2}$
D 2

2. A straight line is 4·5 cm long. $\frac{2}{5}$ of the line is

A 0·4 cm
B 1·8 cm
C 2 cm
D 0·18 cm

3. The mean of four numbers is 12. The mean of three of the numbers is 13. What is the fourth number?

A 9
B 12·5
C 7
D 1

4. How many cubes of edge 3 cm are needed to fill a box with internal dimensions 12 cm by 6 cm by 6 cm?

A 8
B 18
C 16
D 24

For questions **5** to **7** use the diagram below.

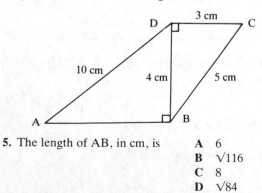

5. The length of AB, in cm, is

A 6
B $\sqrt{116}$
C 8
D $\sqrt{84}$

6. The sine of angle DCB is

A 0·8
B 1·25
C 0·6
D 0·75

7. The tangent of angle CBD is

A 0·6
B 0·75
C 1·333
D 1·6

8. The value of 4865·355 correct to 2 significant figures is

A 4865·36
B 4865·35
C 4900
D 49

9. What values of y satisfy the inequality $4y - 1 < 0$?

A $y < 4$
B $y < -\frac{1}{4}$
C $y > \frac{1}{4}$
D $y < \frac{1}{4}$

10. The area of a circle is 100π cm². The radius, in cm, of the circle is

A 50
B 10
C $\sqrt{50}$
D 5

11. If $f(x) = x^2 - 3$, then $f(3) - f(-1) =$

A 5
B 10
C 8
D 9

12. In the triangle BE is parallel to CD. What is x?

A $6\frac{2}{3}$
B 6
C $7\frac{1}{2}$
D $5\frac{3}{4}$

13. The cube root of 64 is

A 2
B 4
C 8
D 16

14. Given $a + b = 10$
and $a - b = 4$
then $2a - 5b =$

A 0
B -1
C 1
D 3

15. Given $16^x = 4^4$, what is x?

A -2
B $-\frac{1}{2}$
C $\frac{1}{2}$
D 2

16. What is the area, in m², of a square with each side 0·02 m long?

A 0·0004
B 0·004
C 0·04
D 0·4

17. I start with x, then square it,   **A** $(3x)^2 - 4$
multiply by 3 and finally   **B** $(3x - 4)^2$
subtract 4. The final result is   **C** $3x^2 - 4$
                              D $3(x - 4)^2$

18. How many prime numbers are   **A**   1
there between 50 and 60?   **B**   2
                              C   3
                              D   4

19. What are the coordinates of   **A** $(-2, 2)$
the point $(2, -2)$ after   **B** $(2, -2)$
reflection in the line $y = -x$?   **C** $(-2, -2)$
                              D $(2, 2)$

20. The area of a circle is 36π cm^2.   **A** 6π
The circumference, in cm, is   **B** 18π
                              C $12\sqrt{\pi}$
                              D 12π

21. The gradient of the line   **A** $\frac{2}{3}$
$2x - 3y = 4$ is   **B** $1\frac{1}{2}$

                              C $-\frac{4}{3}$

                              D $-\frac{3}{4}$

22. When all three sides of a   **A**   3
triangle are trebled in length,   **B**   6
the area is increased by a   **C**   9
factor of   **D**   27

23. $a = \sqrt{\left(\dfrac{m}{x}\right)}$ **A** $a^2 m$
                           B $a^2 - m$
  $x =$ **C** $\dfrac{m}{a^2}$

                              D $\dfrac{a^2}{m}$

24. A coin is tossed three times.   **A** $\frac{1}{3}$
The probability of getting   **B** $\frac{1}{6}$
three 'heads' is   **C** $\frac{1}{8}$

                              D $\frac{1}{16}$

25. A triangle has sides of length   **A**   12
5 cm, 5 cm and 6 cm. What is   **B**   15
the area, in cm^2?   **C**   18
                              D   20

Fig. A

Letter Post

Weight not over	1st class	2nd class		Weight not over	1st class	2nd class
60g	18p	13p		500g	92p	70p
100g	26p	20p		600g	£1.15	85p
150g	32p	24p		700g	£1.35	£1.00
200g	40p	30p		750p	£1.45	£1.05
250g	48p	37p		800g	£1.55	Not admissible over 750g
300g	56p	43p		900g	£1.70	
350g	64p	49p		1000g	£1.85	
400g	72p	55p		Each extra 250g or part thereof 45p		
450g	82p	62p				

Fig. C

Car facts

[1] Second figure is with rear seat fully folded

	price	likely discount	engine capacity	engine power	doors	length	width	luggage capacity [1]	sill height	fuel tank	fuel consumption overall	steady 70mph	simulated city	insurance group	time to cover ¼ mile	time to reach 60mph	30 to 50mph in 4th gear	top speed
	£	%	cc	bhp		metres	metres	litres	cm	litres	mpg	mpg	mpg		sec	sec	sec	mph
SALOONS																		
Audi 80 1.8GL	9,224	15	1781	90	4	4.41	1.69	390	67	68	31.2	39	16.9	5	18.1	11.0	7.9	109
Honda Accord EX 2.0	8,750	2	1955	105	4	4.54	1.70	445	66	60	27.8	36	15.8	6/7	17.9	11.2	9.0	112
Mazda 626 2.0GLX	8,124	14	1998	102	4	4.43	1.69	440/645	64	60	25.9	33	15.3	6	18.3	11.6	9.2	106
Toyota Camry 2.0GLi	9,659	12	1995	106	4	4.44	1.69	400	84	55	30.1	34	17.0	5	18.5	11.8	10.1	105
ESTATES																		
Austin Montego 2.0HL	8,987	14	1994	105	4	4.47	1.71	480/860	56	50	29.3	36	17.2	5	18.9	12.0	10.6	104
Citroën BX19 TRS	8,903	14	1905	105	4	4.24	1.68	390/785	40	52	28.5	37	14.6	5	17.9	10.6	7.6	108
Fiat Regata 100S Weekend	8,590	13	1585	99	4	4.27	1.65	480/795	52	55	29.1	37	15.2	5	19.0	12.6	12.1	109
Nissan Bluebird 2.0GL	8,845	14	1973	105	4	4.37	1.69	480/725	59	60	28.1	35	16.8	6	18.5	12.0	10.5	105

Fig. E

London to Harrow, Watford, Chesham, Amersham and Aylesbury

Fig. F

Marylebone	...	≉ 1910	...	...	...	...	...	≉ 2010	...	...	...	...	...	≉ 2110	...
Baker Street	1850	↓	1905	1920	1933	1935	1950	↓	2005	2020	2033	2035	2050	↓	2105
Finchley Road	1856	↓	1911	1926	1939	1941	1956	↓	2011	2026	2039	2041	2056	↓	2111
Wembley Park	1902	↓	1917	1932	↓	1947	2002	↓	2017	2032	↓	2047	2102	↓	2117
Preston Road	1904	↓	1919	1934	↓	1949	2004	↓	2019	2034	↓	2049	2104	↓	2119
Northwick Park	1907	↓	1922	1937	↓	1952	2007	↓	2022	2037	↓	2052	2107	↓	2122
Harrow-on-the-Hill	1909	1922	1924	1939	1949	1954	2009	2022	2024	2039	2049	2054	2109	2122	2124
North Harrow	1912	↓	1927	1942	↓	1957	2012	↓	2027	2042	↓	2057	2112	↓	2127
Pinner	1914	↓	1929	1944	↓	1959	2014	↓	2029	2044	↓	2059	2114	↓	2129
Northwood Hills	1917	↓	1932	1947	↓	2002	2017	↓	2032	2047	↓	2102	2117	↓	2132
Northwood	1920	↓	1935	1950	↓	2005	2020	↓	2035	2050	↓	2105	2120	↓	2135
Moor Park	1923	1930	1938	1953	1957	2008	2023	2030	2038	2053	2057	2108	2123	2130	2138
Croxley	1927	↓	1942	1957	↓	2012	2027	↓	2042	2057	↓	2112	2127	↓	2142
Watford	1933	↓	1947	2003	↓	2017	2033	↓	2047	2103	↓	2117	2133	↓	2147
Rickmansworth	—	1934	—	—	2001	—	—	2034	—	—	2101	—	—	2134	—
Chorleywood	...	1938	...	...	2005	...	...	2038	...	...	2105	...	...	2138	...
Chesham dep	...	*1928c*	...	...	*1958c*	...	...	*2028c*	...	...	*2058c*	...	...	*2128c*	...
Chalfont & Latimer	...	1943	...	...	2009	...	...	2043	...	...	2109	...	...	2143	...
Chesham arr	...	*1954c*	...	...	*2019c*	...	...	*2054c*	...	...	*2119c*	...	...	*2154c*	...
Amersham	...	1948	...	...	2013	...	...	2048	...	...	2113	...	...	2148	...
Great Missenden	...	1955	...	...	—	...	...	2055	...	...	—	...	...	2155	...
Wendover	...	2002	...	...	...	...	...	2102	...	...	...	...	...	2202	...
Stoke Mandeville	...	2006	...	...	...	...	...	2106	...	...	...	...	...	2206	...
Aylesbury	...	2010	...	...	...	...	...	2110	...	...	...	...	...	2210	...

Car Ferry Tariffs

Motorist Fares/Vehicle Rates for Single Journeys

	DOVER –CALAIS				FOLKESTONE –BOULOGNE				
	Tariff E	Tariff D	Tariff C	Tariff B	Tariff E	Tariff D	Tariff C	Tariff B	Tariff A
MOTORIST FARES (driver and accompanying passengers)	£	£	£	£	£	£	£	£	£
Adult	10.00	10.00	10.00	10.00	10.00	10.00	10.00	10.00	10.00
Child (4 but under 14 years)	5.00	5.00	5.00	5.00	5.00	5.00	5.00	5.00	5.00
VEHICLE RATES Cars, Motorised Caravans, Minibuses and Three-wheeled Vehicles									
Up to 4.00m in length	16.00	24.00	33.00	43.00	16.00	21.00	33.00	43.00	50.00
Up to 4.50m in length	16.00	30.00	42.00	52.00	16.00	27.00	42.00	52.00	60.00
Up to 5.50m in length	16.00	34.00	50.00	60.00	16.00	31.00	50.00	60.00	70.00
Over 5.50m: each additional metre (or part thereof)	9.00	9.00	9.00	9.00	9.00	9.00	9.00	9.00	9.00

Fig. G

FUEL SHARE OF FINAL USER ENERGY MARKET (EXCLUDING TRANSPORT)

	1976	1977	1978	1979	1980	1981	1982	1983	1984	1985
COAL	23.3	22.3	20.7	20.8	17.8	18.3	18.4	18.4	16.4	18.7
GAS	30.0	31.0	33.0	34.4	39.0	40.3	41.3	42.4	44.2	44.5
ELECTRICITY	16.3	16.3	16.7	16.8	17.8	18.0	18.0	18.5	19.5	19.3
OIL	30.4	30.4	29.6	28.2	25.4	23.4	22.3	20.7	19.9	17.5

Fig. H

Flying to FARO Airport	Mimosa		Buganvilia		Elma		
Departure airport	Gatwick		Gatwick		Gatwick		★ Child Reductions (see page 3) ★
Number of nights	7	14	7	14	7	14	
Departure day/approx. time	Sat 1545		Sat 1545		Sat 1545		
Arrive back day/approx. time	Sat 2210		Sat 2210		Sat 2210		
First departure	5 Apr		5 Apr		3 May		
Last departure	18 Oct	11 Oct	18 Oct	11 Oct	18 Oct	11 Oct	
Holiday number	H3561		H3554		H3559		
DEPARTURES BETWEEN 1 Apr-23 Apr **Bargain**	150	184	150	184	—	—	**50%**
24 Apr-8 May **Bargain**	140	174	140	164	131	156	
9 May-15 May **Bargain**	150	184	150	175	141	166	
16 May-26 May	172	221	170	195	161	186	**25%**
27 May-5 Jun **Bargain**	161	231	155	179	147	171	
6 Jun-19 Jun **Bargain**	181	251	165	199	157	191	
20 Jun-26 Jun	201	263	175	209	167	201	
27 Jun-3 Jul	205	264	178	213	169	204	
4 Jul-17 Jul	208	265	181	213	172	206	
18 Jul-10 Aug	222	289	195	227	186	219	**15%**
11 Aug-25 Aug	209	278	182	211	174	204	
26 Aug-7 Sep	204	273	177	211	168	203	
8 Sep-28 Sep	195	252	172	207	163	198	
29 Sep-5 Oct	172	213	171	205	162	196	**40%**
6 Oct-17 Oct **Bargain**	152	190	151	182	142	174	
18 Oct-24 Oct	170	—	170	—	160	—	
Transfer airport/hotel approx.	45 mins		45 mins		45 mins		—

Fig. I

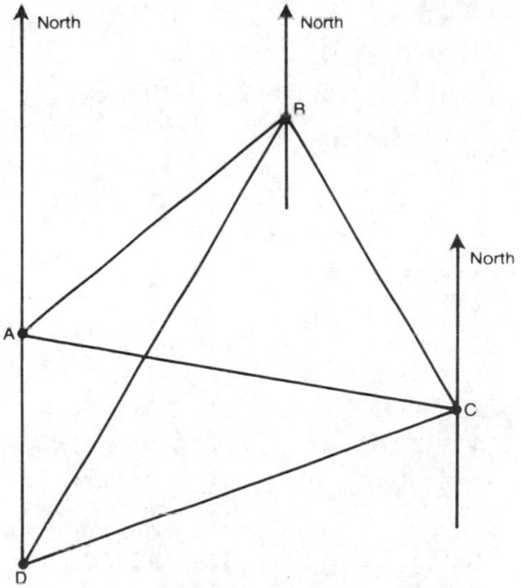

Fig. J

Answers

PART 1

page 1 **Exercise 1**

1. 7·91	**2.** 22·22	**3.** 7·372	**4.** 0·066	**5.** 466·2
6. 1·22	**7.** 1·67	**8.** 1·61	**9.** 16·63	**10.** 24·1
11. 26·7	**12.** 3·86	**13.** 0·001	**14.** 1·56	**15.** 0·0288
16. 2·176	**17.** 0·02	**18.** 0·0001	**19.** 7·56	**20.** 0·7854
21. 360	**22.** 34 000	**23.** 18	**24.** 0·74	**25.** 2·34
26. 1620	**27.** 8·8	**28.** 1200	**29.** 0·00175	**30.** 13·2
31. 200	**32.** 0·804	**33.** 0·8	**34.** 0·077	**35.** 0·0009
36. 0·01	**37.** 184	**38.** 20	**39.** 0·099	**40.** 3

page 2 **Exercise 2**

1. 20   **2.** 256; 65536

3. $6^2 + 7^2 + 42^2 = 43^2$; $x^2 + (x + 1)^2 + [x(x + 1)]^2 = (x^2 + x + 1)^2$

4. (a) $54 \times 9 = 486$ (b) $57 \times 8 = 456$ (c) Three possible answers

5. 32   **8.** 21   **11.** 37

page 3 **Exercise 3**

1. $1\frac{11}{20}$	**2.** $\frac{11}{24}$	**3.** $1\frac{1}{2}$	**4.** $\frac{5}{12}$	**5.** $\frac{4}{15}$
6. $\frac{1}{10}$	**7.** $\frac{8}{15}$	**8.** $\frac{5}{42}$	**9.** $\frac{15}{26}$	**10.** $\frac{5}{12}$
11. $4\frac{1}{2}$	**12.** $1\frac{2}{3}$	**13.** $\frac{23}{40}$	**14.** $\frac{3}{40}$	**15.** $1\frac{7}{8}$
16. $1\frac{1}{12}$	**17.** $1\frac{5}{6}$	**18.** $2\frac{5}{8}$	**19.** $6\frac{1}{10}$	**20.** $9\frac{1}{10}$
21. $1\frac{9}{26}$	**22.** $\frac{1}{9}$	**23.** $\frac{2}{3}$	**24.** $5\frac{1}{4}$	**25.** $10\frac{2}{5}$

26. (a) $\frac{1}{2}, \frac{7}{12}, \frac{2}{3}$ (b) $\frac{2}{3}, \frac{3}{4}, \frac{5}{6}$ (c) $\frac{1}{3}, \frac{5}{8}, \frac{17}{24}, \frac{3}{4}$ (d) $\frac{5}{6}, \frac{8}{9}, \frac{11}{12}$

27. (a) $\frac{1}{2}$ (b) $\frac{3}{4}$ (c) $\frac{17}{24}$ (d) $\frac{7}{18}$ (e) $\frac{3}{10}$ (f) $\frac{5}{12}$

29. 9   **30.** same

page 3 **Exercise 4**

1. 0·25	**2.** 0·4	**3.** 0·8	**4.** 0·75	**5.** 0·5
6. 0·375	**7.** 0·9	**8.** 0·625	**9.** 0·41$\dot{6}$	**10.** 0·1$\dot{6}$
11. 0·$\dot{6}$	**12.** 0·8$\dot{3}$	**13.** 0·$\dot{2}$85 71$\dot{4}$	**14.** 0·$\dot{4}$28571$\dot{1}$	**15.** 0·$\dot{4}$
16. 0·4$\dot{5}$	**17.** 1·2	**18.** 2·625	**19.** 2·$\dot{3}$	**20.** 1·7
21. 2·1875	**22.** 2·$\dot{2}$85 71$\dot{4}$	**23.** 2·$\dot{8}$57 14$\dot{2}$	**24.** 3·19	**25.** $\frac{1}{5}$
26. $\frac{7}{10}$	**27.** $\frac{1}{4}$	**28.** $\frac{9}{20}$	**29.** $\frac{9}{25}$	**30.** $\frac{13}{25}$
31. $\frac{1}{8}$	**32.** $\frac{5}{8}$	**33.** $\frac{21}{25}$	**34.** $2\frac{7}{20}$	**35.** $3\frac{19}{20}$
36. $1\frac{1}{20}$	**37.** $3\frac{1}{5}$	**38.** $\frac{27}{100}$	**39.** $\frac{7}{1000}$	**40.** $\frac{11}{100000}$
41. 0·58	**42.** 1·42	**43.** 0·65	**44.** 1·61	**45.** 0·07
46. 0·16	**47.** 3·64	**48.** 0·60	**49.** $\frac{4}{15}$, 0·33, $\frac{1}{3}$	**50.** $\frac{2}{7}$, 0·3, $\frac{4}{9}$

51. $\frac{7}{11}$, 0·705, 0·71   **52.** $\frac{5}{18}$, 0·3, $\frac{4}{13}$

page 4 **Exercise 5**

1. 3, 11, 19, 23, 29, 31, 37, 47, 59, 61, 67, 73
2. (a) 4, 8, 12, 16, 20   (b) 6, 12, 18, 24, 30   (c) 10, 20, 30, 40, 50
   (d) 11, 22, 33, 44, 55 (e) 20, 40, 60, 80, 100
3. 12 and 24   **4.** 15
5. (a) 1, 2, 3, 6   (b) 1, 3, 9   (c) 1, 2, 5, 10
   (d) 1, 3, 5, 15   (e) 1, 2, 3, 4, 6, 8, 12, 24 (f) 1, 2, 4, 8, 16, 32
6. (a) rational   (b) rational   (c) irrational   (d) rational
   (e) rational   (f) irrational   (g) rational   (h) irrational
   (i) irrational   (j) rational   (k) rational   (l) irrational
7. (a) Yes. Divide by 2, 3, 5, 7, 11, 13. (i.e. prime numbers $< \sqrt{263}$)
   (b) No
   (c) Prime numbers $< \sqrt{1147}$

page 4 **Exercise 6**

1. 18, 22   2. 30, 37   3. 63, 55   4. $-7, -12$ 5. 21, 27
6. 2, -5 7. 16, 22   8. 16, 32   9. 25, 15   10. $-4, -10$
11. $-4, -3$ 12. $7\frac{1}{2}, 3\frac{3}{4}$ 13. $\frac{1}{3}, \frac{1}{9}$ 14. $2, \frac{2}{3}$ 15. 32, 47
16. 840, 6720   17. 5, -1 18. 2, 1

page 4 **Exercise 7**

1. 3, 5, 7, 9, 11   2. 11, 16, 21, 26, 31   3. 12, 10, 8, 6, 4
4. 5, $6\frac{1}{2}$, 8, $9\frac{1}{2}$, 11   5. 3, 6, 12, 24, 48   6. 64, 32, 16, 8, 4
7. 2, 4, 16, 256, 65536   8. 0, 1, 2, 5, 26   9. 3, 4, 3, 4, 3
10. 2, $\frac{1}{2}$, 2, $\frac{1}{2}$, 2   11. 1, 4, 9, 16   12. 0, 3, 8, 15
13. 2, 8, 18, 32   14. 3, 5, 7, 9   15. 2, 6, 12, 20
16. 2, 4, 8, 16   17. 6, 12, 20, 30   18. 6, 12, 24, 48
19. $3n$ 20. $5n$ 21. n^2
22. 2^{n-2} 23. 3^n 24. $n(n + 2)$

page 5 **Exercise 8**

1. (a) 8   (b) 8·17   (c) 8·17   2. (a) 20   (b) 19·6   (c) 19·62
3. (a) 20   (b) 20·0   (c) 20·04   4. (a) 1   (b) 0·815   (c) 0·81
5. (a) 311   (b) 311   (c) 311·14   6. (a) 0   (b) 0·275   (c) 0·28
7. (a) 0   (b) 0·00747   (c) 0·01   8. (a) 16   (b) 15·6   (c) 15·62
9. (a) 900   (b) 900   (c) 900·12   10. (a) 4   (b) 3·56   (c) 3·56
11. (a) 5   (b) 5·45   (c) 5·45   12. (a) 21   (b) 21·0   (c) 20·96
13. (a) 0   (b) 0·0851   (c) 0·09   14. (a) 1   (b) 0·515   (c) 0·52
15. (a) 3   (b) 3·07   (c) 3·07   16. 5·7   17. 0·8   18. 11·2
19. 0·1   20. 0·0   21. 11·1   22. (i) 10·5   (ii) 4·3
23. (i) 13, 11   (ii) 3, 1   (iii) 0·846, 0·6   24. (a) 56 cm^2 (b) 28 cm^2 25. 3·30, 2·87

page 5 **Exercise 9**

1. 70·56   2. 118·958   3. 451·62   4. 33678·8   5. 0·6174
6. 1068   7. 19·53   8. 18914·4   9. 38·72   10. 0·00979
11. 2·4   12. 11   13. 41   14. 8·9   15. 4·7
16. 56   17. 0·0201   18. 30·1   19. 1·3   20. 0·31
21. 210·21   22. 294   23. 282·131   24. 35   25. 242

page 6   *Exercise 10*

1. 4×10^3	**2.** 5×10^2	**3.** 7×10^4	**4.** 6×10	**5.** $2 \cdot 4 \times 10^3$
6. $3 \cdot 8 \times 10^2$	**7.** $4 \cdot 6 \times 10^4$	**8.** $4 \cdot 6 \times 10$	**9.** 9×10^5	**10.** $2 \cdot 56 \times 10^3$
11. 7×10^{-3}	**12.** 4×10^{-4}	**13.** $3 \cdot 5 \times 10^{-3}$	**14.** $4 \cdot 21 \times 10^{-1}$	**15.** $5 \cdot 5 \times 10^{-5}$
16. 1×10^{-2}	**17.** $5 \cdot 64 \times 10^5$	**18.** $1 \cdot 9 \times 10^7$	**19.** $1 \cdot 1 \times 10^9$	**20.** $1 \cdot 67 \times 10^{-24}$
21. $5 \cdot 1 \times 10^8$	**22.** $2 \cdot 5 \times 10^{-10}$	**23.** $6 \cdot 023 \times 10^{23}$	**24.** 3×10^{10}	**25.** £$3 \cdot 6 \times 10^6$

page 6   *Exercise 11*

1. $1 \cdot 5 \times 10^7$	**2.** 3×10^8	**3.** $2 \cdot 8 \times 10^{-2}$	**4.** 7×10^{-9}	**5.** 2×10^6
6. 4×10^{-6}	**7.** 9×10^{-2}	**8.** $6 \cdot 6 \times 10^{-8}$	**9.** $3 \cdot 5 \times 10^{-7}$	**10.** 10^{-16}
11. 8×10^9	**12.** $7 \cdot 4 \times 10^{-7}$	**13.** c, a, b	**14.** 13	**15.** 16

16. (i) $8 \cdot 75 \times 10^2$, $3 \cdot 75 \times 10^2$ (ii) 10^8, $4 \cdot 29 \times 10^7$

17. 50 min   **18.** 6×10^2 **19.** (a) $20 \cdot 5$ s   (b) $6 \cdot 3 \times 10^{91}$ years

page 7   *Exercise 12*

1. $1 : 3$	**2.** $1 : 6$	**3.** $1 : 50$	**4.** $1 : 1 \cdot 6$	**5.** $1 : 0 \cdot 75$
6. $1 : 0 \cdot 375$	**7.** $1 : 25$	**8.** $1 : 8$	**9.** $2 \cdot 4 : 1$	**10.** $2 \cdot 5 : 1$
11. $0 \cdot 8 : 1$	**12.** $0 \cdot 02 : 1$	**13.** £15, £25	**14.** £36, £84	**15.** 140 m, 110 m

16. £18, £27, £72   **17.** 15 kg, 75 kg, 90 kg   **18.** 46 min, 69 min, 69 min

19. £39   **20.** 18 kg, 36 kg, 54 kg, 72 kg   **21.** £400, £1000, £1000, £1600

22. $5 : 3$	**23.** £200	**24.** $3 : 7$	**25.** $\frac{1}{7}x$	**26.** 6
27. 12	**28.** £120	**29.** 300 g	**30.** 625	

page 8   *Exercise 13*

1. £1·68	**2.** £84	**3.** 6 days	**4.** $2\frac{1}{2}$ litres	**5.** 60 km
6. 119 g	**7.** £68·40	**8.** $2\frac{1}{4}$ weeks	**9.** 80p	**10.** (a) 12; (b) 2100
11. 4	**12.** 5·6 days	**13.** £175	**14.** 540°	**15.** £1·20
16. 190 m	**17.** 1250	**18.** 14 min	**19.** 11·2 h	**20.** 10 h
21. 57·1 min	**22.** 243 kg	**23.** 12 days		

page 9   *Exercise 14*

1. (a) Fr 218   (b) $121·80   (c) Ptas 36 400   (d) DM 6·3   (e) Lire 5244   (f) $1·566

2. (a) £45·87   (b) £1436·78   (c) £1·79   (d) £10·34   (e) £219·30   (f) £5·22

3. £1·80   **4.** Spain by £1·20   **5.** £20 000

6. Germany £12 400; USA £14 300; Britain £15 000; Belgium £17 200, France £17 800

7. 3·25 Swiss francs = £1   **8.** DM 1690   **9.** £2197·46   **10.** $62\frac{1}{2}$p

page 9   *Exercise 15*

1. (a) 70 m   (b) 16 m   (c) 3·55 m   (d) 108·5 m

2. (a) 5 cm   (b) 3·5 cm   (c) 0·72 cm   (d) 2·86 cm

3. (a) 450 000 cm   (b) 4500 m   (c) 4·5 km

4. 12·3 km   **5.** 4·71 km   **6.** 50 cm   **7.** 640 cm   **8.** 5·25 cm

page 10   *Exercise 16*

1. 40 m by 30 m; 12 cm²; 1200 m²   **2.** 1 m², 6 m²   **3.** 0·32 km²   **4.** 5 m²

5. 150 km²   **6.** 1200 hectares   **7.** 240 cm²   **8.** 1 : 50 000

page 10   ***Exercise 17***

1. (a) $\frac{3}{5}$ (b) $\frac{6}{25}$ (c) $\frac{7}{20}$ (d) $\frac{1}{50}$
2. (a) 25%   (b) 10%   (c) $87\frac{1}{2}$%   (d) $33\frac{1}{3}$%   (e) 72%   (f) 31%
3. (a) 0·36   (b) 0·28   (c) 0·07   (d) 0·134   (e) 0·6   (f) 0·875
4. (a) 45%; $\frac{1}{2}$; 0·6   (b) 4%; $\frac{6}{16}$; 0·38   (c) 11%; 0·111; $\frac{1}{9}$ (d) 0·3; 32%; $\frac{1}{3}$
5. (a) 85%   (b) 77·5%   (c) 23·75%   (d) 56%   (e) 10%   (f) 37·5%

page 11   ***Exercise 18***

1. (a) £15   (b) £900   (c) $2·80   (d) 125
2. £32   **3.** 13·2p   **4.** 52·8 kg
5. (a) £1·02   (b) £21·58   (c) £2·22   (d) £0·53
6. £243·28   **7.** £26 182   **8.** 96·8%   **9.** 77·5%
10. £71·48   **11.** 200 kg   **12.** 29 000   **13.** 500 cm   **14.** £6·30
15. 400 kg   **16.** 325   **17.** £35·25   **18.** £8425·60

page 12   ***Exercise 19***

1. £49   **2.** £136   **3.** 14 000   **4.** 5500   **5.** 22 000
6. 12 000   **7.** 62p   **8.** 72p

page 12   ***Exercise 20***

1. (a) 25%, profit   (b) 25%, profit   (c) 10%, loss   (d) 20%, profit
    (e) 30%, profit   (f) 7·5%, profit   (g) 12%, loss   (h) 54%, loss
2. 28%   **3.** $44\frac{4}{9}$%   **4.** 46·9%   **5.** 12%   **6.** $5\frac{1}{3}$%
7. (a) £50   (b) £450   (c) £800   (d) £12·40
8. £500   **9.** £12   **10.** £5   **11.** 60p   **12.** £220
13. 14·3%   **14.** 20%   **15.** 8 : 11   **16.** 21%   **17.** 20%

page 13   ***Exercise 21***

1. 7·15%   **2.** 4·23%   **3.** 12·9%   **4.** 0·177%   **5.** 6·15%
6. 0·0402%   **7.** 1·55%   **8.** 2·1%, 5·5%; Katy   **9.** 8%

page 14   ***Exercise 22***

1. (a) £2180   (b) £2376·20   (c) £2590·06
2. (a) £5550   (b) £6838·16   (c) £8425·29
3. £13 108   **4.** £5657   **5.** No. Should be > £193.
6. (a) £14 033   (b) £734   (c) £107 946
7. 8   **8.** 11   **9.** (b) $x = 9$ (c) 9 years

page 14   ***Exercise 23***

1. (a) £1380   (b) 412·50   (c) £3337·50   (d) £3855   (e) £4890·50   (f) £13 875
2. (a) £10 000   (b) £5200   (c) £13 150   (d) £15 965·56   (e) £16 212·50   (f) £20 471·67
3. (a) £183·20   (b) £268
4. 32%   **5.** 29%   **6.** £99·60; £628·80

page 16   ***Exercise 24***

1. (a) $2\frac{1}{2}$ h   (b) $3\frac{1}{8}$ h   (c) 75 s   (d) 4 h
2. (a) 20 m/s   (b) 30 m/s   (c) $83\frac{1}{3}$ m/s   (d) 108 km/h   (e) 79·2 km/h
    (f) 1·2 cm/s   (g) 90 m/s   (h) 25 mph   (i) 0·03 miles per second

3. (a) 75 km/h   (b) 4·52 km/h   (c) 7·6 m/s   (d) 4×10^6 m/s   (e) $2·5 \times 10^8$ m/s
    (f) 200 km/h   (g) 3 km/h
4. (a) 110 000 m   (b) 10 000 m   (c) 56 400 m   (d) 4500 m   (e) 50 400 m
    (f) 80 m   (g) 960 000 m
5. (a) 3·125 h   (b) 76·8 km/h   **6.** (a) 4·45 h   (b) 23·6 km/h
7. 46 km/h
8. (a) 8 m/s   (b) 7·6 m/s   (c) 102·63 s   (d) 7·79 m/s
9. 1230 km/h   **10.** 3 h   **11.** 100 s   **12.** $1\frac{1}{2}$ minutes
13. 600 m   **14.** $53\frac{1}{3}$ s   **15.** 5 cm/s   **16.** 60 s   **17.** 120 mph

page 17   *Exercise 25*

1. 3·041   **2.** 1460   **3.** 0·030 83   **4.** 47·98   **5.** 130·6
6. 0·4771   **7.** 0·3658   **8.** 37·54   **9.** 8·000   **10.** 0·6537
11. 0·037 16   **12.** 34·31   **13.** 0·7195   **14.** 3·598   **15.** 0·2445
16. 2·043   **17.** 0·3798   **18.** 0·7683   **19.** −0·5407   **20.** 0·070 40
21. 2·526   **22.** 0·094 78   **23.** 0·2110   **24.** 3·123   **25.** 2·230
26. 128·8   **27.** 4·268   **28.** 3·893   **29.** 0·6290   **30.** 0·4069
31. 9·298   **32.** 0·1010   **33.** 0·3692   **34.** 1·125   **35.** 1·677
36. 0·9767   **37.** 0·8035   **38.** 0·3528   **39.** 2·423   **40.** 1·639
41. 0·000 465 9   **42.** 0·3934   **43.** −0·7526   **44.** 2·454   **45.** 40 000
46. 0·070 49   **47.** 405 400   **48.** 471·3   **49.** 20 810   **50.** $2·218 \times 10^6$
51. $1·237 \times 10^{-24}$ **52.** 3·003   **53.** 0·035 81   **54.** 47·40   **55.** −1748
56. 0·011 38   **57.** 1757   **58.** 0·026 35   **59.** 0·1651   **60.** 5447
61. 0·006 562   **62.** 0·1330   **63.** 0·4451   **64.** 0·036 16   **65.** 19·43
66. $1·296 \times 10^{-15}$ **67.** $5·595 \times 10^{14}$ **68.** $1·022 \times 10^{-8}$ **69.** 0·019 22   **70.** 0·9613

page 18   *Revision Exercise 1A*

1. (a) 185   (b) 150   (c) 40   (d) $\frac{11}{12}$ (e) $2\frac{4}{5}$ (f) $\frac{2}{5}$

2. 128 cm   **3.** $\frac{2}{5}$ **4.** $\frac{a}{b}$ **5.** (a) 0·0547   (b) 0·055   (c) $5·473 \times 10^{-2}$

6. 1·238   **7.** (a) 3×10^7 (b) $3·7 \times 10^4$ (c) $2·7 \times 10^{13}$
8. (a) £26   (b) 6 : 5   (c) 6   **9.** £75
10. (a) (i) 57·2%   (ii) $87\frac{1}{2}$%   (b) 40%   (c) 80p   **11.** 5%
12. (a) £500   (b) $37\frac{1}{2}$%   **13.** £350   **14.** £150·50, £8910
15. (a) 2·4 km   (b) 1 km^2 **16.** (a) 300 m   (b) 60 cm   (c) 150 cm^2
17. (a) 1 : 50 000   (b) 1 : 4 000 000   **18.** (a) 22%   (b) 20·8%   (c) £240
19. (a) (i) 7 m/s   (ii) 200 m/s   (iii) 5 m/s   (b) (i) 144 km/h   (ii) 2·16 km/h
20. (a) 0·005 m/s   (b) 1·6 s   (c) 172·8 km   **21.** $33\frac{1}{3}$ mph
22. (a) (i) 0·05   (ii) $1\frac{1}{2}$ (b) (i) 12·2   (ii) 38·7   (iii) 0·0387
    (c) (i) 138   (ii) 13·8   (iii) 0·0138
23. (a) 5, 8, 11, 14   (b) 2, 6, 18, 54   (c) 60, 53, 46, 39   (d) 3, 9, 81, 6561
24. $2·3 \times 10^9$
25. (a) 600   (b) 10 000   (c) 3   (d) 60
26. (a) 0·5601   (b) 3·215   (c) 0·6161   (d) 0·4743
27. (a) 0·340   (b) $4·08 \times 10^{-6}$ (c) 64·9   (d) 0·119
28. 33·1%

page 20   *Examination Exercise 1B*

1. (a) 3·05   (b) 3
2. (a) 31   (b) D   (c) B, D or A, E   (d) 34   (e) N, R, T, U

3. (a) 3   (b) 10   (c) 1, 9   (d) 1, 8   (e) 3, 9
     (f) $p = 1, q = 3, r = 9, s = 8, t = 10$
4. (a) (i) £11·76   (ii) £8.99   (iii) £9·60   (iv) £2·10
     (b) (i) 90p   (ii) 54p   (iii) £1·71   (iv) 18p
     (c) £35·78   (d) £32·92
5. 176p   **6.** (a) 3 km   (b) 625 000 m^2
7. 0·4 m^3 **8.** (i) £36100   (ii) £5367
9. (i) £20·52   (ii) £36·90
10. (a) £13130   (b) (i) £280; £313·60; £351·23   (ii) 40·5%
11. (a) 300   (b) 10p   (c) 11p   (d) 31·4%   (e) £141·45
12. (b) 1 m   (c) 1 m   (d) £1250; £1625
     (e) 25 of A = £31 250; 17 of B = £27 625; choose B.

PART 2

page 24   ***Exercise 1***

1. 13	**2.** 211	**3.** −12	**4.** −31	**5.** −66
6. 6·1	**7.** 9·1	**8.** −35	**9.** 18·7	**10.** −9
11. −3	**12.** 3	**13.** −2	**14.** −14	**15.** −7
16. 3	**17.** 181	**18.** −2·2	**19.** 8·2	**20.** 17
21. 2	**22.** −6	**23.** −15	**24.** −14	**25.** −2
26. −12	**27.** −80	**28.** −13·1	**29.** −4·2	**30.** 12·4
31. −7	**32.** 8	**33.** 4	**34.** −10	**35.** 11
36. 4	**37.** −20	**38.** 8	**39.** −5	**40.** −10
41. −26	**42.** −21	**43.** 8	**44.** 1	**45.** −20·2
46. −50	**47.** −508	**48.** −29	**49.** 0	**50.** −21
51. −0·1	**52.** −4	**53.** 6·7	**54.** 1	**55.** −850
56. 4	**57.** 6	**58.** −4	**59.** −12	**60.** −31

page 25   ***Exercise 2***

1. −8	**2.** 28	**3.** 12	**4.** 24	**5.** 18
6. −35	**7.** 49	**8.** −12	**9.** −2	**10.** 9
11. −4	**12.** 4	**13.** −4	**14.** 8	**15.** 70
16. −7	**17.** $\frac{1}{4}$	**18.** $-\frac{3}{5}$	**19.** −0·01	**20.** 0·0002
21. 121	**22.** 6	**23.** −600	**24.** −1	**25.** −20
26. −2·6	**27.** −700	**28.** 18	**29.** −1000	**30.** 640
31. −6	**32.** −42	**33.** −0·4	**34.** −0·4	**35.** −200
36. −35	**37.** −2	**38.** $\frac{1}{2}$	**39.** $-\frac{1}{4}$	**40.** −90

page 25   ***Exercise 3***

1. −10	**2.** 1	**3.** 12	**4.** −28	**5.** −2
6. 16	**7.** −3	**8.** 14	**9.** −28	**10.** 4
11. $-\frac{1}{6}$	**12.** 9	**13.** −30	**14.** 24	**15.** −1
16. −2	**17.** −30	**18.** 7	**19.** 3	**20.** 16
21. 93	**22.** 2400	**23.** 10	**24.** 1	**25.** −4
26. 48	**27.** −1	**28.** 0	**29.** −8	**30.** 170
31. −3	**32.** 1	**33.** 1	**34.** 0	**35.** 15
36. 5	**37.** −2·4	**38.** −180	**39.** 5	**40.** −994
41. 2	**42.** −48	**43.** 60	**44.** −2·5	**45.** −32
46. 0	**47.** −0·1	**48.** −16	**49.** −4·3	**50.** $-\frac{1}{16}$

page 26   ***Exercise 4***

1. 21   **2.** 1·62   **3.** 396   **4.** 650   **5.** 63·8
6. 9×10^{12} **7.** $10\frac{1}{2}$ **8.** 800   **9.** $ac + ab - a^2$ **10.** $r - p + q$
11. 802; $4n + 2$ **12.** $2n + 6$

page 27   ***Exercise 5***

1. 7   **2.** 13   **3.** 13   **4.** 22   **5.** 1
6. -1 **7.** 18   **8.** -4 **9.** -3 **10.** 37
11. 0   **12.** -4 **13.** -7 **14.** -2 **15.** -3
16. -8 **17.** -30 **18.** 16   **19.** -10 **20.** 0
21. 7   **22.** -6 **23.** -2 **24.** -7 **25.** -5
26. 3   **27.** 4   **28.** -8 **29.** -2 **30.** 2
31. 0   **32.** 4   **33.** -4 **34.** -3 **35.** -9
36. 4

page 27   ***Exercise 6***

1. 9   **2.** 27   **3.** 4   **4.** 16   **5.** 36
6. 18   **7.** 1   **8.** 6   **9.** 2   **10.** 8
11. -7 **12.** 15   **13.** -23 **14.** 3   **15.** 32
16. 36   **17.** 144   **18.** -8 **19.** -7 **20.** 13
21. 5   **22.** -16 **23.** 84   **24.** 17   **25.** 6
26. 0   **27.** -25 **28.** -5 **29.** 17   **30.** $-1\frac{1}{2}$
31. 19   **32.** 8   **33.** 19   **34.** 16   **35.** -16
36. 12   **37.** 36   **38.** -12 **39.** 2   **40.** 11
41. -23 **42.** -26 **43.** 5   **44.** 31   **45.** $4\frac{1}{2}$

page 27   ***Exercise 7***

1. -20 **2.** 16   **3.** -42 **4.** -4 **5.** -90
6. -160 **7.** -2 **8.** -81 **9.** 4   **10.** 22
11. 14   **12.** 5   **13.** 1   **14.** $\sqrt{5}$ **15.** 4
16. $-6\frac{1}{2}$ **17.** 54   **18.** 25   **19.** 4   **20.** 312
21. 45   **22.** 22   **23.** 14   **24.** -36 **25.** -7
26. 1   **27.** 901   **28.** -30 **29.** -5 **30.** $7\frac{1}{2}$
31. -7 **32.** $-\frac{3}{13}$ **33.** 7   **34.** -2 **35.** 0
36. $-4\frac{1}{2}$ **37.** 6   **38.** 2   **39.** 26   **40.** -9
41. $3\frac{1}{4}$ **42.** $-\frac{5}{6}$ **43.** 4   **44.** $2\frac{2}{3}$ **45.** $3\frac{1}{4}$
46. $-2\frac{1}{6}$ **47.** -13 **48.** 12   **49.** $1\frac{1}{3}$ **50.** $-\frac{5}{36}$

page 28   ***Exercise 8***

1. $3x + 11y$ **2.** $2a + 8b$ **3.** $3x + 2y$ **4.** $5x + 5$
5. $9 + x$ **6.** $3 - 9y$ **7.** $5x - 2y - x^2$ **8.** $2x^2 + 3x + 5$
9. $-10y$ **10.** $3a^2 + 2a$ **11.** $7 + 7a - 7a^2$ **12.** $5x$
13. $\dfrac{10}{a} - b$ **14.** $\dfrac{5}{x} - \dfrac{5}{y}$ **15.** $\dfrac{3m}{x}$ **16.** $\dfrac{1}{2} - \dfrac{2}{x}$
17. $\dfrac{5}{a} + 3b$ **18.** $-\dfrac{n}{4}$ **19.** $7x^2 - x^3$ **20.** $2x^2$
21. $x^2 + 5y^2$ **22.** $-12x^2 - 4y^2$ **23.** $5x - 11x^2$ **24.** $\dfrac{8}{x^2}$
25. $5x + 2$ **26.** $12x - 7$ **27.** $3x + 4$ **28.** $11 - 6x$
29. $-5x - 20$ **30.** $7x - 2x^2$ **31.** $3x^2 - 5x$ **32.** $x - 4$

33. $5x^2 + 14x$ **34.** $-4x^2 - 3x$ **35.** $5a + 8$ **36.** $a + 9$
37. $ab + 4a$ **38.** $y^2 + y$ **39.** $2x - 2$ **40.** $6x + 3$
41. $x - 4$ **42.** $7x + 5y$ **43.** $4x^2 - 11x$ **44.** $2x^2 + 14x$
45. $3y^2 - 4y + 1$ **46.** $12x + 12$ **47.** $4ab - 3a + 14b$ **48.** $2x - 4$

page 29   **Exercise 9**

1. $x^2 + 4x + 3$ **2.** $x^2 + 5x + 6$ **3.** $y^2 + 9y + 20$ **4.** $x^2 + x - 12$
5. $x^2 + 3x - 10$ **6.** $x^2 - 5x + 6$ **7.** $a^2 - 2a - 35$ **8.** $z^2 + 7z - 18$
9. $x^2 - 9$ **10.** $k^2 - 121$ **11.** $2x^2 - 5x - 3$ **12.** $3x^2 - 2x - 8$
13. $2y^2 - y - 3$ **14.** $49y^2 - 1$ **15.** $9x^2 - 4$ **16.** $6a^2 + 5ab + b^2$
17. $3x^2 + 7xy + 2y^2$ **18.** $6b^2 + bc - c^2$ **19.** $-5x^2 + 16xy - 3y^2$ **20.** $15b^2 + ab - 2a^2$
21. $2x^2 + 2x - 4$ **22.** $6x^2 + 3x - 9$ **23.** $24y^2 + 4y - 8$ **24.** $6x^2 - 10x - 4$
25. $4a^2 - 16b^2$ **26.** $x^3 - 3x^2 + 2x$ **27.** $8x^3 - 2x$ **28.** $3y^3 + 3y^2 - 18y$
29. $x^3 + x^2y + x^2z + xyz$ **30.** $3za^2 + 3zam - 6zm^2$

page 29   **Exercise 10**

1. $x^2 + 8x + 16$ **2.** $x^2 + 4x + 4$ **3.** $x^2 - 4x + 4$ **4.** $4x^2 + 4x + 1$
5. $y^2 - 10y + 25$ **6.** $9y^2 + 6y + 1$ **7.** $x^2 + 2xy + y^2$ **8.** $4x^2 + 4xy + y^2$
9. $a^2 - 2ab + b^2$ **10.** $4a^2 - 12ab + 9b^2$ **11.** $3x^2 + 12x + 12$ **12.** $9 - 6x + x^2$
13. $9x^2 + 12x + 4$ **14.** $a^2 - 4ab + 4b^2$ **15.** $2x^2 + 6x + 5$ **16.** $2x^2 + 2x + 13$
17. $5x^2 + 8x + 5$ **18.** $2y^2 - 14y + 25$ **19.** $10x - 5$ **20.** $-8x + 8$
21. $-10y + 5$ **22.** $3x^2 - 2x - 8$ **23.** $2x^2 + 4x - 4$ **24.** $-x^2 - 18x + 15$

page 30   **Exercise 11**

1. 8   **2.** 9   **3.** 7   **4.** 10   **5.** $\frac{1}{3}$
6. 10   **7.** $1\frac{1}{2}$ **8.** -1 **9.** $-1\frac{1}{2}$ **10.** $\frac{1}{3}$
11. 35   **12.** 130   **13.** 14   **14.** $\frac{2}{3}$ **15.** $3\frac{1}{3}$
16. $-2\frac{1}{2}$ **17.** 3   **18.** $1\frac{1}{8}$ **19.** $\frac{3}{10}$ **20.** $-1\frac{1}{4}$
21. 10   **22.** 27   **23.** 20   **24.** 18   **25.** 28
26. -15 **27.** $\frac{99}{100}$ **28.** 0   **29.** 1000   **30.** $-\frac{1}{1000}$
31. 1   **32.** -7 **33.** -5 **34.** $1\frac{1}{6}$ **35.** 1
36. 2   **37.** -5 **38.** -3 **39.** $-1\frac{1}{2}$ **40.** 2
41. 1   **42.** $3\frac{1}{2}$ **43.** 2   **44.** -1 **45.** $10\frac{2}{3}$
46. $1 \cdot 1$ **47.** -1 **48.** 2   **49.** $2\frac{1}{2}$ **50.** $1\frac{1}{3}$

page 30   **Exercise 12**

1. $-1\frac{1}{2}$ **2.** 2   **3.** $-\frac{2}{5}$ **4.** $-\frac{1}{3}$ **5.** $1\frac{2}{3}$
6. 6   **7.** $-\frac{2}{5}$ **8.** $-3\frac{1}{5}$ **9.** $\frac{1}{2}$ **10.** -4
11. 18   **12.** 5   **13.** 4   **14.** 3   **15.** $2\frac{3}{4}$
16. $-\frac{7}{22}$ **17.** $\frac{1}{4}$ **18.** 1   **19.** 4   **20.** -11
21. $-7\frac{1}{3}$ **22.** $1\frac{1}{4}$ **23.** -5 **24.** 6   **25.** 3
26. 6   **27.** 2   **28.** 3   **29.** 4   **30.** 3
31. $10\frac{1}{2}$ **32.** 5   **33.** 2   **34.** -1 **35.** -17
36. $-2\frac{9}{10}$ **37.** $2\frac{10}{21}$ **38.** $\frac{1}{3}$ **39.** 14   **40.** 15

page 31   **Exercise 13**

1. $\frac{1}{4}$
2. -3
3. 4
4. $-7\frac{2}{3}$
5. -43
6. 11
7. $-\frac{1}{2}$
8. 0
9. 1
10. $-1\frac{2}{3}$
11. $\frac{1}{4}$
12. 0
13. $-\frac{6}{7}$
14. $1\frac{9}{17}$
15. $1\frac{22}{23}$
16. $\frac{2}{11}$
17. $10, 8, 6$
18. $13, 12, 5$
19. $10, 8, 6$
20. $13, 12, 5$
21. $5, 4, 3$
22. $13, 12, 5$
23. 4 cm
24. 5 m
25. 4

page 32   **Exercise 14**

1. $\frac{1}{3}$
2. $\frac{1}{5}$
3. $1\frac{2}{3}$
4. -3
5. $\frac{5}{11}$
6. -2
7. 6
8. $3\frac{3}{4}$
9. -7
10. $-7\frac{2}{3}$
11. 2
12. 3
13. 4
14. -2
15. -3
16. 3
17. $1\frac{5}{7}$
18. $4\frac{4}{5}$
19. 10
20. 24
21. 2
22. 3
23. 5
24. -4
25. $6\frac{3}{4}$
26. -3
27. 0
28. 3
29. 0
30. 1
31. 2
32. 3
33. 4
34. $\frac{3}{5}$
35. $1\frac{1}{8}$
36. -1
37. 1
38. 1
39. $\frac{1}{4}$
40. $-\frac{1}{3}$
41. $\frac{9}{10}$
42. 1
43. 2
44. $-\frac{1}{7}$
45. 2
46. 3

page 33   **Exercise 15**

1. $91, 92, 93$
2. $21, 22, 23, 24$
3. $57, 59, 61$
4. $506, 508, 510$
5. $12\frac{1}{2}$
6. $12\frac{1}{2}$
7. $11\frac{2}{3}$
8. $8\frac{1}{3}, 41\frac{2}{3}$
9. $1\frac{1}{4}, 13\frac{3}{4}$
10. $3\frac{1}{3}$ cm
11. 12 cm
12. $20°$
13. 5 cm
14. 7 cm
15. $18\frac{1}{2}, 27\frac{1}{2}$
16. $20°, 60°, 100°$
17. $45°, 60°, 75°$
18. 5
19. $6, 8$
20. $12, 24, 30$
21. $5, 15, 8$
22. $59\frac{2}{3}$ kg, $64\frac{2}{3}$ kg, $72\frac{2}{3}$ kg
23. $24, 22, 15$
24. $48, 12$
25. $40, 8$
26. 6
27. $168\cdot84$ cm^2
28. 14
29. 45p, 31p
30. £$21\cdot50$

page 34   **Exercise 16**

1. £3700
2. 3
3. 8
4. $1\frac{3}{7}$ m
5. $80°, 100°$
6. $30°, 60°, 90°, 120°, 150°, 270°$
7. $26, 58$
8. 2 km
9. 8 km
10. 400 m
11. 21
12. 23
13. £3600
14. 15
15. 2 km
16. $6, 7, 8, 9$
17. $2, 3, 4, 5$

page 36   **Exercise 17**

1. $x = 2, y = 1$
2. $x = 4, y = 2$
3. $x = 3, y = 1$
4. $x = -2, y = 1$
5. $x = 3, y = 2$
6. $x = 5, y = -2$
7. $x = 2, y = 1$
8. $x = 5, y = 3$
9. $x = 3, y = -1$
10. $a = 2, b = -3$
11. $a = 5, b = \frac{1}{4}$
12. $a = 1, b = 3$
13. $m = \frac{1}{2}, n = 4$
14. $w = 2, x = 3$
15. $x = 6, y = 3$
16. $x = \frac{1}{2}, z = -3$
17. $m = 1\frac{15}{17}, n = \frac{11}{17}$
18. $c = 1\frac{16}{23}, d = 2\frac{12}{23}$

page 36   **Exercise 18**

1. 1
2. -3
3. -2
4. 15
5. -12
6. -3
7. -2
8. -11
9. -21
10. 1
11. 0
12. 15
13. -10
14. 3
15. 6
16. -11
17. 2
18. 5
19. -19
20. -4
21. x
22. $-3x$
23. $4x$
24. $4y$
25. $9y$
26. $3x$
27. $-8x$
28. $4x$
29. $2x$
30. $3y$

page 37   **Exercise 19**

1. $x = 2, y = 4$
2. $x = 1, y = 4$
3. $x = 2, y = 5$
4. $x = 3, y = 7$
5. $x = 5, y = 2$
6. $a = 3, b = 1$
7. $x = 1, y = 3$
8. $x = 1, y = 3$

9. $x = -2, y = 3$ **10.** $x = 4, y = 1$ **11.** $x = 1, y = 5$ **12.** $x = 0, y = 2$
13. $x = \frac{5}{7}, y = 4\frac{3}{7}$ **14.** $x = 1, y = 2$ **15.** $x = 2, y = 3$ **16.** $x = 4, y = -1$
17. $x = 3, y = 1$ **18.** $x = 1, y = 2$ **19.** $x = 2, y = 1$ **20.** $x = -2, y = 1$
21. $x = 1, y = 2$ **22.** $a = 4, b = 3$ **23.** $x = -23, y = -78$ **24.** $x = 3, y = \frac{1}{2}$
25. $x = 4, y = 3$ **26.** $x = 5, y = -2$ **27.** $x = \frac{1}{3}, y = -2$ **28.** $x = 5\frac{5}{14}, y = \frac{2}{7}$
29. $x = 3, y = -1$ **30.** $x = 5, y = 0\cdot2$

page 37   *Exercise 20*

1. $5\frac{1}{2}, 9\frac{1}{2}$ **2.** $6, 3$ or $2\frac{2}{5}, 5\frac{2}{5}$ **3.** $4, 10$ **4.** $a = 2, c = 7$
5. $m = 4, c = -3$ **6.** $a = 1, b = -2$ **7.** $m = 1p, w = 3p$ **8.** TV £200, video £450
9. $7, 3$ **10.** white 2 oz, brown $3\frac{1}{2}$ oz   **11.** 120 m, 240 m
12. 150 m, 350 m   **13.** $2p \times 15, 5p \times 25$ **14.** $10p \times 14, 50p \times 7$ **15.** 20
16. man £50, woman £70   **17.** current 4 m/s, kipper 10 m/s
18. $\frac{5}{7}$ **19.** $\frac{3}{5}$ **20.** boy 10, mouse 3   **21.** $4, 7$
22. $y = 3x - 2$ **23.** walks 4 m/s, runs 5 m/s   **24.** £1 $\times$ 15, £5 $\times$ 5
25. $36, 9$ **26.** wind $4\frac{1}{2}$ knots, submarine $20\frac{1}{2}$ knots   **27.** $a = 1, b = 2, c = 5$
28. $y = 2x^2 - 3x + 5$ **29.** $y = x^2 + 3x + 4$ **30.** $y = x^2 + 2x - 3$

page 39   *Exercise 21*

1. $x(x + 5)$ **2.** $x(x - 6)$ **3.** $x(7 - x)$ **4.** $y(y + 8)$
5. $y(2y + 3)$ **6.** $2y(3y - 2)$ **7.** $3x(x - 7)$ **8.** $2a(8 - a)$
9. $3c(2c - 7)$ **10.** $3x(5 - 3x)$ **11.** $7y(8 - 3y)$ **12.** $x(a + b + 2c)$
13. $x(x + y + 3z)$ **14.** $y(x^2 + y^2 + z^2)$ **15.** $ab(3a + 2b)$ **16.** $xy(x + y)$
17. $2a(3a + 2b + c)$ **18.** $m(a + 2b + m)$ **19.** $2k(x + 3y + 2z)$ **20.** $a(x^2 + y + 2b)$
21. $xk(x + k)$ **22.** $ab(a^2 + 2b)$ **23.** $bc(a - 3b)$ **24.** $ae(2a - 5e)$
25. $ab(a^2 + b^2)$ **26.** $x^2y(x + y)$ **27.** $2xy(3y - 2x)$ **28.** $3ab(b^2 - a^2)$
29. $a^2b(2a + 5b)$ **30.** $ax^2(y - 2z)$ **31.** $2ab(x + b + a)$ **32.** $yx(a + x^2 - 2yx)$

page 39   *Exercise 22*

1. $(a + b)(x + y)$ **2.** $(a + b)(y + z)$ **3.** $(x + y)(b + c)$ **4.** $(x + y)(h + k)$
5. $(x + y)(m + n)$ **6.** $(a + b)(h - k)$ **7.** $(a + b)(x - y)$ **8.** $(m + n)(a - b)$
9. $(h + k)(s + t)$ **10.** $(x + y)(s - t)$ **11.** $(a - b)(x - y)$ **12.** $(x - y)(s - t)$
13. $(a - x)(s - y)$ **14.** $(h - b)(x - y)$ **15.** $(m - n)(a - b)$ **16.** $(x - z)(k - m)$
17. $(2a + b)(x + 3y)$ **18.** $(2a + b)(x + y)$ **19.** $(2m + n)(h - k)$ **20.** $(m - n)(2h + 3k)$
21. $(2x + y)(3a + b)$ **22.** $(2a - b)(x - y)$ **23.** $(x^2 + y)(a + b)$ **24.** $(m - n)(s + 2t^2)$

page 40   *Exercise 23*

1. $(x + 2)(x + 5)$ **2.** $(x + 3)(x + 4)$ **3.** $(x + 3)(x + 5)$ **4.** $(x + 3)(x + 7)$
5. $(x + 2)(x + 6)$ **6.** $(y + 5)(y + 7)$ **7.** $(y + 3)(y + 8)$ **8.** $(y + 5)(y + 5)$
9. $(y + 3)(y + 12)$ **10.** $(a + 2)(a - 5)$ **11.** $(a + 3)(a - 4)$ **12.** $(z + 3)(z - 2)$
13. $(x + 5)(x - 7)$ **14.** $(x + 3)(x - 8)$ **15.** $(x - 2)(x - 4)$ **16.** $(y - 2)(y - 3)$
17. $(x - 3)(x - 5)$ **18.** $(a + 2)(a - 3)$ **19.** $(a + 5)(a + 9)$ **20.** $(b + 3)(b - 7)$
21. $(x - 4)(x - 4)$ **22.** $(y + 1)(y + 1)$ **23.** $(y - 7)(y + 4)$ **24.** $(x - 5)(x + 4)$
25. $(x - 20)(x + 12)$ **26.** $(x - 15)(x - 11)$ **27.** $(y + 12)(y - 9)$ **28.** $(x - 7)(x + 7)$
29. $(x - 3)(x + 3)$ **30.** $(x - 4)(x + 4)$

page 40   *Exercise 24*

1. $(2x + 3)(x + 1)$ **2.** $(2x + 1)(x + 3)$ **3.** $(3x + 1)(x + 2)$ **4.** $(2x + 3)(x + 4)$
5. $(3x + 2)(x + 2)$ **6.** $(2x + 5)(x + 1)$ **7.** $(3x + 1)(x - 2)$ **8.** $(2x + 5)(x - 3)$
9. $(2x + 7)(x - 3)$ **10.** $(3x + 4)(x - 7)$ **11.** $(2x + 1)(3x + 2)$ **12.** $(3x + 2)(4x + 5)$
13. $(3x - 2)(x - 3)$ **14.** $(y - 2)(3y - 5)$ **15.** $(4y - 3)(y - 5)$ **16.** $(2y + 3)(3y - 1)$

17. $(2x-5)(3x-6)$ **18.** $(5x+2)(2x+1)$ **19.** $(6x-1)(x-3)$ **20.** $(4x+1)(2x-3)$
21. $(6x+5)(2x-1)$ **22.** $(16x+3)(x+1)$ **23.** $(2a-1)(2a-1)$ **24.** $(x+2)(12x-7)$
25. $(x+3)(15x-1)$ **26.** $(8x+1)(6x+5)$ **27.** $(16x-3)(4x+1)$ **28.** $(15x-1)(8x+5)$
29. $(3x-1)(3x+1)$ **30.** $(2a-3)(2a+3)$

page 40   **Exercise 25**

1. $(y-a)(y+a)$ **2.** $(m-n)(m+n)$ **3.** $(x-t)(x+t)$ **4.** $(y-1)(y+1)$
5. $(x-3)(x+3)$ **6.** $(a-5)(a+5)$ **7.** $(x-\frac{1}{2})(x+\frac{1}{2})$ **8.** $(x-\frac{1}{3})(x+\frac{1}{3})$
9. $(2x-y)(2x+y)$ **10.** $(a-2b)(a+2b)$ **11.** $(5x-2y)(5x+2y)$ **12.** $(3x-4y)(3x+4y)$
13. $\left(x-\frac{y}{2}\right)\left(x+\frac{y}{2}\right)$ **14.** $(3m-\frac{2}{3}n)(3m+\frac{2}{3}n)$ **15.** $(4t-\frac{2}{5}s)(4t+\frac{2}{5}s)$ **16.** $\left(2x-\frac{z}{10}\right)\left(2x+\frac{z}{10}\right)$
17. $x(x-1)(x+1)$ **18.** $a(a-b)(a+b)$ **19.** $x(2x-1)(2x+1)$ **20.** $2x(2x-y)(2x+y)$
21. $3x(2x-y)(2x+y)$ **22.** $2m(3m-2n)(3m+2n)$ **23.** $5(x-\frac{1}{2})(x+\frac{1}{2})$
24. $2a(5a-3b)(5a+3b)$ **25.** $3y(2x-z)(2x+z)$ **26.** $4ab(3a-b)(3a+b)$
27. $2a^3(5a-2b)(5a+2b)$ **28.** $9xy(2x-5y)(2x+5y)$ **29.** 161 **30.** 404
31. 4400 **32.** 2421 **33.** 4329 **34.** $0{\cdot}75$ **35.** $4{\cdot}8$ **36.** -2469 **37.** $0{\cdot}0761$
38. $-10\,900$ **39.** $53{\cdot}6$ **40.** $0{\cdot}000\,005$

page 41   **Exercise 26**

1. $-3, -4$ **2.** $-2, -5$ **3.** $3, -5$ **4.** $2, -3$ **5.** $2, 6$
6. $-3, -7$ **7.** $2, 3$ **8.** $5, -1$ **9.** $-7, 2$ **10.** $-\frac{1}{2}, 2$
11. $\frac{2}{3}, -4$ **12.** $1\frac{1}{2}, -5$ **13.** $\frac{2}{3}, 1\frac{1}{2}$ **14.** $\frac{1}{4}, 7$ **15.** $\frac{3}{5}, -\frac{1}{2}$
16. $7, 8$ **17.** $\frac{5}{6}, \frac{1}{2}$ **18.** $7, -9$ **19.** $-1, -1$ **20.** $3, 3$
21. $-5, -5$ **22.** $7, 7$ **23.** $-\frac{1}{3}, \frac{1}{2}$ **24.** $-1\frac{1}{4}, 2$ **25.** $13, -5$
26. $-3, \frac{1}{6}$ **27.** $\frac{1}{10}, -2$ **28.** $1, 1$ **29.** $\frac{2}{9}, -\frac{1}{4}$ **30.** $-\frac{1}{4}, \frac{3}{5}$

page 41   **Exercise 27**

1. $0, 3$ **2.** $0, -7$ **3.** $0, 1$ **4.** $0, \frac{1}{3}$ **5.** $4, -4$
6. $7, -7$ **7.** $\frac{1}{2}, -\frac{1}{2}$ **8.** $\frac{2}{3}, -\frac{2}{3}$ **9.** $0, -1\frac{1}{2}$ **10.** $0, 1\frac{1}{2}$
11. $0, 5\frac{1}{2}$ **12.** $\frac{1}{4}, -\frac{1}{4}$ **13.** $\frac{1}{2}, -\frac{1}{2}$ **14.** $0, \frac{5}{8}$ **15.** $0, \frac{1}{12}$
16. $0, 6$ **17.** $0, 11$ **18.** $0, 1\frac{1}{2}$ **19.** $0, 1$ **20.** $0, 4$
21. $0, 3$ **22.** $\frac{1}{2}, -\frac{1}{2}$ **23.** $1\frac{1}{3}, -1\frac{1}{3}$ **24.** $3, -3$ **25.** $0, 2\frac{2}{5}$
26. $\frac{1}{3}, -\frac{1}{3}$ **27.** $0, \frac{1}{4}$ **28.** $0, \frac{1}{6}$ **29.** $\frac{1}{4}, -\frac{1}{4}$ **30.** $0, \frac{1}{5}$

page 42   **Exercise 28**

1. $-\frac{1}{2}, -5$ **2.** $-\frac{2}{3}, -3$ **3.** $-\frac{1}{2}, -\frac{2}{3}$ **4.** $\frac{1}{3}, 3$ **5.** $\frac{2}{5}, 1$
6. $\frac{1}{3}, 1\frac{1}{2}$ **7.** $-0{\cdot}63, -2{\cdot}37$ **8.** $-0{\cdot}27, -3{\cdot}73$ **9.** $0{\cdot}72, 0{\cdot}28$ **10.** $6{\cdot}70, 0{\cdot}30$
11. $0{\cdot}19, -2{\cdot}69$ **12.** $0{\cdot}85, -1{\cdot}18$ **13.** $0{\cdot}61, -3{\cdot}28$ **14.** $-1\frac{2}{3}, 4$ **15.** $-1\frac{1}{2}, 5$
16. $3{\cdot}56, -0{\cdot}56$ **17.** $0{\cdot}16, -3{\cdot}16$ **18.** $-\frac{1}{2}, 2\frac{1}{3}$ **19.** $-\frac{1}{3}, -8$ **20.** $1\frac{2}{3}, -1$
21. $2{\cdot}28, 0{\cdot}22$ **22.** $-0{\cdot}35, -5{\cdot}65$ **23.** $-\frac{2}{3}, \frac{1}{2}$ **24.** $-0{\cdot}58, 2{\cdot}58$ **25.** $-2{\cdot}69, 0{\cdot}19$
26. $0{\cdot}22, -1{\cdot}55$ **27.** $-0{\cdot}37, 5{\cdot}37$ **28.** $-\frac{5}{6}, 1\frac{3}{4}$ **29.** $-\frac{7}{9}, 1\frac{1}{4}$ **30.** $1\frac{2}{5}, -2\frac{1}{4}$
31. $-4, 1\frac{1}{2}$ **32.** $-3, 1\frac{2}{3}$ **33.** $-2, 1\frac{2}{3}$ **34.** $-3\frac{1}{2}, \frac{1}{5}$ **35.** $-3, \frac{4}{5}$
36. $-8\frac{1}{2}, 11$

page 42   **Exercise 29**

1. $-3, 2$ **2.** $-3, -7$ **3.** $-\frac{1}{2}, 2$ **4.** $1, 4$ **5.** $-1\frac{2}{3}, \frac{1}{2}$
6. $-0{\cdot}39, -4{\cdot}28$ **7.** $-0{\cdot}16, 6{\cdot}16$ **8.** 3 **9.** $2, -1\frac{1}{3}$ **10.** $-3, -1$
11. $0{\cdot}66, -22{\cdot}66$ **12.** $-7, 2$ **13.** $\frac{1}{4}, 7$ **14.** $-\frac{1}{2}, \frac{3}{5}$ **15.** $0, 3\frac{1}{2}$

16. $-\frac{1}{4}, \frac{1}{4}$ **17.** $-2\cdot77, 1\cdot27$ **18.** $-\frac{2}{3}, 1$ **19.** $-\frac{1}{2}, 2$ **20.** 0, 3
21. (a) -1 (b) $0\cdot6258$ (c) $0\cdot5961$ (d) $0\cdot2210$

page 43   **Exercise 30**

1. 8, 11   **2.** 11, 13   **3.** 12 cm   **4.** 6 cm   **5.** $x = 11$
6. 10 cm × 24 cm   **7.** 8 km north, 15 km east   **8.** 12 eggs   **9.** 13 eggs
10. 4   **11.** 2, 5   **12.** $\frac{40}{x}$ h, $\frac{40}{x-2}$ h, 10 km/h   **13.** 4 km/h
14. 20 mph   **15.** 5 mph   **16.** 157 km   **17.** $x = 2$ **18.** $x = 3$
19. $\frac{3}{4}$ **20.** 9 cm or 13 cm

page 44   **Revision Exercise 2A**

1. (a) $-2\frac{1}{2}$ (b) $2\frac{2}{3}$ (c) 0, -5 (d) 2, -2 (e) $-5, 2\frac{2}{3}$
2. (a) 14   (b) 18   (c) 28
3. (a) $(2x - y)(2x + y)$ (b) $2(x + 3)(x + 1)$ (c) $(2 - 3k)(3m + 2n)$ (d) $(2x + 1)(x - 3)$
4. (a) $x = 3, y = -2$ (b) $m = 1\frac{1}{2}, n = -3$ (c) $x = 7, y = \frac{1}{2}$ (d) $x = -1, y = -2$
5. (a) 8   (b) 140   (c) 29   (d) 42   (e) 6   (f) -6
6. (a) $2x - 21$ (b) $(1 - 2x)(2a - 3b)$ (c) 23
7. (a) 1   (b) $10\frac{1}{2}$ (c) 0, $3\frac{1}{2}$ (d) $-3, -2$ (e) 12
8. (a) $z(z - 4)(z + 4)$ (b) $(x^2 + 1)(y^2 + 1)$ (c) $(2x + 3)(x + 4)$ **9.** $\frac{7}{8}$
10. (a) $c = 5, d = -2$ (b) $x = 2, y = -1$ (c) $x = 9, y = -14$ (d) $s = 5, t = -3$
11. (a) $\frac{1}{2}, -\frac{1}{2}$ (b) $\frac{7}{11}$ (c) 3   (d) 0, 5
12. (a) $1\cdot78, -0\cdot28$ (b) $1\cdot62, -0\cdot62$ (c) $0\cdot87, -1\cdot54$ (d) $1\cdot54, -4\cdot54$
13. (a) $x = 9$ (b) $x = 10$
14. (a) 2   (b) -3 (c) 36   (d) 0   (e) 36   (f) 4
15. speed = 5 mph   **16.** 8 cm × 6·5 cm   **17.** (a) $-2, 4$ (b) 16   (c) $6\cdot19, 0\cdot81$
18. $-\frac{1}{5}, 3$ **19.** 8   **20.** $x = 13$ **21.** 21   **22.** 18
23. 6 cm   **24.** -4

page 46   **Examination Exercise 2B**

1. £1·50   **2.** (i) $£(C + 3A)$ (ii) $T = C + nA$
3. (a) $2x; y - 2$ (b) $3p^2 + 5pq - 2q^2$ (c) $(x - 5)(x + 4); 3(k - 8)(k + 8)$
     (d) $4\frac{5}{6}; 7$ (e) $x = -13, y = -11$
4. (i) 10; 14; 46   (ii) $s = 2w + 6$ (iii) 206   (iv) 310
5. (a) $3x$ (b) £500   (c) £1400
6. (b) 58   (c) 48   (d) $W = 2R + 4$ (e) 44
     (f) $5\frac{1}{3}$ (g) Impossible
7. (a) (i) $6x + 12$ (ii) 8   (b) $-0\cdot73, 2\cdot73$
8. (i) 104   (iii) '29 square'
9. (a) $-30; -30$ (b) $-8, 3; 0, -5$
10. (ii) 45 mph

11. (a) $\frac{30}{x}$ (b) $\frac{30}{x + 0\cdot5}$ (c) $x = 2\frac{1}{2}$; 12 minutes

12. (a) (i) $x + 4h$ (ii) $x(x + 4h)$ (b) $\frac{1}{4}\left(\frac{2500}{x} - x\right)$

     (c) 24·03   (d) 24 mm × 104 mm

PART 3

page 50   *Exercise 1*

1. $10 \cdot 2 \text{ m}^2$ **2.** 22 cm^2 **3.** 103 m^2 **4.** 9 cm^2 **5.** 31 m^2
6. 6000 cm^2 **7.** 26 m^2 **8.** 18 cm^2 **9.** 20 m^2 **10.** 13 m
11. 15 cm **12.** 56 m **13.** $8 \text{ m, } 10 \text{ m}$ **14.** 12 cm **15.** 2500
16. 6 square units   **17.** 12 square units   **18.** 1849   **20.** 1100 m

page 51   *Exercise 2*

1. $48 \cdot 3 \text{ cm}^2$ **2.** $28 \cdot 4 \text{ cm}^2$ **3.** $66 \cdot 4 \text{ m}^2$ **4.** $3 \cdot 07 \text{ cm}^2$ **5.** $18 \cdot 2 \text{ cm}^2$
6. $12 \cdot 3 \text{ cm}^2$ **7.** $2 \cdot 78 \text{ cm}^2$ **8.** $36 \cdot 4 \text{ m}^2$ **9.** $62 \cdot 4 \text{ m}^2$ **10.** $30 \cdot 4 \text{ m}^2$
11. $44 \cdot 9 \text{ cm}^2$ **12.** $0 \cdot 277 \text{ m}^2$ **13.** 63 m^2 **14.** $70 \cdot 7 \text{ m}^2$ **15.** 14 m^2
16. $65 \cdot 8 \text{ cm}^2$ **17.** $18 \cdot 1 \text{ cm}^2$ **18.** $8 \cdot 03 \text{ m}^2$ **19.** 14 m^2 **20.** $52 \cdot 0 \text{ cm}^2$
21. 124 cm^2 **22.** $69 \cdot 8 \text{ m}^2$ **23.** $57 \cdot 1 \text{ cm}^2$ **24.** $10 \cdot 7 \text{ cm}$ **25.** $50 \cdot 9°$
26. $4 \cdot 10 \text{ m}$ **27.** $4 \cdot 85 \text{ m}$ **28.** $7 \cdot 23 \text{ cm}$ **29.** $60°; 23 \cdot 4 \text{ cm}^2$
30. 292 **31.** 110 cm^2 **32.** (a) $\dfrac{360°}{n}$ (b) $\dfrac{n}{2} \sin \dfrac{360°}{n}$
33. $18 \cdot 7 \text{ cm}$

page 53   *Exercise 3*

1. (a) $31 \cdot 4 \text{ cm}$ (b) $78 \cdot 5 \text{ cm}^2$ **2.** (a) $18 \cdot 8 \text{ cm}$ (b) $28 \cdot 3 \text{ cm}^2$
3. (a) $51 \cdot 4 \text{ cm}$ (b) 157 cm^2 **4.** (a) $26 \cdot 6 \text{ cm}$ (b) $49 \cdot 1 \text{ cm}^2$
5. (a) $26 \cdot 3 \text{ cm}$ (b) $33 \cdot 3 \text{ cm}^2$ **6.** (a) $25 \cdot 0 \text{ cm}$ (b) $38 \cdot 5 \text{ cm}^2$
7. (a) $35 \cdot 7 \text{ cm}$ (b) $21 \cdot 5 \text{ cm}^2$ **8.** (a) $50 \cdot 3 \text{ cm}$ (b) 174 cm^2
9. (a) $22 \cdot 0 \text{ cm}$ (b) $10 \cdot 5 \text{ cm}^2$ **10.** (a) $75 \cdot 4 \text{ cm}$ (b) 412 cm^2
11. (a) $25 \cdot 1 \text{ cm}$ (b) $25 \cdot 1 \text{ cm}^2$ **12.** (a) $18 \cdot 8 \text{ cm}$ (b) $12 \cdot 6 \text{ cm}^2$

page 54   *Exercise 4*

1. $2 \cdot 19 \text{ cm}$ **2.** $30 \cdot 2 \text{ m}$ **3.** $2 \cdot 65 \text{ km}$ **4.** $9 \cdot 33 \text{ cm}$ **5.** $14 \cdot 2 \text{ mm}$
6. $497\,000 \text{ km}^2$ **7.** $21 \cdot 5 \text{ cm}^2$ **8.** (a) $40 \cdot 8 \text{ m}^2$ (b) 6
9. $30;$ (a) 1508 cm^2 (b) 508 cm^2 **10.** 5305 **11.** 29
12. 970 **13.** (a) 80 (b) 7 **14.** $5 \cdot 39 \text{ cm } (\sqrt{29})$
15. (a) $33 \cdot 0 \text{ cm}$ (b) $70 \cdot 9 \text{ cm}^2$ **16.** (a) 98 cm^2 (b) $14 \cdot 0 \text{ cm}^2$
17. $1:3:5$ **18.** 796 m^2 **19.** $57 \cdot 5°$ **20.** Yes
21. $1 \cdot 716 \text{ cm}$

page 56   *Exercise 5*

1. (a) $2 \cdot 09 \text{ cm}; 4 \cdot 19 \text{ cm}^2$ (b) $7 \cdot 85 \text{ cm}; 39 \cdot 3 \text{ cm}^2$
    (c) $8 \cdot 20 \text{ cm}; 8 \cdot 20 \text{ cm}^2$
2. $31 \cdot 9 \text{ cm}^2$ **3.** $31 \cdot 2 \text{ cm}^2$
4. (a) $7 \cdot 07 \text{ cm}^2$ (b) $19 \cdot 5 \text{ cm}^2$
5. (a) $85 \cdot 9°$ (b) $57 \cdot 3°$ (c) $6 \cdot 25 \text{ cm}$
6. (a) 12 cm (b) $30°$ **7.** (a) $3 \cdot 98 \text{ cm}$ (b) $74 \cdot 9°$
8. (a) $30°$ (b) $10 \cdot 5 \text{ cm}$ **9.** (a) 18 cm (b) $38 \cdot 2°$
10. (a) 10 cm (b) $43 \cdot 0°$
11. (a) $6 \cdot 14 \text{ cm}$ (b) $27 \cdot 6 \text{ m}$ (c) $28 \cdot 6 \text{ cm}^2$ **12.** $15 \cdot 1 \text{ km}^2$

page 58   *Exercise 6*

1. (a) 14·5 cm   (b) 72·6 cm²   (c) 24·5 cm²   (d) 48·1 cm²
2. (a) 5·08 cm²   (b) 82·8 m²   (c) 5·14 cm²
3. (a) 60°, 9·06 cm²   (b) 106·3°, 11·2 cm²
4. 3 cm   **5.** 3·97 cm   **6.** 13·5 cm², 405 cm³
7. 130 cm²; 184 cm²   **8.** 459 cm², 651 cm²   **9.** 19·6 cm²
10. 0·313 r^2 **11.** (a) 8·37 cm   (b) 54·5 cm   (c) 10·4 cm
12. 81·2 cm²

page 59   *Exercise 7*

1. (a) 30 cm³   (b) 168 cm³   (c) 110 cm³   (d) 94·5 cm³
     (e) 754 cm³   (f) 283 cm³
2. (a) 503 cm³   (b) 760 m³   (c) 12·5 cm³
3. 3·98 cm   **4.** 6·37 cm   **5.** 1·89 cm   **6.** 5·37 cm
7. 9·77 cm   **8.** 7·38 cm   **9.** 1273 cm   **10.** 4·24 litres
11. 106 cm/s   **12.** 1570 cm³, 12·57 kg   **13.** 3 : 4   **14.** cubes by 77 cm³
15. No   **16.** 1·19 cm   **17.** 53 times   **18.** 191 cm

page 61   *Exercise 8*

1. 20·9 cm³   **2.** 524 cm³   **3.** 4189 cm³   **4.** 101 cm³   **5.** 268 cm³
6. 4·19x^3 cm³   **7.** 0·004 19 m³   **8.** 3 cm³   **9.** 93·3 cm³   **10.** 48 cm³
11. 92·4 cm³   **12.** 262 cm³   **13.** 235 cm³   **14.** 415 cm³   **15.** 5 m
16. 2·43 cm   **17.** 23·9 cm   **18.** 6 cm   **19.** 3·72 cm   **20.** 1933 g
21. 106 s   **22.** (a) 125   (b) 2744   (c) $2·7 \times 10^7$
23. (a) 0·36 cm   (b) 0·427 cm   **24.** (a) 6·69 cm   (b) 39·1 cm
25. 10$\frac{2}{3}$ cm³ **26.** 1·05 cm³   **27.** 488 cm³   **28.** 4·19 cm³   **29.** 53·6 cm³
30. 74·5 cm³   **31.** 4·24 cm   **32.** 123 cm³   **33.** 54 400 cm³
34. (a) 16π   (b) 8 cm   (c) 6 cm   **35.** 471 cm³   **36.** 2720 cm³
37. 943 cm³   **38.** 5050 cm³

page 63   *Exercise 9*

1. (a) 36π cm²   (b) 72π cm²   (c) 60π cm²   (d) 2·38π m²
     (e) 400π m²   (f) 65π cm²   (g) 192π mm²   (h) 10·2π cm²
     (i) 0·000 4π m²   (j) 98π cm², 147π cm²
2. 1·64 cm   **3.** 2·12 cm   **4.** 3·46 cm
5. (a) 3 cm   (b) 4 cm   (c) 3 cm   (d) 0·2 m   (e) 6 cm   (f) 2·5 cm   (g) 6 cm
6. 303 cm²   **7.** £1178   **8.** £3870   **9.** 94·0 cm³
10. 44·6 cm²   **11.** 675 cm²   **12.** $1·62 \times 10^8$ years   **13.** 377 cm²
14. 20 cm, 10 cm   **15.** 71·7 cm²   **16.** 147 cm²

page 65   *Revision Exercise 3A*

1. (a) 14 cm²   (b) 54 cm²   (c) 50 cm²   (d) 18 m²
2. (a) 56·5 m, 254 m²   (b) 10·8 cm   (c) 3·99 cm
3. (a) 9π cm²   (b) 8 : 1   **4.** 3·43 cm², 4·57 cm²
5. (a) 12·2 cm   (b) 61·1 cm²
6. (a) 11·2 cm   (b) 10·3 cm   (c) 44·7 cm²   (d) 31·5 cm²   (e) 13·2 cm²
7. 103·1°   **8.** 9·95 cm   **9.** (a) 904 cm³   (b) 5·76 cm
10. 8·06 cm   **11.** 99·5 cm³   **12.** 333 cm³, 201 cm³   **13.** 4 cm
14. (a) 15·6 cm² (b) 93·5 cm³   (c) 3741 cm³   **15.** 0·370 cm   **16.** 104 cm²
17. 5·14 cm²   **18.** 68p   **19.** 25   **20.** 20 cm²

page 66   *Examination Exercise 3B*

1. (i) 154 m^2 (ii) 42 m^2 (iii) 3360 g
2. (a) 197·9 cm^3 (b) 1357 mm^3 (c) 145
3. (i) 500 s   (ii) 9·42 × 10^8 km   (iii) 29·87 km/s
4. (a) 36°   (b) 5·71 cm   (c) 3·71 cm   (d) 10·6 cm^2 (e) 106 cm^2
5. (a)(i) 306 cm   (ii) 7138 cm^2 (b) 28·6%
6. (a)(i) 6·16 cm^2 (ii) 1·44 cm^3 (b) 851 cm^2
7. (i) 50 cm   (ii) $\frac{1}{3}\pi \times 10^6$ cm^3 (iii) 1 : 8
8. (a) 54·5 cm^3 (b) 0·007 54 m^3 (c) 2·01 h
9. (i) 10 cm   (ii) 8   (iii) 2·5 cm, 1·25 cm   (iv) 4096
10. (a)(i) 10π (ii) 12π (b)(i) $\frac{\theta}{180}\pi r$ (ii) $\frac{\theta}{180}\pi\,(r+12)$
   (c)(i) 60   (ii) 30

PART 4

page 70   *Exercise 1*

1. 95°   **2.** 49°   **3.** 100°   **4.** 77°   **5.** 129°   **6.** 95°
7. $a = 30°$ **8.** $e = 30°, f = 60°$ **9.** 110°   **10.** $x = 54°$ **11.** $a = 40°$
12. $a = 36°, b = 72°, c = 144°, d = 108°$ **13.** 105°   **14.** $a = 30°, b = 120°, c = 150°$
15. $x = 20°, y = 140°$ **16.** $a = 120°, b = 34°, c = 26°$ **17.** $a = 68°, b = 54°$
18. 25°   **19.** 44°   **20.** $a = 30°, b = 60°, c = 150°, d = 120°$
21. $a = 10°, b = 80°$ **22.** $e = 71°, f = 21°$ **23.** 144°   **24.** 70°
25. 41°, 66°   **26.** 46°, 122°   **27.** 36°

page 71   *Exercise 2*

1. $a = 72°, b = 108°$ **2.** $x = 60°, y = 120°$ **3.** $(2n - 4) \times 90°$ **4.** 110°
5. 60°   **6.** $128\frac{4}{7}°$ **7.** 15   **8.** 12
9. 9   **10.** 18   **11.** 12   **12.** 36°

page 72   *Exercise 3*

1. $a = 116°, b = 64°, c = 64°$ **2.** $a = 64°, b = 40°$
3. $x = 68°$ **4.** $a = 40°, b = 134°, c = 134°$
5. $m = 69°, y = 65°$ **6.** $t = 48°, u = 48°, v = 42°$
7. $a = 118°, b = 100°, c = 62°$ **8.** $a = 34°, b = 76°, c = 70°, d = 70°$
9. 72°, 108°

page 73   *Exercise 4*

1. 10 cm   **2.** 4·12 cm   **3.** 4·24 cm   **4.** 9·90 cm   **5.** 8·72 cm
6. 5·66 cm   **7.** 6·63 cm   **8.** 5 cm   **9.** 17 cm   **10.** 4 cm
11. 9·85 cm   **12.** 7·07 cm   **13.** 3·46 m   **14.** 40·3 km   **15.** 13·6 cm
16. 6·34 m   **17.** 4·58 cm   **18.** 84·9 km   **19.** 24 cm   **20.** 9·80 cm
21. 5, 4, 3; 13, 12, 5; 25, 24, 7; 41, 40, 9; 61, 60, 11   **22.** $x = 4$ m, 20·6 m
23. 9·49 cm   **24.** 18·5 km

page 75   **Exercise 5**

1. (a) 1, 1   (b) 1, 1   (c) 2, 2   (d) 2, 2
   (e) 4, 4   (f) 0, 2   (g) 5, 5   (h) 0, 1
   (i) 1, 1   (j) 0, 2   (k) 0, 2   (l) 0, 2
   (m) ∞, ∞   (n) 0, 4

4. square 4, 4; rectangle 2, 2; parallelogram 0, 2; rhombus 2, 2; trapezium 0, 1; kite, 1, 1; equilateral triangle 3, 3; regular hexagon 6, 6

5. 34°, 56°   6. 35°, 35°   7. 72°, 108°, 80°   8. 40°, 30°, 110°   9. 116°, 32°, 58°
10. 55°, 55°   11. 26°, 26°, 77°   12. 52°, 64°, 116°   13. 70°, 40°, 110°   14. 54°, 72°, 36°
15. 60°, 15°, 75°, 135°

page 77   **Exercise 6**

1. $a = 2\frac{1}{2}$ cm, $e = 3$ cm   2. $x = 6$ cm, $y = 10$ cm   3. $x = 12$ cm, $y = 8$ cm
4. $m = 10$ cm, $a = 16\frac{2}{3}$ cm   5. $y = 6$ cm   6. $x = 4$ cm, $w = 1\frac{1}{2}$ cm
7. $e = 9$ cm, $f = 4\frac{1}{2}$ cm   8. $x = 13\frac{1}{3}$ cm, $y = 9$ cm   9. $m = 6$ cm, $n = 6$ cm
10. $m = 5\frac{1}{3}$ cm, $z = 4\frac{4}{5}$ cm   11. $v = 5\frac{1}{3}$ cm, $w = 6\frac{2}{3}$ cm   12. No
13. 2 cm, 6 cm   14. 16 m
15. (a) Yes (b) No   (c) No   (d) Yes   (e) Yes   (f) No   (g) No   (h) Yes
18. 0·618; 1·618 : 1

page 78   **Exercise 7**

1. A and G; B and E

page 80   **Exercise 8**

1. 16 cm^2 2. 27 cm^2 3. $11\frac{1}{4}$ cm^2 4. $14\frac{1}{2}$ cm^2 5. 128 cm^2 6. 12 cm^2
7. 8 cm   8. 18 cm   9. $4\frac{1}{2}$ cm   10. $7\frac{1}{2}$ cm   11. $2\frac{1}{2}$ cm   12. 6 cm
13. (a) $16\frac{2}{3}$ cm^2 (b) $10\frac{2}{3}$ cm^2 14. (a) 25 cm^2 (b) 21 cm^2 15. 8 cm^2 16. 6 cm
17. 24 cm^2 18. (a) $1\frac{4}{5}$ cm   (b) 3 cm   (c) 3 : 5   (d) 9 : 25   19. 150
20. 360   21. Less (for the same weight)

page 82   **Exercise 9**

1. 480 cm^3 2. 540 cm^3 3. 160 cm^3 4. 4500 cm^3 5. 81 cm^3
6. 11 cm^3 7. 16 cm^3 8. $85\frac{1}{3}$ cm^3 9. 4 cm   10. 21 cm
11. 4·6 cm   12. 9 cm   13. 6·6 cm   14. $4\frac{1}{2}$ cm   15. $168\frac{3}{4}$ cm^3
16. 106·3 cm^3 17. 12 cm   18. (a) 2 : 3   (b) 8 : 27   19. 8 : 125
20. $x_1^3 : x_2^3$ 21. 54 kg   22. 240 cm^2 23. $9\frac{3}{8}$ litres   24. $2812\frac{1}{2}$ cm^2

page 85   **Exercise 10**

1. $a = 27°$, $b = 30°$ 2. $c = 20°$, $d = 45°$ 3. $c = 58°$, $d = 41°$, $e = 30°$
4. $f = 40°$, $g = 55°$, $h = 55°$ 5. $a = 32°$, $b = 80°$, $c = 43°$ 6. $x = 34°$, $y = 34°$, $z = 56°$
7. 43°   8. 92°   9. 42°
10. $c = 46°$, $d = 44°$ 11. $e = 49°$, $f = 41°$ 12. $g = 76°$, $h = 52°$
13. 48°   14. 32°   15. 22°
16. $a = 36°$, $x = 36°$

page 87   **Exercise 11**

1. $a = 94°$, $b = 75°$ 2. $c = 101°$, $d = 84°$ 3. $x = 92°$, $y = 116°$ 4. $c = 60°$, $d = 45°$
5. 37°   6. 118°   7. $e = 36°$, $f = 72°$ 8. 35°
9. 18°   10. 90°   11. 30°   12. $22\frac{1}{2}°$

13. $n = 58°$, $t = 64°$, $w = 45°$ **14.** $a = 32°$, $b = 40°$, $c = 40°$
15. $a = 18°$, $c = 72°$ **16.** 55°   **17.** $e = 41°$, $f = 41°$, $g = 41°$
18. 8°   **19.** $x = 30°$, $y = 115°$ **20.** $x = 80°$, $z = 10°$

page 88   **Exercise 12**

1. $a = 18°$ **2.** $x = 40°$, $y = 65°$, $z = 25°$ **3.** $c = 30°$, $e = 15°$
4. $f = 50°$, $g = 40°$ **5.** $h = 70°$, $k = 40°$, $i = 40°$ **6.** $m = 108°$, $n = 36°$
7. $x = 50°$, $y = 68°$ **8.** $a = 74°$, $b = 32°$ **9.** $e = 36°$
10. $k = 63°$, $m = 54°$ **11.** $k = 50°$, $m = 50°$, $n = 80°$, $p = 80°$
12. $n = 16°$, $p = 46°$ **13.** (a) 24°   (b) 78°   (c) 48°
14. (a) p (b) $2p$ (c) $90 - 2p$ **15.** $x = 70°$, $y = 20°$, $z = 55°$
16. (b) $2a$, $180 - 3a$ **19.** 55°, 60°, 65°   **20.** (a) 64°   (b) $180 - 2x$

page 91   **Exercise 13**

2. 3   **3.** 4   **4.** 10 cm

page 91   **Exercise 14**

1. 93°   **2.** 36°   **7.** 7·8 cm
8. (a) 7·2 cm   (b) 5·2 cm   (c) 12·2 cm   (d) 8·2 cm
10. 5·0 cm   **13.** 10·4 cm   **14.** 3·5 cm   **15.** 6·6 cm
16. 4·9 cm, 7·75 cm

page 93   **Exercise 16**

1. (a), (b), (d)   **2.** (a) $a = 4$ cm, $x = 4$ cm, $y = 6$ cm   (b) 240 cm^3
3. (a) $a = 10$ cm, $b = 6$ cm, $c = 10$ cm, $d = 10$ cm   (b) 64 cm^3
4. (a) 168 mm^2 (b) 16800 mm^3
5. $a = \sqrt{2}$, $b = \sqrt{2}$, $c = \sqrt{3}$, $x = \sqrt{2}$, $y = \sqrt{3}$

page 94   **Revision Exercise 4A**

2. 80°   **3.** (a) 30°   (b) $22\frac{1}{2}°$ (c) 12
4. (a) 40°   (b) 100°   **5.** 4·12 cm
6. (i) 3 cm   (ii) 5·66 cm   **7.** (c) $2\frac{4}{5}$ cm   **8.** (b) 6 cm
9. $3\frac{2}{3}$ cm, $1\frac{1}{11}$ cm   **10.** 6 cm   **11.** (c) $5\frac{1}{3}$ cm   **12.** 250 cm^3
13. (a) $3\frac{1}{3}$ cm   (b) 1620 cm^3 **14.** (a) 1 m^2 (b) 1000 cm^3
15. (a) 50°   (b) 128°   (c) $c = 50°$, $d = 40°$ (d) $x = 10°$, $y = 40°$
16. (a) 55°   (b) 45°   **17.** (b) 3·6 cm

page 96   **Examination Exercise 4B**

1. (a) 23°   (b) 23°
2. (a) 12   (b) (i) EF   (ii) DE   (iii) A, E   (c) (i) 10 cm   (ii) $10\sqrt{2}$ cm
3. 47°   **5.** (b) (i) 94 cm^2 (ii) 376 cm^2 (c) Depends on net
6. (a) 60°   (b) 30°   (c) 15°
7. 1·125 m   **8.** 348·8 cm^3
9. (a) 4 cm   (b) 864 cm^3 (c) 3888 cm^2 (d) 1728 cm^2
10. (a) $\sqrt{(2x + 1)}$ (b) 41, 40, 9; 61, 60, 11
12. (i) 112 m   (iii) 49 m

PART 5

page 98   *Exercise 1*

1. $\frac{5}{7}$

2. $\frac{7}{8}$

3. $5y$

4. $\frac{1}{2}$

5. $4x$

6. $\frac{x}{2y}$

7. 2

8. $\frac{a}{2}$

9. $\frac{2b}{3}$

10. $\frac{a}{5b}$

11. a

12. $\frac{7}{8}$

13. $\frac{5+2x}{3}$

14. $\frac{3x+1}{x}$

15. $\frac{32}{25}$

16. $\frac{4+5a}{5}$

17. $\frac{3}{4-x}$

18. $\frac{b}{3+2a}$

19. $\frac{5x+4}{8x}$

20. $\frac{2x+1}{y}$

21. $\frac{x+2y}{3xy}$

22. $\frac{6-b}{2a}$

23. $\frac{2b+4a}{b}$

24. $x-2$

page 99   *Exercise 2*

1. $\frac{x+2}{x-3}$

2. $\frac{x}{x+1}$

3. $\frac{x+4}{2(x-5)}$

4. $\frac{x+5}{x-2}$

5. $\frac{x+3}{x+2}$

6. $\frac{x+5}{x-2}$

7. $\frac{x^2+1}{x^2}$

8. $\frac{1+4x}{2}$

9. $2x^2-1$

10. $2x-1$

11. $12x+1$

12. $\frac{1-4x}{2}$

13. $\frac{3x^2-1}{2x}$

14. $\frac{6x+2}{12x-3}$

15. $\frac{3x^2+1}{x^2+2}$

16. $\frac{x+2}{x}$

page 99   *Exercise 3*

1. $\frac{3}{5}$

2. $\frac{3x}{5}$

3. $\frac{3}{x}$

4. $\frac{4}{7}$

5. $\frac{4x}{7}$

6. $\frac{4}{7x}$

7. $\frac{7}{8}$

8. $\frac{7x}{8}$

9. $\frac{7}{8x}$

10. $\frac{5}{6}$

11. $\frac{5x}{6}$

12. $\frac{5}{6x}$

13. $\frac{23}{20}$

14. $\frac{23x}{20}$

15. $\frac{23}{20x}$

16. $\frac{1}{12}$

17. $\frac{x}{12}$

18. $\frac{1}{12x}$

19. $\frac{5x+2}{6}$

20. $\frac{7x+2}{12}$

21. $\frac{9x+13}{10}$

22. $\frac{1-2x}{12}$

23. $\frac{2x-9}{15}$

24. $\frac{-3x-12}{14}$

25. $\frac{3x+1}{x(x+1)}$

26. $\frac{7x-8}{x(x-2)}$

27. $\frac{8x+9}{(x-2)(x+3)}$

28. $\frac{4x+11}{(x+1)(x+2)}$

29. $\frac{-3x-17}{(x+3)(x-1)}$

30. $\frac{11-x}{(x+1)(x-2)}$

page 100   *Exercise 4*

1. $2\frac{1}{2}$

2. 3

3. $\frac{B}{A}$

4. $\frac{T}{N}$

5. $\frac{K}{M}$

6. $\frac{4}{y}$

7. $\frac{C}{B}$

8. $\frac{D}{4}$

9. $\frac{T+N}{9}$

10. $\frac{B-R}{A}$

11. $\frac{R+T}{C}$

12. $\frac{N-R^2}{L}$

13. $\frac{R-S^2}{N}$

14. 2

15. -7

16. $T-A$

17. $S-B$

18. $N-D$

19. $M-B$

20. $L-D^2$

21. $T-N^2$

22. $N+M-L$

23. $R-S-Z$

24. 7

25. $A+R$

26. $E+A$

27. $F+B$

28. F^2+B^2

29. $A+B+D$

30. A^2+E

31. $L + B$ **32.** $N + T$ **33.** 2   **34.** $4\frac{1}{2}$ **35.** $\dfrac{N - C}{A}$

36. $\dfrac{L - D}{B}$ **37.** $\dfrac{F - E}{D}$ **38.** $\dfrac{H + F}{N}$ **39.** $\dfrac{T + Z}{Y}$ **40.** $\dfrac{B + L}{R}$

41. $\dfrac{Q - m}{V}$ **42.** $\dfrac{n + a + m}{t}$ **43.** $\dfrac{s - t - n}{q}$ **44.** $\dfrac{t + s^2}{n}$ **45.** $\dfrac{c - b}{V^2}$

46. $\dfrac{r + 6}{n}$ **47.** $\dfrac{s - d}{m}$ **48.** $\dfrac{t + b}{m}$ **49.** $\dfrac{j - c}{m}$ **50.** 2

51. $2\frac{2}{3}$ **52.** $\dfrac{C - AB}{A}$ **53.** $\dfrac{F - DE}{D}$ **54.** $\dfrac{a - hn}{h}$ **55.** $\dfrac{q + bd}{b}$

56. $\dfrac{n - rt}{r}$ **57.** $\dfrac{b + 4t}{t}$ **58.** $\dfrac{z - St}{S}$ **59.** $\dfrac{s + vd}{v}$ **60.** $\dfrac{g - mn}{m}$

page 101   **Exercise 5**

1. 12   **2.** 10   **3.** BD **4.** TB **5.** RN

6. bm **7.** 26   **8.** $BT + A$ **9.** $AN + D$ **10.** $B^2N - Q$

11. $ge + r$ **12.** $4\frac{1}{2}$ **13.** $\dfrac{DC - B}{A}$ **14.** $\dfrac{pq - m}{n}$ **15.** $\dfrac{vS + t}{r}$

16. $\dfrac{qt + m}{z}$ **17.** $\dfrac{bc - m}{A}$ **18.** $\dfrac{AE - D}{B}$ **19.** $\dfrac{nh + f}{e}$ **20.** $\dfrac{qr - b}{g}$

21. 4   **22.** -2 **23.** 2   **24.** $A - B$ **25.** $C - E$

26. $D - H$ **27.** $n - m$ **28.** $q - t$ **29.** $s - b$ **30.** $r - v$

31. $m - t$ **32.** 2   **33.** $\dfrac{T - B}{X}$ **34.** $\dfrac{M - Q}{N}$ **35.** $\dfrac{V - T}{M}$

36. $\dfrac{N - L}{R}$ **37.** $\dfrac{v^2 - r}{r}$ **38.** $\dfrac{w - t^2}{n}$ **39.** $\dfrac{n - 2}{q}$ **40.** $\frac{1}{4}$

41. $-\frac{1}{7}$ **42.** $\dfrac{B - DE}{A}$ **43.** $\dfrac{D - NB}{E}$ **44.** $\dfrac{h - bx}{f}$ **45.** $\dfrac{v^2 - Cd}{h}$

46. $\dfrac{NT - MB}{M}$ **47.** $\dfrac{mB + ef}{fN}$ **48.** $\dfrac{TM - EF}{T}$ **49.** $\dfrac{yx - zt}{y}$ **50.** $\dfrac{k^2m - x^2}{k^2}$

page 101   **Exercise 6**

1. $\frac{1}{2}$ **2.** $1\frac{2}{3}$ **3.** $\dfrac{B}{C}$ **4.** $\dfrac{T}{X}$ **5.** $\dfrac{M}{B}$

6. $\dfrac{n}{m}$ **7.** $\dfrac{v}{t}$ **8.** $\dfrac{n}{\sin 20°}$ **9.** $\dfrac{7}{\cos 30°}$ **10.** $\dfrac{B}{x}$

11. $6\frac{2}{3}$ **12.** $\dfrac{ND}{B}$ **13.** $\dfrac{HM}{N}$ **14.** $\dfrac{et}{b}$ **15.** $\dfrac{vs}{m}$

16. $\dfrac{mb}{t}$ **17.** $1\frac{1}{2}$ **18.** $3\frac{1}{3}$ **19.** $\dfrac{B - DC}{C}$ **20.** $\dfrac{Q + TC}{T}$

21. $\dfrac{V + TD}{D}$ **22.** $\dfrac{L}{MB}$ **23.** $\dfrac{N}{BC}$ **24.** $\dfrac{m}{cd}$ **25.** $\dfrac{tc - b}{t}$

26. $\dfrac{xy - z}{x}$ **27.** 1   **28.** $\frac{5}{6}$ **29.** $\dfrac{A}{C - B}$ **30.** $\dfrac{V}{H - G}$

31. $\dfrac{r}{n + t}$ **32.** $\dfrac{b}{q - d}$ **33.** $\dfrac{m}{t + n}$ **34.** $\dfrac{b}{d - h}$ **35.** $\dfrac{d}{C - e}$

36. $\dfrac{m}{r - e^2}$ **37.** $\dfrac{n}{b - t^2}$ **38.** $\dfrac{d}{mn - b}$ **39.** $\dfrac{M - Nq}{N}$ **40.** $\dfrac{Y + Tc}{T}$

41. $\dfrac{N - 2MP}{2M}$ **42.** $\dfrac{B - 6Ac}{6A}$ **43.** $\dfrac{K}{(C - B)M}$ **44.** $\dfrac{z}{y(y + z)}$ **45.** $\dfrac{m^2}{n - p}$

46. $\dfrac{q}{w - t}$

page 102   **Exercise 7**

1. 4   **2.** 24   **3.** 11   **4.** $B^2 - A$ **5.** $D^2 - C$

6. $H^2 + E$ **7.** $\dfrac{c^2 - b}{a}$ **8.** $a^2 + m$ **9.** $\dfrac{b^2 + t}{g}$ **10.** $b - r^2$

11. $d - t^2$ **12.** $b^2 + d$ **13.** $n - c^2$ **14.** $b - f^2$ **15.** $c - g^2$

16. $\dfrac{M - P^2}{N}$ **17.** $\dfrac{D - B}{A}$ **18.** $A^4 + D$ **19.** $\pm\sqrt{g}$ **20.** ± 4

21. $\pm\sqrt{B}$ **22.** $\pm\sqrt{(B - A)}$ **23.** $\pm\sqrt{(M + A)}$ **24.** $\pm\sqrt{(b - a)}$ **25.** $\pm\sqrt{(C - m)}$

26. $\pm\sqrt{(d - n)}$ **27.** $\pm\sqrt{\dfrac{n}{m}}$ **28.** $\pm\sqrt{\dfrac{b}{a}}$ **29.** $\dfrac{at}{z}$ **30.** $\pm\sqrt{\left(\dfrac{m + t}{a}\right)}$

31. $\pm\sqrt{(a - n)}$ **32.** $\pm\sqrt{40}$ **33.** $\pm\sqrt{(B^2 + A)}$ **34.** $\pm\sqrt{(x^2 - y)}$ **35.** $\pm\sqrt{(t^2 - m)}$

36. 8   **37.** $\dfrac{M^2 - A^2 B}{A^2}$ **38.** $\dfrac{M}{N^2}$ **39.** $\dfrac{N}{B^2}$ **40.** $a - b^2$

41. $\pm\sqrt{(a^2 - t^2)}$ **42.** $\pm\sqrt{(m - x^2)}$ **43.** $\dfrac{4}{\pi^2} - t$ **44.** $\dfrac{B^2}{A^2} - 1$ **45.** $\pm\sqrt{\left(\dfrac{C^2 + b}{a}\right)}$

46. $\pm\sqrt{\left(\dfrac{b^2 + a^2 x}{a^2}\right)}$ **47.** $\pm\sqrt{(x^2 - b)}$ **48.** $\pm\sqrt{(c - b)a}$ **49.** $\dfrac{c^2 - b^2}{a}$ **50.** $\pm\sqrt{\left(\dfrac{m}{a + b}\right)}$

page 103   **Exercise 8**

1. $3\frac{2}{3}$ **2.** 3   **3.** $\dfrac{D - B}{2N}$ **4.** $\dfrac{E + D}{3M}$ **5.** $\dfrac{2b}{a - b}$

6. $\dfrac{e + c}{m + n}$ **7.** $\dfrac{3}{x + k}$ **8.** $\dfrac{C - D}{R - T}$ **9.** $\dfrac{z + x}{a - b}$ **10.** $\dfrac{nb - ma}{m - n}$

11. $\dfrac{d + xb}{x - 1}$ **12.** $\dfrac{a - ab}{b + 1}$ **13.** $\dfrac{d - c}{d + c}$ **14.** $\dfrac{M(b - a)}{b + a}$ **15.** $\dfrac{n^2 - mn}{m + n}$

16. $\dfrac{m^2 + 5}{2 - m}$ **17.** $\dfrac{2 + n^2}{n - 1}$ **18.** $\dfrac{e - h^2}{b - a}$ **19.** $\dfrac{3x}{a + x}$ **20.** $\dfrac{e - c}{a - d}$

21. $\dfrac{d}{a - b - c}$ **22.** $\dfrac{ab}{m + n - a}$ **23.** $\dfrac{s - t}{b - a}$ **24.** $2x$ **25.** $\dfrac{v}{3}$

26. $\dfrac{a(b + c)}{b - 2a}$ **27.** $\dfrac{5x}{3}$ **28.** $-\dfrac{4z}{5}$ **29.** $\dfrac{mn}{p^2 - m}$ **30.** $\dfrac{mn + n}{4 + m}$

page 103   **Exercise 9**

1. $-\left(\dfrac{by + c}{a}\right)$ **2.** $+\sqrt{\left(\dfrac{e^2 + ab}{a}\right)}$ **3.** $\dfrac{n^2}{m^2} + m$ **4.** $\dfrac{a - b}{1 + b}$ **5.** $3y$

6. $\dfrac{a}{e^2 + c}$ **7.** $-\left(\dfrac{a + lm}{m}\right)$ **8.** $\dfrac{t^2 g}{4\pi^2}$ **9.** $\dfrac{4\pi^2 d}{t^2}$ **10.** $\pm\sqrt{\dfrac{a}{3}}$

11. $\pm\sqrt{\left(\dfrac{t^2 e - ba}{b}\right)}$ **12.** $\dfrac{1}{a^2 - 1}$ **13.** $\dfrac{a + b}{x}$ **14.** $\pm\sqrt{(x^4 - b^2)}$ **15.** $\dfrac{c - a}{b}$

16. $\dfrac{a^2 - b}{a + 1}$ **17.** $\pm\sqrt{\left(\dfrac{G^2}{16\pi^2} - T^2\right)}$ **18.** $-\left(\dfrac{ax + c}{b}\right)$ **19.** $\dfrac{1 + x^2}{1 - x^2}$ **20.** $\pm\sqrt{\left(\dfrac{a^2 m}{b^2} + n\right)}$

21. $\dfrac{P - M}{E}$ **22.** $\dfrac{RP - Q}{R}$ **23.** $\dfrac{z - t^2}{x}$ **24.** $(g - e)^2 - f$ **25.** $\dfrac{4np + me^2}{mn}$

page 104   **Exercise 10**

1. (a) $S = ke$ (b) $v = kt$ (c) $x = kz^2$ (d) $y = k\sqrt{x}$ (e) $T = k\sqrt{L}$
   (f) $C = kr$ (g) $A = kr^2$ (h) $V = kr^3$
2. (a) 9   (b) $2\frac{2}{3}$ 3. (a) 35   (b) 11
4. (a) 75   (b) 4

5.
x	1	3	4	$5\frac{1}{2}$
z	4	12	16	22

6.
r	1	2	4	$1\frac{1}{2}$
V	4	32	256	$13\frac{1}{2}$

7.
h	4	9	25	$2\frac{1}{4}$
w	6	9	15	$4\frac{1}{2}$

8. (a) 18   (b) 2   9. (a) 42   (b) 4   10. 333 N/cm^2
11. 180 m; 2 s   12. 675 J; $\sqrt{\frac{4}{3}}$ cm   13. 4 cm; 49 h   14. $15\frac{5}{8}$ h   15. 9000 N; 25 m/s
16. $15^4 : 1$ (50625 : 1)

page 105   **Exercise 11**

1. (a) $x = \dfrac{k}{y}$ (b) $s = \dfrac{k}{t^2}$ (c) $t = \dfrac{k}{\sqrt{q}}$.   (d) $m = \dfrac{k}{w}$ (e) $z = \dfrac{k}{t^2}$
2. (a) 1   (b) 4   3. (a) $2\frac{1}{2}$ (b) $\frac{1}{2}$
4. (a) 36   (b) ± 4 5. (a) $1\cdot 2$ (b) ± 2
6. (a) 16   (b) ± 10 7. (a) 6   (b) 16
8. (a) $\frac{1}{2}$ (b) $\frac{1}{20}$

9.
y	2	4	1	$\frac{1}{4}$
z	8	4	16	64

10.
t	2	5	20	10
v	25	4	$\frac{1}{4}$	1

11.
x	1	4	256	36
r	12	6	$\frac{3}{4}$	2

12. (a) 6   (b) 50   13. (a) $0\cdot 36$ (b) 6   14 $k = 100$, $n = 3$

	x	1	2	4	10
	z	100	$12\frac{1}{2}$	$1\cdot 5625$	$\frac{1}{10}$

15. $k = 12$, $n = 2$

v	1	4	36	10000
y	12	6	2	$\frac{3}{25}$

16. $2\cdot 5$ m^3; 200 N/m^2 17. 3 h; 48 men
18. 2 days; 200 days   19. 6 cm

page 107   **Exercise 12**

1. 3^4 2. $4^2 \times 5^3$ 3. 3×7^3 4. $2^3 \times 7$ 5. 10^{-3}
6. $2^{-2} \times 3^{-3}$ 7. $15^{\frac{1}{2}}$ 8. $3^{\frac{1}{3}}$ 9. $10^{\frac{1}{3}}$ 10. $5^{\frac{1}{2}}$
11. x^7 12. y^{13} 13. z^4 14. z^{100} 15. m
16. e^{-5} 17. y^2 18. w^6 19. y 20. x^{10}
21. 1   22. w^{-5} 23. w^{-5} 24. x^7 25. a^8
26. k^3 27. 1   28. x^{29} 29. y^2 30. x^6
31. z^4 32. t^{-4} 33. $4x^6$ 34. $16y^{10}$ 35. $6x^4$
36. $10y^5$ 37. $15a^4$ 38. $8a^3$ 39. 3   40. $4y^2$
41. $\frac{5}{2}y$ 42. $32a^4$ 43. $108x^5$ 44. $4z^{-3}$ 45. $2x^{-4}$
46. $\frac{5}{2}y^5$ 47. 1   48. $21w^{-3}$ 49. $2n^4$ 50. $2x$

page 108   **Exercise 13**

1. 27   2. 1   3. $\frac{1}{9}$ 4. 25   5. 2
6. 4   7. 9   8. 2   9. 27   10. 3
11. $\frac{1}{3}$ 12. $\frac{1}{2}$ 13. 1   14. $\frac{1}{5}$ 15. 10
16. 8   17. 32   18. 4   19. $\frac{1}{9}$ 20. $\frac{1}{8}$
21. 18   22. 10   23. 1000   24. $\frac{1}{1000}$ 25. $\frac{1}{9}$
26. 1   27. $1\frac{1}{2}$ 28. $\frac{1}{25}$ 29. $\frac{1}{10}$ 30. $\frac{1}{4}$
31. $\frac{1}{4}$ 32. 100 000   33. 1   34. $\frac{1}{32}$ 35. $0\cdot 1$

36. 0·2   **37.** 1·5   **38.** 1   **39.** 9   **40.** $1\frac{1}{2}$

41. $\frac{3}{10}$ **42.** 64   **43.** $\frac{1}{100}$ **44.** $1\frac{2}{3}$ **45.** $\frac{1}{100}$

46. 1   **47.** 100   **48.** 6   **49.** 750   **50.** −7

page 108   Exercise 14

1. $25x^4$ **2.** $49y^6$ **3.** $100a^2b^2$ **4.** $4x^2y^4$ **5.** $2x$

6. $\frac{1}{9y}$ **7.** x^2 **8.** $\frac{x^2}{2}$ **9.** 1   **10.** $\frac{2}{x}$

11. $36x^4$ **12.** $25y$ **13.** $16x^2$ **14.** $27y$ **15.** 25

16. 1   **17.** 49   **18.** 1   **19.** $8x^6y^3$ **20.** $100x^2y^6$

21. $\frac{3x}{2}$ **22.** $\frac{2}{x}$ **23.** x^3y^5 **24.** $12x^3y^2$ **25.** $10y^4$

26. $3x^3$ **27.** $x^3y^2z^4$ **28.** x **29.** $3y$ **30.** $27x^{\frac{3}{2}}$

31. $10x^3y^5$ **32.** $32x^2$ **33.** $\frac{5}{2}x^2$ **34.** $\frac{9}{x^2}$ **35.** $2a^2$

36. $a^3b^3c^6$ **37.** (a) 2^5 (b) 2^7 (c) 2^6 (d) 2^0

38. (a) 3^{-3} (b) 3^{-4} (c) 3^{-1} (d) 3^{-2} **39.** 16

40. $\frac{1}{4}$ **41.** $\frac{1}{6}$ **42.** 1   **43.** $16\frac{1}{8}$ **44.** $\frac{3}{8}$

45. $\frac{1}{4}$ **46.** $\frac{5}{256}$ **47.** $1\frac{1}{16}$ **48.** 0   **49.** $\frac{1}{4}$

50. $\frac{1}{4}$ **51.** 3   **52.** 4   **53.** −1   **54.** −2

55. 3   **56.** 3   **57.** 1   **58.** $\frac{1}{5}$ **59.** 0

60. −4   **61.** 2   **62.** −5   **63.** 1   **64.** $\frac{1}{18}$

65. (a) 3·60   (b) 5·44

page 109   Exercise 15

1. (a) 2·828   (b) 4·243   (c) 6·245   (d) 2·762   (e) 10·54   (f) 20·32

2. (a) 2·080   (b) 3·037   (c) 4·121   (d) 0·9872   (e) 10·16

3. (a) 2·036   (b) 3·041   (c) 10·28   (d) 1·933

4. 4·85   **5.** 3·28   **6.** 0·076923   **7.** 9·472   **8.** 1·781

page 110   Exercise 16

1. (a) 24, 12, 42, 18, 60, 24, 78   (b) 54, 62, 7, 38, 42, 46, 50

page 111   Exercise 17

1. $<$ **2.** $>$ **3.** $>$ **4.** $=$ **5.** $<$

6. $<$ **7.** $=$ **8.** $>$ **9.** $<$ **10.** $>$

11. $<$ **12.** $>$ **13.** $>$ **14.** $>$ **15.** $=$

16. F   **17.** F   **18.** T   **19.** F   **20.** F

21. T   **22.** T   **23.** F   **24.** F   **25.** $x > 13$

26. $x < -1$ **27.** $x < 12$ **28.** $x \leqslant 2\frac{1}{2}$ **29.** $x > 3$ **30.** $x \geqslant 8$

31. $x < \frac{1}{4}$ **32.** $x \geqslant -3$ **33.** $x < -8$ **34.** $x < 4$ **35.** $x > -9$

36. $x < 8$ **37.** $x > 3$ **38.** $x \geqslant 1$ **39.** $x < 1$ **40.** $x > 2\frac{1}{3}$

page 112   Exercise 18

1. $x > 5$ **2.** $x \leqslant 3$ **3.** $x > 6$ **4.** $x \geqslant 4$

5. $x < 1$ **6.** $x < -3$ **7.** $x > 0$ **8.** $x > 4$

9. $x > 2$ **10.** $x < -3$ **11.** $1 < x < 4$ **12.** $-2 \leqslant x \leqslant 5$

13. $1 \leqslant x < 6$ **14.** $0 \leqslant x < 5$ **15.** $-1 \leqslant x \leqslant 7$ **16.** $-2 < x < 2$

17. $-4 < x < 4$ **18.** $x < -1$ **19.** $x \geqslant 3$ or $x \leqslant -3$ **20.** $0 < x < 4$

288 **Answers Part 5**

21. $5 \leqslant x \leqslant 9$ **22.** $-1 < x < 4$ **23.** $1 \leqslant x \leqslant 6$ **24.** $\frac{1}{2} < x < 8$
25. $-8 < x < 2$ **26.** $\{1, 2, 3, 4, 5, 6\}$ **27.** $\{7, 11, 13, 17, 19\}$ **28.** $\{2, 4, 6, 8, 10\}$
29. $\{4, 9, 16, 25, 36, 49\}$ **30.** $\{5, 10\}$ **31.** $\{-4, -3, -2, -1\}$ **32.** $\{2, 3, 4, \ldots 12\}$
33. $\{1, 4, 9\}$ **34.** $\{2, 3, 5, 7, 11\}$ **35.** $\{2, 4, 6, \ldots 18\}$ **36.** $n = 5$
37. $x = 7$ **38.** $y = 5$ **39.** $4 < z < 5$ **40.** $4 < p < 5$
41. $\frac{1}{2}$ (or other values) **42.** $1, 2, 3, \ldots 14$ **43.** 19 **44.** $\frac{1}{2}$ (or other values)
45. 19 **47.** 17

page 113   **Exercise 19**

1. $x \leqslant 3$ **2.** $y \geqslant 2\frac{1}{2}$ **3.** $1 \leqslant x \leqslant 6$ **4.** $x < 7, y < 5$
5. $y \geqslant x$ **6.** $x + y < 10$ **7.** $x < 8, y > -2, y < x$
8. $x \geqslant 0, y \geqslant x - 1, x + y \leqslant 7$ **9.** $y \geqslant 0, y \leqslant x + 2, x + y \leqslant 6$
28. (a) maximum value $= 26$ at $(6, 5)$ (b) minimum value $= 12$ at $(3, 3)$
29. (a) maximum value $= 25$ at $(8, 3)$ (b) minimum value $= 9$ at $(7, 2)$
30. (a) maximum value $= 40$ at $(20, 0)$ (b) maximum value $= 112$ at $(14, 8)$

page 114   **Revision Exercise 5A**

1. (a) $\dfrac{9x}{20}$ (b) $\dfrac{7}{6x}$ (c) $\dfrac{5x - 2}{6}$ (d) $\dfrac{5x + 23}{(x - 1)(x + 3)}$

2. (a) $(x - 2)(x + 2)$ (b) $\dfrac{3}{x + 2}$

3. (a) $s = t(r + 3)$ (b) $r = \dfrac{s - 3t}{t}$ (c) $t = \dfrac{s}{r + 3}$

4. (a) $z = x - 5y$ (b) $m = \dfrac{11}{k + 3}$ (c) $z = \dfrac{T^2}{C^2}$

5. (a) 50 (b) 50 **6.** (a) 16 (b) ± 4
7. (a) (i) 3 (ii) 4 (iii) $\frac{1}{4}$ (b) (i) 4 (ii) 0
8. (a) 9, 10 (b) 2, 3, 4, 5 **9.** $\dfrac{t^2}{k^2} - 5$ **10.** $\dfrac{z + 2}{z - 3}$

11. (a) $\frac{3}{5}$ (b) $\dfrac{k(1 - y)}{y}$ **13.** (a) $1\frac{5}{6}$ (b) $0 \cdot 09$

14. 21 **15.** (a) $\dfrac{5 + a^2}{2 - a}$ (b) $-\left(\dfrac{cz + b}{a}\right)$ (c) $\dfrac{a^2 + 1}{a^2 - 1}$

16. (a) $\dfrac{7}{2x}$ (b) $\dfrac{3a + 7}{a^2 - 4}$ (c) $\dfrac{x - 8}{x(x + 1)(x - 2)}$ **17.** $p = \dfrac{10t^2}{s}$

18. $x = 2 \cdot 19$ (to 3 S.F.) **19.** $y \geqslant 2, x + y \leqslant 6, y \leqslant 3x$
20. $x \geqslant 0, y \geqslant x - 2, x + y \leqslant 7$ **22.** (a) 512 (b) 6 h (c) 2^{21}

page 116   **Examination Exercise 5B**

1. (a) 3 (b) $\dfrac{4\pi^2 l}{T^2}$ **2.** 8

3. (a) $h = \dfrac{2d^2}{3}$ (b) (i) $x \geqslant 2$ (ii) $x < 6$ (iii) 2, 3, 4, 5 (c) 6, 7
4. (i) 4 (ii) $\frac{1}{2}$
5. (i) $-\frac{1}{2}$ (ii) $\frac{4}{3}$ (iii) $\frac{4}{25}$ (iv) $x = \dfrac{64}{y^2}$

6. (a) $x - 2$ (b) $x - 2 = \dfrac{1}{x}$ (c) $2 \cdot 333333; 2 \cdot 428571$ (d) $2 \cdot 41$

7. (a) 4, 49, 484   (b) 4·472136

8. (a) 0, 1, 3, 6, 10   (b) 1, 2, 3, 4   (c) $\frac{1}{2}(n-1)n$; n (e) 99

9. (i) (a) 1·008 m³   (b) $V = 2x^2(2x + 0·2)$

   (ii) (a) 0·594444; 0·607485; 0·577436   (b) 0·59761; 0·59863; 0·59819   (c) 0·598

PART 6

page 120   *Exercise 2*

1. 4·54	**2.** 3·50	**3.** 3·71	**4.** 6·62	**5.** 8·01
6. 31·9	**7.** 45·4	**8.** 4·34	**9.** 17·1	**10.** 13·2
11. 38·1	**12.** 3·15	**13.** 516	**14.** 79·1	**15.** 5·84
16. 2·56	**17.** 18·3	**18.** 8·65	**19.** 11·9	**20.** 10·6
21. 119	**22.** 10·1	**23.** 3·36 cm	**24.** 4·05 cm	**25.** 4·10 cm
26. 11·7 cm	**27.** 9·48 cm	**28.** 5·74 cm	**29.** 9·53 cm	**30.** 100 m
31. 56·7 m	**32.** 16·3 cm	**33.** 0·952 cm	**34.** 8·27 m	

page 121   *Exercise 3*

1. 5, 5·55	**2.** 13·1, 27·8	**3.** 34·6, 41·3	**4.** 20·4, 11·7	**5.** 94·1, 94·1
6. 15·2, 10, 6·43	**7.** 4·26	**8.** 3·50	**9.** 26·2	**10.** 8·82

11. (a) 17·4 cm   (b) 11·5 cm   (c) 26·5 cm

12. (a) 6·82 cm   (b) 6·01 cm   (c) 7·31 cm

page 122   *Exercise 4*

1. 36.9°	**2.** 44·4°	**3.** 48·2°	**4.** 60°	**5.** 36·9°
6. 50·2°	**7.** 29·0°	**8.** 56·4°	**9.** 38·9°	**10.** 43·9°
11. 41·8°	**12.** 39·3°	**13.** 60·3°	**14.** 50·5°	**15.** 13·6°
16. 34·8°	**17.** 60·0°	**18.** 42·0°	**19.** 36·9°	**20.** 51·3°
21. 19·6°	**22.** 17·9°	**23.** 32·5°	**24.** 59·6°	**25.** 54·8°
26. 46·3°				

page 124   *Exercise 5*

1. 19·5°	**2.** 4·1 m	**3.** (a) 26·0 km	(b) 23·4 km	
4. (a) 88·6 km	(b) 179·3 km	**5.** 4·1 m	**6.** 8·6 m	
7. (a) 484 km	(b) 858 km	(c) 986 km, 060·6°	**8.** 954 km, 133°	**9.** 56·3°
10. 35·5°	**11.** 71·6°	**12.** 91·8°	**13.** 180 m	**14.** 36·4°
15. 10·3 cm	**16.** 9·51 cm	**17.** 71·1°	**18.** 67·1 m	**19.** 138 m
20. 83·2 km	**21.** 60°	**22.** 13·9 cm	**23.** Yes	
24. 11·1 m; 11·1 s; 222 m		**25.** 4·4 m	**26.** 3·13 m	**27.** 25·6 cm

page 125   *Exercise 6*

1. 97.3 m	**2.** 88.7 n miles	**3.** 103 km
4. 99 km; 023½°	**5.** 9180 km/h; 255°	**6.** 11 km

page 126   *Exercise 7*

1. (a) 13 cm   (b) 13·6 cm   (c) 17·1°

2. (a) 4·04 m   (b) 38·9°   (c) 11·2 m   (d) 19·9°

3. (a) 8·49 cm   (b) 8·49 cm   (c) 10·4 cm   (d) 35·3°   (e) 35·3°

4. (a) 10 m (b) 7·81 m (c) 9·43 m (d) 70·2°
5. (a) 14·1 cm (b) 18·7 cm (c) 69·3° (d) 29·0° (e) 41·4°
6. (a) 4·47 m (b) 7·48 m (c) 63·4° (d) 74·5° (e) 53·3°
7. 10·8 cm; 21·8°
8. (a) $h \tan 65°$ (b) $h \tan 57°$ (c) 22·7 m
9. 22·6 m **10.** 55·0 m **11.** 7·26 m **12.** 43·3°

page 128   **Exercise 8**

1. 6·38 m **2.** 12·5 m **3.** 5·17 cm **4.** 40·4 cm **5.** 7·81 m, 7·10 m
6. 3·55 m, 6·68 m **7.** 8·61 cm **8.** 9·97 cm **9.** 8·52 cm **10.** 15·2 cm
11. 35·8° **12.** 42·9° **13.** 32·3° **14.** 37·8° **15.** 35·5°, 48·5°
16. 68·8°, 80·0° **17.** 64·6° **18.** 34·2° **19.** 50·6° **20.** 39·1°
21. 39·5° **22.** 21·6°

page 130   **Exercise 9**

1. 6·24 **2.** 6·05 **3.** 5·47 **4.** 9·27 **5.** 10·1
6. 8·99 **7.** 5·87 **8.** 4·24 **9.** 11·9 **10.** 154
11. 25·2° **12.** 78·5° **13.** 115·0° **14.** 111·1° **15.** 24·0°
16. 92·5° **17.** 99·9° **18.** 38·2° **19.** 137·8° **20.** 34·0°
21. 60·2° **22.** 8·72 **23.** 1·40 **24.** 7·38

page 131   **Exercise 10**

1. 6·0 cm **2.** 10·8 m **3.** 35·6 km **4.** 25·2 m **5.** 38·6°, 48·5°, 92·9°
6. 40·4 m **7.** 9·8 km; 085·7° **8.** (a) 29·6 km (b) 050·5°
9. (a) 10·8 m (b) 72·6° (c) 32·6°
10. 378 km, 048·4° **11.** (a) 62·2° (b) 2·33 km
12. 9·64 m **13.** 8·6°
17. (a) 17·5°, 162·5° (b) 41·5°, 318·5° (c) 68°, 248°

page 132   **Revision Exercise 6A**

1. (a) 45·6° (b) 58·0° (c) 3·89 cm (d) 33·8 m
2. (a) 1·75 (b) 60·3° **3.** (a) 12·7 cm (b) 5·92 cm (c) 36·1°
4. 5·39 cm **5.** (a) 220° (b) 295°
6. 0·335 m **7.** (a) 6·61 cm (b) 12·8 cm (c) 5·67 cm
8. (a) 86·9 cm (b) 53·6 cm (c) 133 cm **9.** 52·4 m
10. (a) 14·1 cm (b) 35·3° (c) 35·3°
11. (a) 6·63 cm (b) 41·8° **12.** (a) 11·3 cm (b) 8·25 cm (c) 55·6°
13. 45·2 km, 33·6 km **14.** 73·4° **15.** 8·76 m, 9·99 m
16. 0·539 **17.** 4·12 cm, 9·93 cm **18.** 26·4°

page 134   **Examination Exercise 6B**

1. (i) 42·1° (ii) 6·02
2. (a) 36 m (b) 27 m (c) 14 m (d) 480 m²
3. (i) 51° (ii) 20·8 cm
4. (a) 5·2 km (b) 3 km (c) 3·10
5. (b) 4·60 km (c) 1·95 km (d) Semi-circle with AB as diameter
(e) 037°
6. (i) 59·1 cm (ii) 49·8°
7. (a) 99 km (b) 15°, 075° (d) (i) 16·4 km (ii) 79·6 km
8. (ii) 124 km (iii) 024°

 9. (a) 170 m   (b) 51·3°; 41·5°
10. (b) 174 cm   (c) No. Only 95·5 cm
11. (a) 32·5 m   (c) (i) 25·38 m   (ii) 24°
12. (a) 23·6 m   (b) 249 m   (c) 318 m   (d) 2·83 hectares

PART 7

page 138   *Exercise 1*

11. (0, 0), (1, 4), (1·6, 1·6)   **12.** (0, 1), $(2\frac{1}{4}, 1)$, $(4\frac{1}{2}, 10)$
13. (−2, −6), (1·25, 3·75), (4·5, 0·5)   **14.** (−1·5, 1·5)(0·67, 8), (3·5, 8), (3·5, −3·5)
15. (4, −2), (0·33, 5·33), (−2·28, −5·14)   **16.** (−2, 3), (0·6, 8·2), (2·5, 2·5), (1·33, 1·33)
17. (a) £560   (b) 2400 miles   **18.** (a) 1·53 kg (b) 7·1 h
19. (a) £188   (b) 158 km/h   (c) £210   **20.** (a) £4315   (b) 26 000 miles (c) £380

page 139   *Exercise 2*

 1. $1\frac{1}{2}$ **2.** 2   **3.** 3   **4.** $1\frac{1}{2}$ **5.** $\frac{1}{2}$
 6. $-\frac{1}{6}$ **7.** −7   **8.** −1   **9.** 4   **10.** −4
11. 5   **12.** $-1\frac{3}{7}$ **13.** 6   **14.** 0   **15.** 0
16. infinite   **17.** infinite   **18.** −8   **19.** $5\frac{1}{3}$ **20.** 0
21. $\dfrac{b-d}{a-c}$ **22.** $\dfrac{n+b}{m-a}$ **23.** $\dfrac{2f}{a}$ **24.** −4   **25.** 0
26. $-\dfrac{6d}{c}$ **27.** (a) $-1\frac{1}{5}$ (b) $\frac{1}{10}$ (c) $\frac{4}{5}$
28. (a) infinite   (b) $-\frac{3}{10}$ (c) $\frac{3}{10}$ **29.** $3\frac{1}{2}$
30. (a) $\dfrac{n+4}{2m-3}$ (b) $n - -4$ (c) $m = 1\frac{1}{2}$

page 140   *Exercise 3*

 1. 1, 3   **2.** 1, −2   **3.** 2, 1   **4.** 2, −5   **5.** 3, 4
 6. $\frac{1}{2}$, 6   **7.** 3, −2   **8.** 2, 0   **9.** $\frac{1}{4}$, −4   **10.** −1, 3
11. −2, 6   **12.** −1, 2   **13.** −2, 3   **14.** −3, −4   **15.** $\frac{1}{2}$, 3
16. $-\frac{1}{3}$, 3   **17.** 4, −5   **18.** $1\frac{1}{2}$, −4   **19.** 10, 0   **20.** 0, 4

page 140   *Exercise 4*

 1. $y = 3x + 7$ **2.** $y = 2x - 9$ **3.** $y = -x + 5$ **4.** $y = 2x - 1$
 5. $y = 3x + 5$ **6.** $y = -x + 7$ **7.** $y = \frac{1}{2}x - 3$ **8.** $y = 2x - 3$
 9. $y = 3x - 11$ **10.** $y = -x + 5$ **11.** $y = \frac{1}{3}x - 4$

page 141   *Exercise 5*

 1. 44   **2.** 4·2 km   **4.** $m = 2, c = 3$ **5.** $n = -2·5, k = 15$
 6. $m = 0·2, c = 3$

page 143   *Exercise 7*

 1. (a) 4   (b) 8   (c) 10·6
 2. (a) 3   (b) −5   (c) 1·5
 3. (a) 7·25   (b) −2   (c) −0·8, 3·8   **15.** (a) 0·75   (b) 1·23
16. (a) 3·13   (b) 3·35   **17.** (a) −2·45   (b) 1·4

18. (a) 5　　　　　(b) 10·1　　　　　(c) −1·25
20. (a) 245　　　　(b) 41　　　　　(c) $25 < x < 67$

page 144　　*Exercise 8*

1. (a) 10·7 cm^2　　(b) 1·7 cm × 5·3 cm　　(c) 12·25 cm^2　　(d) 3·5 cm × 3·5 cm
　　(e) square
2. 15 m × 30 m　　**3.** (a) 2·5 s　　(b) 31·3 m　　(c) $2 < t < 3$
4. (a) 108 m/s　　(b) 1·4 s　　(c) $2·3 < t < 3·6$　　**6.** 3·3
11. (a) 1　　　　(b) 2·7　　　　**12.** (a) 0, 2·9　　(b) −0·65, 1·35, 5·3

page 145　　*Exercise 9*

1. (a) 180　　　　(b) $C = 0·2x + 35$
2. (a) 6 gallons　　(b) 40 mpg; 30 mpg　　(c) $33\frac{1}{3}$ mpg; $5\frac{1}{2}$ gallons
3. (a) 2000　　　(b) 270　　　　(c) $1·6 \leqslant x \leqslant 2·4$
4. (a) Yes　　　(b) No　　　　(c) About £250 − £270

page 147　　*Exercise 10*

1. (a) −0·4, 2·4　　(b) −0·8, 3·8　　(c) −1, 3　　　(d) −0·4, 2·4
2. −0·3, 3·3　　**3.** 0·6, 3·4　　**4.** 0·3, 3·7
5. (a) $y = 3$　　(b) $y = -2$　　(c) $y = x + 4$　　(d) $y = x$　　(e) $y = 6$
6. (a) $y = 6$　　(b) $y = 0$　　(c) $y = 4$　　(d) $y = 2x$　　(e) $y = 2x + 4$
7. (a) $y = -4$　　(b) $y = 2x$　　(c) $y = x - 2$　　(d) $y = -3$　　(e) $y = 2$
8. (a) $y = 5$　　(b) $y = 2x$　　(c) $y = 0·2$　　(d) $y = 3 - x$　　(e) $y = 3$
9. (a) $y = 0$　　(b) $y = -2\frac{1}{2}$　　(c) $y = -8x$　　(d) $y = -3$　　(e) $y = -5\frac{1}{2}x$
10. (a) −1·65, 3·65　(b) −1·3, 2·3　　(c) −1·45, 3·45
11. (a) 1·7, 5·3　　(b) 0·2, 4·8　　**12.** (a) −3·3, 0·3　　(b) −4·6, −0·4
13. (a) −2·35, 0·85　(b) −2·8, 1·8　**14.** (a) (i) −0·4, 2·4　　(ii) −0·5, 2　　(b) $-1·25 < x < 2·75$
15. (a) 3·35　　(b) 2·4, 7·6　　(c) 4·25　　　**16.** (a) ±3·74　　(c) ±2·83
17. (a) 1·75　　(b) 0, ±1·4　　**18.** (a) $1·6 < x < 7·4$　　(b) 6·9
19. (a) 40°, 140°　(b) 30°, 150°
20. (a) (i) 16°, 111°　(ii) 153°　　(b) 2·24　　(c) 63°
21. (a) 2·6　　(b) 0·45　　(c) 0·64　　(d) 5·66
22. (a) −1·62, 0·62　　(b) $-\frac{1}{2}$, 1
23. (a) −0·77, 4　　(b) $0·6 < x < 2·8$

page 150　　*Exercise 11*

1. 7·47　　　　**2.** 13·4　　　　**3.** 23·5
4. (b) 38·5　　(c) less　　　**5.** (a) 58　　　(b) greater

page 151　　*Exercise 12*

1. (a) 45 min　　(b) 0915　　(c) 60 km/h　　(d) 100 km/h　　(e) 57·1 km/h
2. (a) 0915　　(b) 64 km/h　　(c) 37·6 km/h　　(d) 47 km　　(e) 80 km/h
3. (b) 1105　　**4.** (b) 1242　　**5.** (b) 1235　　**6.** $1\frac{1}{8}$ h　　**7.** 1 h
8. (a) (i) B　　(ii) A　　(b) 8 s to 18 s　　(c) About 15 s
　　(d) About 9 s　　　　(e) B　　　　(f) A

page 153　　*Exercise 13*

1. (a) $1\frac{1}{2}$ m/s^2　　(b) 675 m　　(c) $11\frac{1}{4}$ m/s
2. (a) 600 m　　(b) 20 m/s　　(c) 225 m　　(d) −2 m/s^2
3. (a) 600 m　　(b) $387\frac{1}{2}$ m　　(c) 0 m/s^2

4. (a) 20 m/s (b) 750 m
5. (a) 8 s (b) 496 m (c) 12·4 m/s
6. (a) 30 m/s (b) $-2\frac{1}{7}$ m/s^2 (c) 20 s
7. (a) 15 m/s (b) $2\frac{1}{4}$ m/s^2 **8.** (a) 40 m/s (b) 10 s
9. (a) 50 m/s (b) 20 s **10.** (a) 20 m/s (b) 20 s

page 154 **Exercise 14**

1. 225 m **2.** 120 m **3.** 1 km **4.** 10 s
5. 60 s **6.** 88 yards **7.** 55 yards **8.** 1·39 km
9. 500 m **10.** Yes. Stopping distance = 46·5 m
11. 94375 m **12.** (a) 0·75 m/s^2 (b) 680 m
13. (a) 0·35 m/s^2 (b) 260 m

page 156 **Revision Exercise 7A**

1. (a) $y = x - 7$ (b) $y = 2x + 5$ (c) $y = -2x + 10$ (d) $y = \dfrac{x+1}{2}$
2. (a) 2 (b) 1 (c) $-3\frac{1}{2}$ (d) 0 (e) 10
3. (a) 2, -7 (b) -4, 5 (c) $\frac{1}{2}$, 4 (d) $-\frac{1}{2}$, 5 (e) -2, 12
 (f) $-\frac{2}{3}$, 8
4. A : $y = 6$; B : $y = \frac{1}{2}x - 3$; C : $y = 10 - x$; D : $y = 3x$
5. A : $4y = 3x - 16$; B : $2y = x - 8$; C : $2y + x = 8$; D : $4y + 3x = 16$
6. (a) $y = 2x - 3$ (b) $y = 3x + 4$ (c) $y = 10 - x$ (d) $y = 7$
7. (a) A $(0, -8)$, $(4, 0)$ (b) 2 (c) $y = 2x - 8$
8. 25 sq. units **9.** -3 **10.** 219
12. (a) $y = 3x$ (b) $y = 0$ (c) $y = 11 - x$ (d) $y = 5x$
13. (a) 1·56, -2·56 (b) ± 2·24 (c) ± 2·65
14. (a) 0·84, 4·15 (b) 0·65 $< x <$ 3·85 (c) 3·3
15. (a) 9·2 (b) 0·6 (c) 1·4 (d) 1·65
16. (a) 0·3 m/s^2 (b) 1050 m (c) 40 s
17. (a) 30 m/s (b) 600 m

page 158 **Examination Exercise 7B**

1. (ii) 3·75 **2.** (b) (i) -8·75 (ii) $x = 1$, $x = 3$·55
3. (d) $x = 1$·4, $x = 3$·2 **4.** (d) 1·41 ± 0·02 (e) 1·15 $\leqslant x \leqslant 1$·3 $(\pm 0$·02)
5. (c) (i) 32 mpg (ii) 53 mph **6.** (i) 10 m/s (ii) 6 m/s
7. (b) (i) -1·25 m/s^2 (ii) 68 m **8.** (b) 1125 am
9. (a) 124·8 m (b) 10 s (c) 15 m/s
10. (a) (ii) 275 feet (b) (ii) 40 mph (iii) More than 50 mph
11. (ii) £1·32
12. (b) 100 (± 2)
 (c) Alcohol level = $170 - 10t$, where t = number of hours after arrest
 (d) Between 6·30 and 8·30 on Sunday

13. (b) (i) 44·5 s (ii) 92·5 s (c) $-\dfrac{49}{1800}$ (d) 0, 274, 116

PART 8

page 163   *Exercise 1*

1. $\begin{pmatrix} 2 & 4 \\ 4 & 2 \end{pmatrix}$ **2.** $\begin{pmatrix} 1 & 6 & -1 \\ 7 & -10 & 6 \end{pmatrix}$ **3.** —   **4.** $\begin{pmatrix} -4 & 2 \\ 0 & 0 \end{pmatrix}$

5. $(8 \quad 10)$ **6.** $\begin{pmatrix} 0 & 15 \\ 3 & -6 \end{pmatrix}$ **7.** $\begin{pmatrix} 0 & -3 \\ 1 & 0 \\ -9 & -5 \end{pmatrix}$ **8.** $\begin{pmatrix} 4 & 3 \\ 7 & 6 \end{pmatrix}$

9. —   **10.** $\begin{pmatrix} 6 & 7 \\ 5 & 0 \end{pmatrix}$ **11.** $\begin{pmatrix} -1 & 9 \\ -2 & -1 \\ 25 & 10 \end{pmatrix}$ **12.** $\begin{pmatrix} 1 & -5\frac{1}{2} \\ \frac{1}{2} & 4 \end{pmatrix}$

13. $\begin{pmatrix} -1 & 12 \\ 4 & 7 \end{pmatrix}$ **14.** $\begin{pmatrix} 15 & 20 \\ -4 & -9 \end{pmatrix}$ **15.** $\begin{pmatrix} 5 & -10 \\ 2 & 7 \end{pmatrix}$ **16.** $\begin{pmatrix} 3 & 14 \\ -2 & 9 \end{pmatrix}$

17. $\begin{pmatrix} 12 \\ 13 \end{pmatrix}$ **18.** $\begin{pmatrix} 5 \\ 13 \end{pmatrix}$ **19.** $\begin{pmatrix} 14 & 1 \\ -32 & -13 \end{pmatrix}$ **20.** $\begin{pmatrix} 8 & -27 \\ 23 & -2 \end{pmatrix}$

21. $\begin{pmatrix} 8 & -27 \\ 23 & -2 \end{pmatrix}$ **22.** —   **23.** —   **24.** $\begin{pmatrix} 16 & 20 \\ 4 & 5 \\ 12 & 15 \end{pmatrix}$

25. $\begin{pmatrix} 30 & 40 \\ -8 & -18 \end{pmatrix}$ **26.** $\begin{pmatrix} 7 \\ 36 \end{pmatrix}$ **27.** $\begin{pmatrix} 12 & 15 \\ 4 & 5 \end{pmatrix}$ **28.** (17)

29. $\begin{pmatrix} -18 & 14 & -6 \\ 26 & -26 & 8 \end{pmatrix}$ **30.** $\begin{pmatrix} 1 & -6 \\ 18 & 13 \end{pmatrix}$ **31.** $\begin{pmatrix} -107 & -84 \\ 252 & 61 \end{pmatrix}$ **32.** —

33. $\begin{pmatrix} -9 & 13 & -17 \\ 3 & -4 & 5 \\ 0 & -7 & 14 \end{pmatrix}$ **34.** $\begin{pmatrix} 59 \\ -21 \end{pmatrix}$ **35.** $\begin{pmatrix} 1 & 5 & 1 \\ 3 & -11 & 0 \\ 22 & -20 & 7 \end{pmatrix}$ **36.** $\begin{pmatrix} 45 & -140 \\ -28 & 101 \end{pmatrix}$

37. $x = 6, y = 3, z = 0$
38. $x = 4, y = 5, z = 7, w = -5, v = 0$
39. $a = 4, b = 9, c = 15, d = 2$
40. $x = 1, y = 4$
41. $m = 5, n = -\frac{1}{3}$
42. $p = 3, q = -1$
43. $x = 1, y = 2, z = -1, w = -2$
44. $y = 2, z = -1, x = 1, w = -2$
45. $a = -3, e = 4, k = 2$
46. $m = 3, n = 5, p = 3, q = 3$
47. $x = 2\frac{2}{3}$
48. $k = \pm 1$
49. (a) $k = 2$ (b) $m = 4$
50. (a) $n = 3$ (b) $q = 9$

Exercise 2   *page 164*

1. $\begin{array}{c} \\ A \\ B \\ C \end{array} \begin{array}{ccc} A & B & C \\ \end{array} \begin{pmatrix} 0 & 1 & 1 \\ 1 & 0 & 2 \\ 1 & 2 & 0 \end{pmatrix}$

2. $\begin{array}{c} \\ A \\ B \\ C \end{array} \begin{array}{ccc} A & B & C \\ \end{array} \begin{pmatrix} 0 & 1 & 0 \\ 1 & 0 & 2 \\ 0 & 2 & 0 \end{pmatrix}$

3. $\begin{array}{c} \\ A \\ B \\ C \\ D \end{array} \begin{array}{cccc} A & B & C & D \\ \end{array} \begin{pmatrix} 0 & 1 & 0 & 1 \\ 1 & 0 & 2 & 0 \\ 0 & 2 & 0 & 2 \\ 1 & 0 & 2 & 0 \end{pmatrix}$

4. $\begin{array}{c} \\ A \\ B \\ C \end{array} \begin{array}{ccc} A & B & C \\ \end{array} \begin{pmatrix} 0 & 1 & 3 \\ 1 & 0 & 1 \\ 3 & 1 & 0 \end{pmatrix}$

5. $\begin{pmatrix} 0 & 1 & 2 & 0 \\ 1 & 0 & 0 & 2 \\ 2 & 0 & 0 & 1 \\ 0 & 2 & 1 & 0 \end{pmatrix} \begin{pmatrix} 5 & 0 & 0 & 4 \\ 0 & 5 & 4 & 0 \\ 0 & 4 & 5 & 0 \\ 4 & 0 & 0 & 5 \end{pmatrix}$

6. $\begin{pmatrix} 0 & 2 & 2 & 0 \\ 2 & 0 & 0 & 1 \\ 2 & 0 & 0 & 2 \\ 0 & 1 & 2 & 0 \end{pmatrix} \begin{pmatrix} 8 & 0 & 0 & 6 \\ 0 & 5 & 6 & 0 \\ 0 & 6 & 8 & 0 \\ 6 & 0 & 0 & 5 \end{pmatrix}$

page 164 **Exercise 3**

1. (150   160) **2.** Team C, 34 points **3.** £60 000

page 165 **Exercise 4**

1. $\begin{pmatrix} 1 & -1 \\ -3 & 4 \end{pmatrix}$ **2.** $\begin{pmatrix} 5 & -2 \\ -2 & 1 \end{pmatrix}$ **3.** $\frac{1}{2}\begin{pmatrix} 2 & -4 \\ -1 & 3 \end{pmatrix}$ **4.** $\frac{1}{3}\begin{pmatrix} 1 & -2 \\ -1 & 5 \end{pmatrix}$

5. $\frac{1}{2}\begin{pmatrix} 2 & 2 \\ 1 & 2 \end{pmatrix}$ **6.** $\frac{1}{5}\begin{pmatrix} 2 & 3 \\ 1 & 4 \end{pmatrix}$ **7.** $\frac{1}{8}\begin{pmatrix} 3 & -1 \\ 2 & 2 \end{pmatrix}$ **8.** $\frac{1}{6}\begin{pmatrix} 4 & 3 \\ -2 & 0 \end{pmatrix}$

9. $\frac{1}{5}\begin{pmatrix} -3 & 2 \\ -1 & -1 \end{pmatrix}$ **10.** no inverse **11.** $\frac{1}{14}\begin{pmatrix} 4 & 2 \\ -1 & 3 \end{pmatrix}$ **12.** $-\frac{1}{5}\begin{pmatrix} 1 & -1 \\ -2 & -3 \end{pmatrix}$

13. $-\frac{1}{5}\begin{pmatrix} -4 & 3 \\ -1 & 2 \end{pmatrix}$ **14.** $\frac{1}{7}\begin{pmatrix} 1 & 0 \\ 5 & 7 \end{pmatrix}$ **15.** $-\frac{1}{6}\begin{pmatrix} -4 & -1 \\ 2 & 2 \end{pmatrix}$ **16.** $\frac{1}{2}\begin{pmatrix} 3 & -4 \\ -1 & 2 \end{pmatrix}$

17. $-\frac{1}{2}\begin{pmatrix} 1 & 0 \\ -3 & -2 \end{pmatrix}$ **18.** $\begin{pmatrix} -1 & 3 \\ 0 & -3 \end{pmatrix}$ **19.** $\begin{pmatrix} 3 & 2 \\ 1 & 3 \end{pmatrix}$ **20.** $\begin{pmatrix} 2 & -3 \\ 2 & 4 \end{pmatrix}$

21. $\frac{2}{3}\begin{pmatrix} 1 & -1 \\ 2 & 1 \end{pmatrix}$ **22.** (a) $\begin{pmatrix} 5 & -11 \\ -1 & 3 \end{pmatrix}$ (b) $\frac{1}{2}\begin{pmatrix} 1 & 3 \\ 0 & 2 \end{pmatrix}$ (c) $\frac{1}{2}\begin{pmatrix} 3 & 1 \\ 1 & 1 \end{pmatrix}$

23. $\begin{pmatrix} 1 & -3 \\ 4 & 0 \end{pmatrix}$ **24.** $\begin{pmatrix} 4 \\ 3 \end{pmatrix}$ **25.** $4; 1; 1\frac{1}{2}$ **26.** $x = -2$

27. (a) 42 (b) 14

page 166 **Exercise 6**

1. (c) (i) (−6, 8)   (ii) (6, −4)   (iii) (8, 6)
2. (c) (i) (8, 8)   (ii) (8, −6)   (iii) (−8, 6)
3. (c) (i) (3, −1)   (ii) (4, 2)   (iii) (−1, 1)
4. (b) (i) $y = 1$ (ii) $y = x$ (iii) $y = -x$ (iv) $y = 2$
5. (f) (1, −1), (−3, −1), (−3, −3)
6. (f) (8, −2), (6, −6), (8, −6)

page 168 **Exercise 8**

1. (a) A′(3, −1) B′(6, −1) C′(6, −3) (b) D′(3, −3) E′(3, −6) F′(1, −6)
   (c) P′(−7, −4), O′(−5, −4) R′(−5, −1)
3. (c) (−2, 1), (−2, −1), (1, −2) **4.** (e) (−5, 2), (−5, 6), (−3, 5)
5. (b) (i) 90° anticlockwise, centre (0, 0) (ii) 180°, centre (2, 1)
      (iii) 90° clockwise, centre (2, 0) (iv) 180°, centre ($3\frac{1}{2}$, $2\frac{1}{2}$)
      (v) 90° anticlockwise, centre (6, 1) (vi) 90° clockwise, centre (1, 3)
6. (e) (i) 180°, centre ($\frac{1}{2}$, $\frac{1}{2}$) (ii) 90° anticlockwise, centre (−2, 4)

page 169 **Exercise 9**

1. (a) $\begin{pmatrix} 7 \\ 3 \end{pmatrix}$ (b) $\begin{pmatrix} 0 \\ -9 \end{pmatrix}$ (c) $\begin{pmatrix} 9 \\ 10 \end{pmatrix}$ (d) $\begin{pmatrix} -10 \\ 3 \end{pmatrix}$ (e) $\begin{pmatrix} -1 \\ 13 \end{pmatrix}$

   (f) $\begin{pmatrix} 10 \\ 0 \end{pmatrix}$ (g) $\begin{pmatrix} -9 \\ -4 \end{pmatrix}$ (h) $\begin{pmatrix} -10 \\ 0 \end{pmatrix}$ **2.** (5, 2) **3.** (5, 6)

4. (8, −5) **5.** (0, 6) **6.** (4, −7) **7.** (−3, 4) **8.** (−3, −5)
9. (−1, −8) **10.** (5, 2) **11.** (−2, 1)

page 170 **Exercise 10**

7. (4, 8), (8, 4), (10, 10) **8.** (3, 6), (7, 2), (9, 8) **9.** (1, 1), (10, 4), (4, 7)
10. (1, 4), (7, 8), (11, 2) **11.** (2, 1), +3 **12.** (11, 9), $\frac{1}{2}$

13. $(5, 4), -2$ **14.** $(6, 6), -1$ **15.** $(\frac{1}{2}, 1), (6\frac{1}{2}, 1), (\frac{1}{2}, 5)$
16. $(3, 4), (3, 3), (6, 3)$ **17.** $(3, 7), (1, 7), (3, 2)$ **18.** $(10, 7), (6, 7), (6, 5)$
19. $(6, 5), (3\frac{1}{2}, 5), (3\frac{1}{2}, 3)$

page 171   *Exercise 11*

1. (b) (i)   Rotation 90° clockwise, centre $(0, -2)$
         (ii)   Reflection in $y = x$
         (iii) Translation $\begin{pmatrix} 3 \\ 7 \end{pmatrix}$
         (iv)   Enlargement, scale factor 2, centre $(-5, 5)$
         (v)   Translation $\begin{pmatrix} -7 \\ -3 \end{pmatrix}$
         (vi)   Reflection in $y = x$
2. (a) Rotation 90° clockwise, centre $(4, -2)$
     (b) Translation $\begin{pmatrix} 8 \\ 2 \end{pmatrix}$
     (c) Reflection in $y = x$
     (d) Enlargement, scale factor $\frac{1}{2}$, centre $(7, -7)$
     (e) Rotation 90° anticlockwise, centre $(-8, 0)$
     (f) Enlargement, scale factor 2, centre $(-1, -9)$
     (g) Rotation 90° anticlockwise, centre $(7, 3)$
3. (a) Enlargement, scale factor $1\frac{1}{2}$, centre $(1, -4)$
     (b) Rotation 90° clockwise, centre $(0, -4)$
     (c) Reflection in $y = -x$
     (d) Translation $\begin{pmatrix} 11 \\ 10 \end{pmatrix}$
     (e) Enlargement, scale factor $\frac{1}{2}$, centre $(-3, 8)$
     (f) Rotation 90° anticlockwise, centre $(\frac{1}{2}, 6\frac{1}{2})$
     (g) Enlargement, scale factor 3, centre $(-2, 5)$

page 172   *Exercise 12*

1. $(2, -3)$ **2.** $(5, -1)$ **3.** $(6, 4)$ **4.** $(4, -6)$ **5.** $(0, 0)$
6. $(-6, 4)$ **7.** $(3, -2)$ **8.** $(3, 2)$ **9.** $(-2, 3)$ **10.** $(0, 0)$
11. $(-3, 2)$ **12.** $(-3, -2)$ **13.** $(-3, -2)$ **14.** $(-6, -4)$ **15.** $(6, 0)$
16. $(-2, 3)$ **17.** $(0, 4)$ **18.** $(6, 8)$ **19.** $(3, 2)$ **20.** $(0, 0)$

page 172   *Exercise 13*

1. $T_6, T_5; \mathbf{R}_b$ **2.** $T_3, T_1; \mathbf{M}_x$ **3.** $T_4, T_2; \mathbf{M}_4$ **4.** $T_2, T_8; \mathbf{M}_4$ **5.** $T_7, T_5; \mathbf{M}_4$
6. $T_1, T_6; \mathbf{M}_3$ **7.** $T_4, T_3; \mathbf{M}_3$ **8.** $T_3, T_6; \mathbf{M}_x$ **9.** $T_8, T_1; \mathbf{R}_c$ **10.** $T_8, T_2; \mathbf{R}_a$
11. $T_8, T_5; \mathbf{M}_x$ **12.** $T_8, T_1; \mathbf{R}_b$ **13.** $T_4, T_1; \mathbf{R}_b$ **14.** $T_3, T_8; \mathbf{I}$ **15.** $T_4, T_3; \mathbf{M}_4$
16. $T_6, T_7, T_1; \mathbf{R}_b$ **17.** $T_6, T_2, T_7; \mathbf{R}_a$ **18.** $T_7, T_2, T_4; \mathbf{M}_4$ **19.** $T_5, T_4, T_8; \mathbf{M}_x$ **20.** $T_7, T_2, T_3; \mathbf{M}_y$
21. $T_2, T_8, T_7; \mathbf{I}$ **22.** $T_5, T_8, T_6; \mathbf{I}$ **23.** $T_4, T_6, T_5; \mathbf{M}_4$ **24.** $T_5, T_3, T_8; \mathbf{R}_a$ **25.** No

page 173   *Exercise 14*

1. (a) $(-4, 4)$ (b) $(2, -2)$ (c) $(0, 0)$ (d) $(0, 4)$ (e) $(0, 0)$
2. (a) $(-2, 5)$ (b) $(-4, 0)$ (c) $(2, -2)$ (d) $(1, -1)$
3. (a) reflection in y-axis
     (b) rotation 180°, centre $(-2, 2)$
     (c) rotation 90° clockwise, centre $(2, 2)$
4. rotation 90° anticlockwise, centre $(0, 0)$

(b) translation $\begin{pmatrix} -2 \\ 5 \end{pmatrix}$

(c) rotation 90° anticlockwise, centre $(2, -4)$

(d) rotation 90° anticlockwise, centre $(-\frac{1}{2}, 3\frac{1}{2})$

5. (a) rotation 90° anticlockwise, centre $(2, 2)$

   (b) enlargement, scale factor $\frac{1}{2}$, centre $(8, 6)$

   (c) rotation 90° clockwise, centre $(-\frac{1}{2}, -3\frac{1}{2})$

6. $\mathbf{A}^{-1}$: reflection in $x = 2$

   $\mathbf{B}^{-1}$: B

   $\mathbf{C}^{-1}$: translation $\begin{pmatrix} 6 \\ -2 \end{pmatrix}$

   $\mathbf{D}^{-1}$: D

   $\mathbf{E}^{-1}$: E

   $\mathbf{F}^{-1}$: translation $\begin{pmatrix} -4 \\ -3 \end{pmatrix}$

   $\mathbf{G}^{-1}$: 90° rotation anticlockwise, centre $(0, 0)$

   $\mathbf{H}^{-1}$: enlargement, scale factor 2, centre $(0, 0)$

7. (a) $(4, 0)$ (b) $(-6, -1)$ (c) $(-2, -2)$ (d) $(2, -2)$ (e) $(6, 2)$

8. (a) $(1, -6)$ (b) $(4, -2)$ (c) $(2, 7)$ (d) $(4, -6)$ (e) $(2, -4)$

9. (a) $(-1, -2)$ (b) $(8, 2)$ (c) $(4, -6)$ (d) $(0, -2)$

10. (b) rotation, 180°, centre $(4, 0)$ (c) translation $\begin{pmatrix} 12 \\ -4 \end{pmatrix}$

page 174   **Exercise 15**

2. (a) enlargement : scale factor 2, centre $(0, 0)$ (b) enlargement : scale factor $-\frac{1}{2}$, centre $(0, 0)$

   (c) reflection in $y = -x$ (d) enlargement : scale factor -2, centre $(0, 0)$

3. A : reflection in x-axis   B : reflection in y-axis   C : reflection in $y = x$

   D : rotation, $-90°$, centre $(0, 0)$ E : reflection in $y = -x$ F : rotation, 180°, centre $(0, 0)$

   G : rotation, $+90°$, centre $(0, 0)$ H : identity (no change)

4. (a) ratio $= 4 : 1$ (b) ratio $= 4 : 1$

5. rotation 45° anticlockwise; enlargement scale factor $\sqrt{2}$ $(1 \cdot 41)$

6. rotation $26 \cdot 6°$ clockwise; enlargement scale factor $\sqrt{5}$ $(2 \cdot 24)$

7. (d) rotation 90° clockwise, centre $(0, 0)$

8. $y = 2x$ 9. $y = 3x$

10. (c) $OB = \sqrt{20}$, $OB' = 3\sqrt{20}$ (d) $36 \cdot 9°$

   (e) rotation $36 \cdot 9°$; enlargement scale factor 3

11. (a) $\begin{pmatrix} 0 & -1 \\ 1 & 0 \end{pmatrix}$ (b) $\begin{pmatrix} -1 & 0 \\ 0 & -1 \end{pmatrix}$ (c) $\begin{pmatrix} 0 \cdot 866 & -0 \cdot 5 \\ 0 \cdot 5 & 0 \cdot 866 \end{pmatrix}$

   (d) $\begin{pmatrix} 0 & 1 \\ -1 & 0 \end{pmatrix}$ (e) $\begin{pmatrix} 0 \cdot 5 & -0 \cdot 866 \\ 0 \cdot 866 & 0 \cdot 5 \end{pmatrix}$ (f) $\begin{pmatrix} -0 \cdot 866 & -0 \cdot 5 \\ 0 \cdot 5 & -0 \cdot 866 \end{pmatrix}$ (g) $\begin{pmatrix} 0 \cdot 707 & -0 \cdot 707 \\ 0 \cdot 707 & 0 \cdot 707 \end{pmatrix}$

   (h) $\begin{pmatrix} 0 \cdot 6 & -0 \cdot 8 \\ 0 \cdot 8 & 0 \cdot 6 \end{pmatrix}$

12. (a) $+90°$ (b) $+36 \cdot 9°$ (c) $-60°$ (d) $-53 \cdot 1°$

page 176   **Exercise 16**

1. (a) reflection in $y = x - 1$ (b) reflection in $y = 1$

   (c) rotation $-90°$, centre $(2, -2)$ (d) enlargement, scale factor 3, centre $(2, -1)$

2. $\mathbf{BA} \equiv \mathbf{A}$ then $\mathbf{B}$ 3. $\mathbf{BA} \equiv \mathbf{A}$ then $\mathbf{B}$

5. (a) $(14, 3)$ (b) $m = 3, n = \frac{1}{2}$ (c) $h = 1, k = -2$

6. (a) $(-2, 2)$ (c) $\begin{pmatrix} 2 \\ -2 \end{pmatrix}$ (d) $\begin{pmatrix} 2 & 0 \\ 0 & 2 \end{pmatrix}\begin{pmatrix} x \\ y \end{pmatrix} + \begin{pmatrix} 2 \\ -2 \end{pmatrix}$

7. (a) $x = 2$ (c) $\begin{pmatrix} 4 \\ 0 \end{pmatrix}$ (d) $\begin{pmatrix} -1 & 0 \\ 0 & 1 \end{pmatrix}\begin{pmatrix} x \\ y \end{pmatrix} + \begin{pmatrix} 4 \\ 0 \end{pmatrix}$

8. (a) $\begin{pmatrix} 2 & 0 \\ 0 & 2 \end{pmatrix}\begin{pmatrix} x \\ y \end{pmatrix} + \begin{pmatrix} -1 \\ -3 \end{pmatrix}$ (b) $\begin{pmatrix} 2 & 0 \\ 0 & 2 \end{pmatrix}\begin{pmatrix} x \\ y \end{pmatrix} + \begin{pmatrix} -\frac{1}{2} \\ -1 \end{pmatrix}$ (c) $\begin{pmatrix} 0 & 1 \\ 1 & 0 \end{pmatrix}\begin{pmatrix} x \\ y \end{pmatrix} + \begin{pmatrix} -3 \\ 3 \end{pmatrix}$

(d) $\begin{pmatrix} -1 & 0 \\ 0 & -1 \end{pmatrix}\begin{pmatrix} x \\ y \end{pmatrix} + \begin{pmatrix} 3 \\ 5 \end{pmatrix}$ (e) $\begin{pmatrix} 1 & 0 \\ 0 & -1 \end{pmatrix}\begin{pmatrix} x \\ y \end{pmatrix} + \begin{pmatrix} 0 \\ 2 \end{pmatrix}$ (f) $\begin{pmatrix} 0 & 1 \\ -1 & 0 \end{pmatrix}\begin{pmatrix} x \\ y \end{pmatrix} + \begin{pmatrix} 4 \\ 0 \end{pmatrix}$

page 177   Exercise 17

1. rotation : $+90°$, centre $(0, 0)$

2. reflection in y-axis

3. reflection in $y = -x$

4. reflection in $y = x$

5. enlargement: scale factor 2, centre $(0, 0)$

6. enlargement: scale factor $\frac{1}{2}$, centre $(0, 0)$

7. stretch: parallel to x-axis, scale factor 3

8. shear: x-axis invariant

9. shear: y-axis invariant

10. stretch: parallel to y-axis, scale factor 2

11. enlargement: scale factor -2, centre $(0, 0)$

12. enlargement: scale factor $-\frac{1}{2}$, centre $(0, 0)$

13. $\begin{pmatrix} 0 & -1 \\ 1 & 0 \end{pmatrix}$ **14.** $\begin{pmatrix} 0 & 1 \\ 1 & 0 \end{pmatrix}$ **15.** $\begin{pmatrix} -1 & 0 \\ 0 & 1 \end{pmatrix}$ **16.** $\begin{pmatrix} -1 & 0 \\ 0 & -1 \end{pmatrix}$

17. $\begin{pmatrix} 3 & 0 \\ 0 & 3 \end{pmatrix}$ **18.** $\begin{pmatrix} 0 & -1 \\ -1 & 0 \end{pmatrix}$ **19.** $\begin{pmatrix} -2 & 0 \\ 0 & -2 \end{pmatrix}$ **20.** $\begin{pmatrix} 1 & 0 \\ 0 & -1 \end{pmatrix}$

21. $\begin{pmatrix} 0 & 1 \\ -1 & 0 \end{pmatrix}$ **22.** $\begin{pmatrix} \frac{1}{2} & 0 \\ 0 & \frac{1}{2} \end{pmatrix}$

page 178   Exercise 18

1. $x = 0$ (y-axis)

2. $y = 0$

3. stretch: parallel to x-axis, scale factor 2

4. stretch: parallel to x-axis, scale factor 3

5. stretch: parallel to y-axis, scale factor 2

6. stretch: parallel to x-axis, scale factor $1\frac{1}{2}$

7. shear: x-axis invariant

8. stretch: parallel to x-axis, scale factor -2

9. stretch: parallel to y-axis, scale factor 3

10. stretch: parallel to x-axis, scale factor $\frac{1}{2}$

11. $y = -x$

12. $y = -x$

page 179   Revision Exercise 8A

1. (a) $\begin{pmatrix} 6 & 4 \\ 2 & 8 \end{pmatrix}$ (b) $\begin{pmatrix} 4 & -1 \\ 1 & 2 \end{pmatrix}$ (c) $\begin{pmatrix} 1\frac{1}{2} & 1 \\ \frac{1}{2} & 2 \end{pmatrix}$ (d) $\begin{pmatrix} -3 & 13 \\ -1 & 11 \end{pmatrix}$

(e) $\begin{pmatrix} 1 & 3 \\ 0 & 4 \end{pmatrix}$

2. (a) $\begin{pmatrix} -9 & -1 \\ 5 & 1\frac{1}{3} \end{pmatrix}$ (b) $\begin{pmatrix} 12 & 6 \\ 4 & 2 \end{pmatrix}$ (c) $\begin{pmatrix} 9 & -2 \\ 2 & -7 \end{pmatrix}$

3. (a) (14); $\begin{pmatrix} -1 & -15 \\ 3 & 15 \end{pmatrix}$ (b) $\mathbf{X} = \begin{pmatrix} 1 & 3 \\ 2 & 4 \end{pmatrix}$ **4.** $\frac{1}{13}\begin{pmatrix} 5 & 1 \\ -3 & 2 \end{pmatrix}$

5. $x = 3$; $-\frac{1}{9}\begin{pmatrix} -1 & -2 \\ -3 & 3 \end{pmatrix}$ **6.** $h = 4$, $k = -4$

7. (a) $a = \pm 4$ (b) $a = \pm 3$

8. (a) $(4, -1)$ (b) $(4, 1)$ (c) $(-3, 2)$

9. (a) $A'(-3, -1)$ $B'(1, -1)$ $C'(-3, -7)$ (b) $A'(2, -2)$ $B'(6, -2)$ $C'(2, -8)$

(c) $A'(1, 1)$ $B'(2, 1)$ $C'(1, -\frac{1}{2})$ (d) $A'(4, 2)$ $B'(3, 2)$ $C'(4, 3\frac{1}{2})$

(e) $A'(-2, 2)$ $B'(-6, 2)$ $C'(-2, 8)$

10. (a) reflection in y-axis   (b) reflection in $y = x$ (c) rotation $-90°$, centre $(0, 0)$

(d) reflection in $y = -x$ (e) rotation $180°$, centre $(0, 0)$ (f) rotation $-90°$, centre $(0, 0)$

11. (a) reflection in $x = \frac{1}{2}$ (b) reflection in $y = -x$ (c) rotation $180°$, centre $(1, 1)$

12. (a) $(-1, -3)$ (b) $(-1, 3)$ (c) $(6, 2)$ (d) $(-3, 1)$ (e) $(-2, 6)$ (f) $(0, 2)$

13. (a) $(-1, 2)$ (b) $(1, -2)$ (c) $(10, -2)$ (d) $(6, -2)$ (e) $(-10, 2)$ (f) $(12, 2)$

14. (a) $(-2, 5)$ (b) $(-4, -3)$ (c) rotation $+90°$, centre $(0, 0)$

15. (a) rotation $+90°$ centre $(0, 0)$ (b) reflection in x-axis (c) rotation $180°$, centre $(0, 0)$
(d) rotation $-90°$, centre $(0, 0)$ (e) reflection in $y = -x$

16. (a) reflection in $y = x$ (b) reflection in y-axis
(c) enlargement, scale factor 3, centre $(0, 0)$

17. (a) $\begin{pmatrix} -1 & 0 \\ 0 & -1 \end{pmatrix}$ (b) $\begin{pmatrix} 1 & 0 \\ 0 & -1 \end{pmatrix}$ (c) $\begin{pmatrix} 4 & 0 \\ 0 & 4 \end{pmatrix}$ (d) $\begin{pmatrix} 0 & -1 \\ -1 & 0 \end{pmatrix}$

(e) $\begin{pmatrix} 0 & 1 \\ -1 & 0 \end{pmatrix}$

18. (a) enlargement, scale factor 2, centre $(0, 0)$; translation $\begin{pmatrix} 5 \\ -2 \end{pmatrix}$

(b) $(11, -4)$ (c) $(3, -1)$ (d) $(1, 3)$

19. (a) $\begin{pmatrix} 0 & 1 \\ 1 & 0 \end{pmatrix}$ (b) $\begin{pmatrix} -1 & 0 \\ 0 & 1 \end{pmatrix}$ (c) $\begin{pmatrix} 0 & 1 \\ -1 & 0 \end{pmatrix}$

(d) $\begin{pmatrix} 0 & -1 \\ 1 & 0 \end{pmatrix}$ $\mathbf{AB} \equiv$ rotation $-90°$, centre $(0, 0)$
$\mathbf{BA} \equiv$ rotation $+90°$, centre $(0, 0)$

page 180 ***Examination Exercise 8B***

1. (i) (a) reflection in x-axis (b) rotation $90°$ anticlockwise, centre $(0, 0)$
(c) translation $\begin{pmatrix} 4 \\ 2 \end{pmatrix}$
(ii) reflection in $y = x$

2. (d) (i) translation $\begin{pmatrix} 6 \\ 0 \end{pmatrix}$ (ii) reflection in x-axis

3. (a) $\begin{pmatrix} 6 & 13 \\ -8 & 8 \end{pmatrix}$ (b) $\begin{pmatrix} 10 & 5 \\ 5 & 5 \end{pmatrix}$

4. (a) $(5, 0), (10, 0), (3, 9)$ (c) rotation $37°$ clockwise, centre $(0, 0)$

5. (a) (i) $\begin{pmatrix} 3 \\ 0 \end{pmatrix}$ (ii) $\begin{pmatrix} 0 \\ -2 \end{pmatrix}$

(c) (i) reflection in y axis
(ii) reflection in $y = -x$
(iii) rotation $90°$ anticlockwise, centre $(0, 0)$

6. (b) reflection in $y = -x$ (c) $P'(3, 0), Q'(2, 1)$ (d) reflection in $x + y = 3$

7. (i) $\frac{1}{4}\begin{pmatrix} -4 & 3 \\ -4 & 2 \end{pmatrix}$ (ii) $(3, -1)$ (iii) $\begin{pmatrix} -8 & 6 \\ -8 & 4 \end{pmatrix}$
(iv) enlargement, scale factor 8, centre $(0, 0)$

8. (a) $A'(2, 2), B'(1, 3)$
(c) rotation $45°$ anticlockwise, centre $(0, 0)$; enlargement scale factor $\sqrt{2}$, centre $(0, 0)$
(d) $(4, -2)$

9. (i) $\begin{pmatrix} 2 \\ 0 \end{pmatrix}, \begin{pmatrix} 0 \\ -2 \end{pmatrix}$ (ii) $\begin{pmatrix} 2 & 0 \\ 0 & -2 \end{pmatrix}$ (iii) $(4, -6)$

PART 9

page 184   **Exercise 1**

1. (a) 7   (b) 4   (c) 4   (d) 19   (e) 8
2. (a) 9   (b) 5   (c) 4   (d) 20   (e) 31
3. (a) 8   (b) 3   (c) 3   (d) 2   (e) 17   (f) 0
4. (a) 59   (b) 11   (c) 5   (d) 40   (e) 11   (f) 124
5. (a) 120   (b) 120   (c) 490   (d) 80   (e) 40   (f) 10
    (g) 500

page 185   **Exercise 2**

1. (a) {5, 7}   (b) {1, 2, 3, 4, 5, 6, 7, 8, 9, 11, 13}   (c) 5   (d) 11
    (e) true   (f) •true   (g) false   (h) true
2. (a) {2, 3, 5, 7}   (b) {1, 2, 3, . . . 9}   (c) 4   (d) Ø
    (e) false   (f) true   (g) false   (h) true
3. (a) {2, 4, 6, 8, 10}   (b) {16, 18, 20}   (c) Ø   (d) 15
    (e) 11   (f) 21   (g) false   (h) false   (i) true
    (j) true
4. (a) {1, 3, 4, 5}   (b) {1, 5}   (c) 1   (d) {1, 5}   (e) {1, 3, 5, 10}
    (f) 4   (g) true   (h) false   (i) true
5. (a) 4   (b) 3   (c) {b, d}   (d) {a, b, c, d, e}   (e) 5
    (f) 2
6. (a) 2   (b) 4   (c) {1, 2, 4, 6, 7, 8, 9}   (d) {7, 9}
    (e) {1, 2, 4}   (f) {1, 2, 4, 7, 9}   (g) {1, 2, 4, 6, 8}   (h) {6, 7, 8, 9}
    (i) {1, 2, 4, 7, 9}

page 186   **Exercise 3**

7. (a) $A \cup B$ (b) $A' \cap B$ (c) $(A \cup B)'$ (d) $Y' \cap X$

page 187   **Exercise 4**

1. (a) $10 - x$ (b) $13 - x$ (d) 5   **2.** 9   **3.** 14
4. 3   **5.** 36   **6.** 3   **7.** 11   **8.** 5
9. 28   **10.** $x = 6$; 26   **11.** 34   **12.** (a) 12   (b) $\frac{1}{2}$
13. $x = 10$; 30   **14.** 2
15. (a) All good footballers are Scottish men.   (b) No good footballers are Scottish men.
    (c) There are some good footballers who are Scottish men.
16. (a) The football players are all boys.   (b) The hockey players are all girls.
    (c) There are some people who play both football and hockey.
    (d) There are no boys who play hockey.   (e) $B \cap F = Ø$ (f) $H \cup F = \mathscr{E}$
17. (a) $S \cap T = Ø$ (b) $F \subset T$ (c) $S \cap F \neq Ø$
    (d) All living creatures are either spiders, animals that fly or animals which taste nice.
    (e) Animals which taste nice are all spiders.
18. (a) All tigers who believe in fairies also believe in Eskimos.
    (b) All tigers who believe in fairies or Eskimos are in hospital.
    (c) There are no tigers in hospital who believe in Eskimos.
    (d) $H \subset T$ (e) $T \cap X \neq Ø$
19. (a) There are no good bridge players called Peter.
    (b) All school teachers are either called Peter, are good bridge players or are women.
    (c) There are some women teachers called Peter.
    (d) $W \cap B = Ø$ (e) $B \subset (W \cap P)$

page 190   *Exercise 5*

1. d	**2.** 2c	**3.** 3c	**4.** 3d	**5.** 5d
6. 3c	**7.** −2d	**8.** −2c	**9.** −3c	**10.** −c
11. c + d	**12.** c + 2d	**13.** 2c + d	**14.** 3c + d	**15.** 2c + 2d
16. 2c + 3d	**17.** 2c − d	**18.** 3c − d	**19.** −c + 2d	**20.** −c + 3d
21. −c + d	**22.** −c −2d	**23.** −2c − 2d	**24.** −3c − 6d	**25.** −2c + 3d
26. c + 6d	**27.** $\overrightarrow{QI}$	**28.** $\overrightarrow{QU}$	**29.** $\overrightarrow{QH}$	**30.** $\overrightarrow{QB}$
31. $\overrightarrow{QF}$	**32.** $\overrightarrow{QJ}$	**33.** $\overrightarrow{QZ}$	**34.** $\overrightarrow{QL}$	**35.** $\overrightarrow{QE}$
36. $\overrightarrow{QX}$	**37.** $\overrightarrow{QW}$	**38.** $\overrightarrow{QK}$		

39. (a) −a   (b) a + b   (c) 2a − b   (d) −a + b
40. (a) a + b   (b) a − 2b   (c) −a + b   (d) −a − b
41. (a) −a − b   (b) 3a − b   (c) 2a − b   (d) −2a + b
42. (a) a − 2b   (b) a − b   (c) 2a   (d) −2a + 3b
43. (a) −2a + 2b   (b) 2a − b   (c) 3a − b   (d) −3a + b
44. (a) 2a − c   (b) 2a − c   (c) 3a   (d) a + b + c   (e) −3a − b
45. (a) b − c   (b) 2b + 2c   (c) a + 2b + 2c   (d) −a − b   (e) −a − b + c
46. (a) a + c   (b) −a + c   (c) a + b + c   (d) b − c   (e) −a + 2c

page 191   *Exercise 6*

1. (a) a   (b) −a + b   (c) 2b   (d) −2a   (e) −2a + 2b
   (f) −a + b   (g) a + b   (h) b   (i) −b + 2a   (j) −2b + a
2. (a) a   (b) −a + b   (c) 3b   (d) −2a   (e) −2a + 3b
   (f) $-a + \frac{3}{2}b$ (g) $a + \frac{3}{2}b$ (h) $\frac{3}{2}b$ (i) −b + 2a   (j) −3b + a
3. (a) 2a   (b) −a + b   (c) 2b   (d) −3a   (e) −3a + 2b
   (f) $-\frac{3}{2}a + b$ (g) $\frac{3}{2}a + b$ (h) $\frac{1}{2}a + b$ (i) −b + 3a   (j) −2b + a
4. (a) $\frac{1}{2}a$ (b) −a + b   (c) 4b   (d) $-\frac{3}{2}a$ (e) $-\frac{3}{2}a + 4b$
   (f) $-a + \frac{8}{3}b$ (g) $\frac{1}{2}a + \frac{8}{3}b$ (h) $-\frac{1}{2}a + \frac{8}{3}b$ (i) $\frac{3}{2}a − b$ (j) a − 4b
5. (a) 5a   (b) b − a   (c) $\frac{3}{2}b$ (d) −6a   (e) $\frac{3}{2}b − 6a$
   (f) b − 4a   (g) 2a + b   (h) a + b   (i) 6a − b   (j) $a − \frac{3}{2}b$
6. (a) 4a   (b) b − a   (c) 3b   (d) −5a   (e) 3b − 5a
   (f) $\frac{3}{4}b − \frac{5}{4}a$ (g) $\frac{15}{4}a + \frac{3}{4}b$ (h) $\frac{11}{4}a + \frac{3}{4}b$ (i) 5a − b   (j) a − 3b
7. $\frac{1}{2}s − \frac{1}{2}t$ **8.** $\frac{1}{3}a + \frac{2}{3}b$ **9.** a + c − b   **10.** 2m + 2n
11. (a) b − a   (b) b − a   (c) 2b − 2a   (d) b − 2a   (e) b − 2a
   (f) 2b − 3a
12. (a) y − z   (b) $\frac{1}{2}y − \frac{1}{2}z$ (c) $\frac{1}{2}y + \frac{1}{2}z$ (d) $-x + \frac{1}{2}y + \frac{1}{2}z$ (e) $-\frac{2}{3}x + \frac{1}{3}y + \frac{1}{3}z$
   (f) $\frac{1}{3}x + \frac{1}{3}y + \frac{1}{3}z$

page 194   *Exercise 7*

13. $\begin{pmatrix} -11 \\ 9 \end{pmatrix}$ **14.** $\begin{pmatrix} -1 \\ -4 \end{pmatrix}$ **15.** $\begin{pmatrix} 4 \\ 8 \end{pmatrix}$ **16.** $\begin{pmatrix} 4 \\ 3 \end{pmatrix}$ **17.** $\begin{pmatrix} 8 \\ -7 \end{pmatrix}$

18. $\begin{pmatrix} 18 \\ -1 \end{pmatrix}$ **19.** $\begin{pmatrix} 7 \\ -4 \end{pmatrix}$ **20.** $\begin{pmatrix} -18 \\ -16 \end{pmatrix}$ **21.** $\begin{pmatrix} 13 \\ 19 \end{pmatrix}$ **22.** $\begin{pmatrix} 8 \\ 5 \end{pmatrix}$

23. $\begin{pmatrix} 10 \\ -13 \end{pmatrix}$ **24.** $\begin{pmatrix} 17 \\ 35 \end{pmatrix}$

page 195   *Exercise 8*

1. $\begin{pmatrix} 2 \\ -2 \end{pmatrix}$ **2.** $\begin{pmatrix} 6 \\ -2 \end{pmatrix}$ **3.** (b) $\begin{pmatrix} 0 \\ 3 \end{pmatrix}$; $\begin{pmatrix} -5 \\ -5 \end{pmatrix}$ **4.** (b) $\begin{pmatrix} 4 \\ 2 \end{pmatrix}$; $\begin{pmatrix} -7 \\ 0 \end{pmatrix}$

5. (a) $\begin{pmatrix} 3 \\ -3 \end{pmatrix}$ (b) $\begin{pmatrix} 1\frac{1}{2} \\ -1\frac{1}{2} \end{pmatrix}$ (c) $\begin{pmatrix} 3\frac{1}{2} \\ 3\frac{1}{2} \end{pmatrix}$; M($3\frac{1}{2}$, $3\frac{1}{2}$)

6. (a) $\begin{pmatrix} -1 \\ 6 \end{pmatrix}$ (b) $\begin{pmatrix} -\frac{1}{2} \\ 3 \end{pmatrix}$ (c) $\begin{pmatrix} 5\frac{1}{2} \\ -4 \end{pmatrix}$; M$(5\frac{1}{2}, -4)$

7. (a) (i) $\begin{pmatrix} -6 \\ 3 \end{pmatrix}$ (ii) $\begin{pmatrix} -2 \\ 1 \end{pmatrix}$ (iii) $\begin{pmatrix} 2 \\ 3 \end{pmatrix}$

   (b) (i) $\begin{pmatrix} 0 \\ -9 \end{pmatrix}$ (ii) $\begin{pmatrix} 0 \\ -3 \end{pmatrix}$ (iii) $\begin{pmatrix} -2 \\ 2 \end{pmatrix}$

8. $\begin{pmatrix} 1 \\ -2 \end{pmatrix}$ or $\begin{pmatrix} -1 \\ 2 \end{pmatrix}$ **9.** $\begin{pmatrix} 0 \\ 2 \end{pmatrix}$ or $\begin{pmatrix} 0 \\ -2 \end{pmatrix}$

10. (a) $\mathbf{q} - \mathbf{p}$ (b) $\mathbf{q} + 2\mathbf{p}$ (c) $\mathbf{p} + \mathbf{q}$

11. (a) $\begin{pmatrix} 1 \\ -3 \end{pmatrix}$ (b) $\begin{pmatrix} -1 \\ 3 \end{pmatrix}$ (c) $\begin{pmatrix} 3 \\ 1 \end{pmatrix}$ (d) $\begin{pmatrix} -3 \\ -1 \end{pmatrix}$

page 196   **Exercise 9**

1. 5   **2.** $\sqrt{17}$ **3.** 13   **4.** 3   **5.** 5
6. $\sqrt{45}$ **7.** $\sqrt{74}$ **8.** $\sqrt{208}$ **9.** 10   **10.** $\sqrt{89}$
11. (a) $\sqrt{320}$ (b) no   **12.** (a) $\sqrt{148}$ (b) no   **13.** $\sqrt{29}$
14. $\sqrt{26}$ **15.** $\sqrt{5}$ **16.** (a) 5   (b) $n = \pm 4$
17. (a) 13   (b) $m = \pm 13$ **18.** (a) 5   (b) $p = 0$
19. (a) 9   (b) 6   (c) 5
20. (a) 30   (b) 5   (c) $\sqrt{50}$ (d) 4

page 197   **Exercise 10**

1. (a) $2\mathbf{a}$; $3\mathbf{b}$ (b) $-\mathbf{b} + \mathbf{a}$ (c) $-3\mathbf{b} + 2\mathbf{a}$ (d) $4\mathbf{a} - 3\mathbf{b}$
   (e) $4\mathbf{a} - 6\mathbf{b}$ (f) $\overrightarrow{EC} = 2\overrightarrow{ED}$
2. (a) $2\mathbf{b}$; $\frac{5}{2}\mathbf{a}$ (b) $-\mathbf{a} + \mathbf{b}$ (c) $-\frac{5}{2}\mathbf{a} - 2\mathbf{b}$ (d) $-5\mathbf{a} + 6\mathbf{b}$
   (e) $-\frac{15}{2}\mathbf{a} + 6\mathbf{b}$ (f) $\overrightarrow{XC} = 3\overrightarrow{XY}$
3. (a) $-\mathbf{b} + \mathbf{a}$; $-3\mathbf{b} + 3\mathbf{a}$ (b) $-2\mathbf{b} + \frac{3}{2}\mathbf{a}$ (c) $-\frac{1}{2}\mathbf{a}$; $-2\mathbf{b} + \frac{3}{2}\mathbf{a}$
4. (a) $-\mathbf{a} + \mathbf{b}$; $-\frac{2}{3}\mathbf{a} + \frac{2}{3}\mathbf{b}$ (b) $\frac{1}{2}\mathbf{a}$; $-\frac{1}{6}\mathbf{a} + \frac{2}{3}\mathbf{b}$ (c) $-\frac{1}{4}\mathbf{a} + 2\mathbf{b}$ (d) $\overrightarrow{MX} = 3\overrightarrow{MP}$
5. (a) $-\mathbf{b} + \mathbf{a}$; $-3\mathbf{a} + \mathbf{b}$ (b) $-\frac{3}{2}\mathbf{a} + \frac{1}{2}\mathbf{b}$ (c) $(k - \frac{3}{2})\mathbf{a} + (\frac{1}{2} - k)\mathbf{b}$ (d) $k = \frac{3}{2}$
6. (a) $-\mathbf{a} + \mathbf{b}$ (b) $-\frac{1}{4}\mathbf{a} + \frac{1}{4}\mathbf{b}$ (c) $\frac{3}{4}\mathbf{a} + \frac{1}{4}\mathbf{b}$ (d) $\mathbf{a} + (m - 1)\mathbf{b}$
   (e) $m = \frac{4}{3}$
7. (a) $-\mathbf{c} + \mathbf{d}$ (b) $-\frac{1}{5}\mathbf{c} + \frac{1}{5}\mathbf{d}$ (c) $\frac{4}{5}\mathbf{c} + \frac{1}{5}\mathbf{d}$ (d) $\mathbf{c} + (n - 1)\mathbf{d}$
   (e) $n = \frac{5}{4}$
8. (a) $-\mathbf{a} + \mathbf{b}$; $-\frac{1}{2}\mathbf{a} + \frac{1}{2}\mathbf{b}$; $\frac{1}{2}\mathbf{a} + \frac{1}{2}\mathbf{b}$ (b) $\frac{1}{3}\mathbf{a} + \frac{1}{3}\mathbf{b}$ (c) $-\frac{2}{3}\mathbf{a} + \frac{1}{3}\mathbf{b}$
   (d) $-\mathbf{a} + \frac{1}{2}\mathbf{b}$ (e) $m = \frac{2}{3}$
9. (a) $-\mathbf{a} + \mathbf{b}$ (b) $\frac{1}{2}\mathbf{b}$ (c) $-\mathbf{a} + \mathbf{c}$ (d) $-\frac{1}{2}\mathbf{a} + \frac{1}{2}\mathbf{c}$
   (e) $\frac{1}{2}\mathbf{a} + \frac{1}{2}\mathbf{c}$ (f) $-\frac{1}{2}\mathbf{b} + \frac{1}{2}\mathbf{a} + \frac{1}{2}\mathbf{c}$ (g) $\mathbf{a} + \mathbf{c} = \mathbf{b}$
10. (a) $-\mathbf{b} + \mathbf{a}$ (b) $m\mathbf{a} + (1 - m)\mathbf{b}$ (c) $4\mathbf{a} + 2\mathbf{b}$ (d) $n = \frac{1}{6}, m = \frac{2}{3}$
11. (a) $-\mathbf{c} + \mathbf{d}$; $-\frac{1}{4}\mathbf{c} + \frac{1}{4}\mathbf{d}$; $\frac{3}{4}\mathbf{c} + \frac{1}{4}\mathbf{d}$ (b) $-\mathbf{c} + \frac{1}{2}\mathbf{d}$

   (c) $(1 - h)\,\mathbf{c} + \dfrac{h}{2}\mathbf{d}$ (d) $(1 - h)\,\mathbf{c} + \dfrac{h}{2}\mathbf{d} = k.\frac{3}{4}\mathbf{c} + \dfrac{k}{4}\mathbf{d}$; $h = \frac{2}{5}, k = \frac{4}{5}$

page 199   **Exercise 11**

1. (a) 5, 10, 1   (b) 21, 101, -29
16. (a) $-9, 11, \frac{1}{2}$ (b) $0.8, -2.7, \frac{1}{80}$ (c) 4, 1·2, 36
17. (a) 0   (b) 6   (c) 12
18. (a) 10   (b) $\frac{1}{2}$ (c) 2
19. (a) $\frac{2}{3}, 24, 6$ (b) $0, \sqrt{2}, \sqrt{6}$ (c) $-6, 6, 9\frac{3}{4}$
20. (a) ± 3 (b) ± 3 (c) ± 2 (d) ± 6

21. (a)10, 21   (b) 111, 411, 990, 112
22. (a) 7   (b) 10   (c) 5   (d) 14
   (e) 7   (f) 7
23. (a) 3   (b) 6   (c) 8   (d) 10
24. (a) 11   (b) 17   (c) 7
25. (a) 5   (b) 17   (c) $1\frac{1}{2}$ (d) 3
26. $a = 3, b = 5$ **27.** $a = 2, b = -5$ **28.** $a = 7, b = 1$

page 201   ***Exercise 12***

1. (a) $x \mapsto 4(x + 5)$ (b) $x \mapsto 4x + 5$ (c) $x \mapsto (4x)^2$ (d) $x \mapsto 4x^2$
   (e) $x \mapsto x^2 + 5$ (f) $x \mapsto 4(x^2 + 5)$ (g) $x \mapsto [4(x + 5)]^2$
2. (a) $-2{\cdot}5$ (b) $\pm\sqrt{\frac{5}{3}}$
3. (a) $x \mapsto 2(x - 3)$ (b) $x \mapsto 2x - 3$ (c) $x \mapsto x^2 - 3$ (d) $x \mapsto (2x)^2$
   (e) $x \mapsto (2x)^2 - 3$ (f) $x \mapsto (2x - 3)^2$
4. (a) 2   (b) 11   (c) 6   (d) 2   (e) 1   (f) 64
5. (a) -3 (b) 2   (c) $1\frac{1}{2}$ (d) 5
6. (a) $x \to 2(3x - 1) + 1$ (b) $x \to 3(2x + 1) - 1$ (c) $x \to 2x^2 + 1$
   (d) $x \to (3x - 1)^2$ (e) $x \to 2(3x - 1)^2 + 1$ (f) $x \to 3(2x^2 + 1) - 1$
7. (a) 11   (b) 9   (c) 11   (d) 14   (e) 81   (f) -1
8. (a) 2   (b) 0, 2   (c) $\pm\sqrt{2}$

9. $x \mapsto \dfrac{x + 2}{5}$ **10.** $x \mapsto \dfrac{x}{5} + 2$ **11.** $x \mapsto \dfrac{x}{6} - 2$ **12.** $x \mapsto \dfrac{3x - 1}{2}$

13. $x \mapsto \dfrac{4x}{3} + 1$ **14.** $x \mapsto \dfrac{(x + 6)/2 - 4}{3}$ **15.** $x \mapsto \dfrac{2(x - 10) - 4}{5}$ **16.** $x \mapsto \dfrac{x - 3}{-7}$

17. $x \mapsto \dfrac{3x - 12}{-5}$ **18.** $x \mapsto \dfrac{3(x - 2) - 4}{-1}$ **19.** $x \mapsto \dfrac{4(5x + 3) + 1}{2}$ **20.** $x \mapsto \dfrac{(7x/3) - 10}{-2}$

21. $x \mapsto 4[5(x - 7) - 6]$
22. (a) $\sqrt{x}$ (b) log   (c) $x!$ (d) x^2 (e) $\frac{1}{x}$
   (f) tan   (g) $\frac{1}{x}$ (h) $\sqrt{}$ (i) cos   (j) log or ln
   (k) tan   (l) $x!$ (m) cos   (n) sin   (o) cos
   (p) $\frac{1}{x}$ (q) $x!$ (r) log

page 201   ***Revision Exercise 9A***

1. (a) {5}   (b) {1, 3, 5, 6, 7}   (c) {2, 4, 6, 7, 8}   (d) {2, 4, 8}   (e) {1, 2, 3, 4, 5, 8}   **2.** 32
3. (a)   (b)   (c)

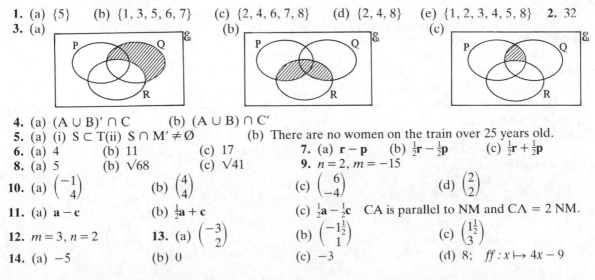

4. (a) $(A \cup B)' \cap C$ (b) $(A \cup B) \cap C'$
5. (a) (i) $S \subset T$(ii) $S \cap M' \neq \emptyset$ (b) There are no women on the train over 25 years old.
6. (a) 4   (b) 11   (c) 17   **7.** (a) $\mathbf{r} - \mathbf{p}$ (b) $\frac{1}{2}\mathbf{r} - \frac{1}{2}\mathbf{p}$ (c) $\frac{1}{2}\mathbf{r} + \frac{1}{2}\mathbf{p}$
8. (a) 5   (b) $\sqrt{68}$ (c) $\sqrt{41}$ **9.** $n = 2, m = -15$

10. (a) $\begin{pmatrix} -1 \\ 4 \end{pmatrix}$ (b) $\begin{pmatrix} 4 \\ 4 \end{pmatrix}$ (c) $\begin{pmatrix} 6 \\ -4 \end{pmatrix}$ (d) $\begin{pmatrix} 2 \\ 2 \end{pmatrix}$

11. (a) $\mathbf{a} - \mathbf{c}$ (b) $\frac{1}{2}\mathbf{a} + \mathbf{c}$ (c) $\frac{1}{2}\mathbf{a} - \frac{1}{2}\mathbf{c}$ CA is parallel to NM and $CA = 2\,NM$.

12. $m = 3, n = 2$ **13.** (a) $\begin{pmatrix} -3 \\ 2 \end{pmatrix}$ (b) $\begin{pmatrix} -1\frac{1}{2} \\ 1 \end{pmatrix}$ (c) $\begin{pmatrix} 1\frac{1}{2} \\ 3 \end{pmatrix}$

14. (a) -5 (b) 0   (c) -3 (d) 8; $ff : x \mapsto 4x - 9$

15. $f^{-1}: x \mapsto \dfrac{(x-4)}{3}$; $h^{-1}: x \mapsto 5x + 2$ (a) 3   (b) $5\frac{1}{3}$
16. (a) 3   (b) 0, 5

page 202   Examination Exercise 9B

1. (a) 3   (b) 7   (c) (i) 15   (ii) 6
(d) Can only use word processor.
2. $\vec{OK} = \frac{1}{2}\mathbf{a}$, $\vec{OL} = 2\mathbf{b}$, $\vec{KB} = \mathbf{b} - \frac{1}{2}\mathbf{a}$, $\vec{AL} = 2\mathbf{b} - \mathbf{a}$; $4:1$
3. $\vec{AC} = \mathbf{x} + \mathbf{y}$, $\vec{AO} = \mathbf{y}$, $\vec{CD} = \mathbf{y} - \mathbf{x}$, $\vec{BF} = \mathbf{y} - 2\mathbf{x}$
4. (b) $\vec{BC} = \begin{pmatrix} 3 \\ -2 \end{pmatrix}$, $\vec{CD} = \begin{pmatrix} -3 \\ -2 \end{pmatrix}$ (d) $\vec{DE} = \begin{pmatrix} -3 \\ 2 \end{pmatrix}$
(e) Continues $B \to C \to D \to E$ and so on.
5. (a) $\frac{2}{5}\mathbf{b}, \frac{3}{5}\mathbf{b}, \mathbf{a} - \mathbf{b}$ (b) $\frac{3}{5}\mathbf{a} - \frac{3}{5}\mathbf{b}, \frac{3}{5}a$
(c) trapezium   (d) $\frac{25}{9}$ (e) 27 cm^2
6. (a) $\frac{1}{3}\mathbf{a} - \mathbf{b}, \frac{1}{5}\mathbf{a} - \frac{3}{5}\mathbf{b}$
(b) $\frac{3}{5}$ (c) (i) 20 cm^2 (ii) 12 cm^2 (iii) 12 cm^2
7. (a) (i) $x^2 - 7$ (ii) $x - 1$ (b) $x = 3$
8. (iii) $\dfrac{2}{\sin x}$, 3·11
9. (a) 16   (b) 2   (d) 32   (e) $x \mapsto 2x + 16$

PART 10

page 206   Exercise 1

1. (a) Squash   (b) 160   (c) 10
3. (a) £3000   (b) £4000   (c) £6000   (d) £11 000
4. red 50°; green 70°; blue 110°; yellow 40°; pink 90°
5. eggs 270°; milk 12°; butter 23·4°; cheese 54°; salt/pepper 0·6°
6. (a) A 60°; B 100°; C 60°; D 140°; E 0°   (b) A 50°; B 75°; C 170°; D 40°; E 25°
(c) A 48·5°; B 76·2°; C 62·3°; D 96·9°; E 76·2°
7. 18°, 54°, 54°, 234°   **8.** 80°, 120°, 160°   **9.** $x = 8$ **10.** 100
11. 21·6°, 1·8°   **12.** (a) 22·5%   (b) $x = 45°$, $y = 114°$
15. *Area* of second apple looks much more than twice area of first.
16. Vertical axis starts at 130.

page 208   Exercise 2

1. (a) mean = 6;   median = 5;   mode = 4.   (b) mean = 9;   median = 7 ; mode = 7.
(c) mean = 6·5;   median = 8;   mode = 9.   (d) mean = 3·5; median = 3·5; mode = 4.
2. (a) mean = 7·82; median = 8;   mode = 8.   (b) mean = 5;   median = 4; mode = 4.
(c) mean = 2·1;   median = 2·5; mode = 4.   (d) mean = $\frac{13}{18}$;   median = $\frac{1}{2}$; mode = $\frac{1}{2}$.
3. 78 kg   **4.** 35·2 cm   **5.** (a) 2   (b) 9
6. (a) 20·4 m   (b) 12·8 m   (c) 1·66 m   **7.** 55 kg   **8.** 12
9. 3·38   **10.** 3·475
11. (a) mean = 3·025; median = 3; mode = 3.   (b) mean = 17·75; median = 17; mode = 17.
(c) mean = 3·38;   median = 4; mode = 4.
12. (a) 5·17   (b) 5   **13.** (i) 9   (ii) 9   (iii) 15
14. (i) 5   (ii) 10   (iii) 10   **15.** 12   **16.** $3\frac{2}{3}$
17. 4·68   **18.** (a) (i) 0·6   (ii) 0   (b) 0·675

19. (a) N (b) mean $= N^2 + 2$; median $= N^2$ (c) 2

20. $\dfrac{ax + by + cz}{a + b + c}$

page 211   *Exercise 3*

1. (a) 45   (b) 28, 60   (c) 32   (d) 35   (e) 25

2. (a) 31   (b) 26, 38   (c) 12   (d) 29   (e) 80

3. (a) 20 kg   (b) 10·5kg   **4.** (a) 80·5 cm   (b) 22 cm

5. (a) 71 s   (b) 20 s   **6.** (a) 45   (b) 14

7. (a) 36·5 g   (b) 20 g   (c) 15

8. (a) 26   (b) 25·8   (c) 26·1

page 213   *Exercise 4*

1. (a) $\frac{1}{13}$ (b) $\frac{1}{2}$ (c) $\frac{1}{52}$ (d) $\frac{3}{52}$

2. (a) $\frac{1}{6}$ (b) $\frac{1}{2}$ (c) $\frac{1}{2}$ (d) $\frac{1}{3}$

3. (a) $\frac{1}{4}$ (b) $\frac{1}{2}$

4. (a) (i) $\frac{3}{5}$ (ii) $\frac{2}{5}$ (b) (i) $\frac{5}{9}$ (ii) $\frac{4}{9}$

5. (a) $\frac{5}{13}$ (b) $\frac{4}{13}$ (c) $\frac{2}{13}$ (d) $\frac{3}{13}$

     (e) $\frac{1}{13}$ (f) 0   (g) $\frac{3}{13}$

6. (a) $\frac{1}{11}$ (b) $\frac{2}{11}$ (c) 0   (d) $\frac{1}{11}$

7. (a) $\frac{13}{49}$ (b) $\frac{3}{49}$ (c) $\frac{10}{49}$ (d) $\frac{1}{49}$

8. (a) $\frac{1}{8}$ (b) $\frac{3}{8}$ (c) $\frac{1}{8}$ (d) $\frac{7}{8}$

9. (a) $\frac{5}{11}$ (b) $\frac{7}{22}$ (c) $\frac{15}{22}$ (d) $\frac{17}{22}$

10. (a) $\frac{1}{2}$ (b) $\frac{3}{25}$ (c) $\frac{9}{100}$ (d) $\frac{3}{25}$

     (e) $\frac{1}{50}$

11. (a) $\frac{1}{12}$ (b) $\frac{1}{36}$ (c) $\frac{5}{18}$ (d) $\frac{1}{6}$

     (e) $\frac{1}{6}$; most likely total $= 7$.

12. (a) $\frac{1}{12}$ (b) $\frac{5}{36}$ (c) $\frac{2}{3}$ (d) $\frac{1}{12}$

     (e) $\frac{1}{36}$

13. (a) 1   (b) 0   (c) 1   (d) 0

14. (a) $\frac{3}{8}$ (b) $\frac{1}{16}$ (c) $\frac{15}{16}$ (d) $\frac{1}{4}$

15. $\frac{5}{999}$ **16.** $\frac{271}{1000}$ **17.** $\frac{x}{12}$, 3

18. (a) $\frac{1}{144}$ (b) $\frac{1}{18}$ (c) $\frac{1}{72}$; head, tail and total of 7.

19. (a) $\frac{1}{216}$ (b) $\frac{1}{72}$ (c) $\frac{1}{8}$ (d) $\frac{5}{108}$

     (e) $\frac{5}{72}$ (f) $\frac{1}{36}$

page 215   *Exercise 5*

1. (a) $\frac{1}{13}, \frac{1}{6}$ (b) $\frac{1}{78}$ **2.** (a) $\frac{1}{2}$ (b) $\frac{1}{2}$ (c) $\frac{1}{4}$

3. (a) $\frac{1}{13}$ (b) $\frac{1}{13}$ (c) $\frac{2}{13}$ **4.** (a) $\frac{1}{78}$ (b) $\frac{1}{104}$ (c) $\frac{1}{24}$

5. (a) $\frac{1}{16}$ (b) $\frac{1}{169}$ (c) $\frac{9}{169}$ (d) $\frac{1}{8}$ (e) $\dfrac{2}{52^2}\left(=\dfrac{1}{1352}\right)$

6. (a) $\frac{1}{16}$ (b) $\frac{25}{144}$ (c) $\frac{5}{18}$ (d) $\frac{25}{72}$

7. (a) $\frac{1}{121}$ (b) $\frac{9}{121}$ (c) $\frac{18}{121}$ **8.** (a) $\frac{1}{288}$ (b) $\frac{1}{72}$ (c) $\frac{1}{32}$

9. (a) $\frac{1}{125}$ (b) $\frac{1}{125}$ (c) $\frac{1}{10000}$ (d) $\frac{3}{500}$ (e) $\frac{3}{500}$

10. (a) $\frac{1}{5}$ (b) $\frac{18}{25}$ (c) $\frac{1}{20}$ (d) $\frac{2}{25}$ (e) $\frac{77}{100}$

11. (a) $\frac{1}{9}$ (b) $\frac{4}{27}$ (c) $\frac{4}{9}$

page 217   *Exercise 6*

1. (a) $\frac{49}{100}$ (b) $\frac{9}{100}$ **2.** (a) $\frac{9}{64}$ (b) $\frac{15}{64}$ **3.** (a) $\frac{7}{15}$ (b) $\frac{1}{15}$

4. (a) $\frac{2}{9}$ (b) $\frac{2}{15}$ (c) $\frac{1}{45}$ (d) $\frac{14}{45}$

5. (a) $\frac{1}{12}$ (b) $\frac{1}{6}$ (c) $\frac{1}{3}$ (d) $\frac{2}{9}$

6. (a) $\frac{1}{216}$ (b) $\frac{125}{216}$ (c) $\frac{25}{72}$ (d) $\frac{91}{216}$

7. (a) $\frac{1}{64}$ (b) $\frac{5}{32}$ (c) $\frac{27}{64}$

8. (a) $\frac{1}{6}$ (b) $\frac{1}{30}$ (c) $\frac{1}{30}$ (d) $\frac{29}{30}$

9. (a) $\frac{27}{64}$ (b) $\frac{1}{64}$ **10.** (a) 6   (b) $\frac{1}{3}$

11. (a) $\frac{1}{64}$ (b) $\frac{27}{64}$ (c) $\frac{9}{64}$ (d) $\frac{27}{64}$; Sum = 1

12. (a) $\frac{3}{20} \times \frac{2}{19} \times \frac{1}{18} \left(= \frac{1}{1140}\right)$ (b) $\frac{1}{4} \times \frac{4}{19} \times \frac{1}{6} \left(= \frac{1}{114}\right)$ (c) $\left(\frac{3}{20} \times \frac{5}{19} \times \frac{12}{18}\right) \times 6$

   (d) $\frac{5}{20} \times \frac{4}{19} \times \frac{3}{18} \times \frac{2}{17}$

13. (a) $\dfrac{1}{10\,000}$ (b) $\dfrac{523}{10000}$ (c) $\dfrac{9^4}{10^4}$

14. (a) $\dfrac{x}{x+y}$ (b) $\dfrac{x(x-1)}{(x+y)(x+y-1)}$ (a) $\dfrac{2xy}{(x+y)(x+y-1)}$ (d) $\dfrac{y(y-1)}{(x+y)(x+y-1)}$

15. (a) $\dfrac{x}{z}$ (b) $\dfrac{x(x-1)}{z(z-1)}$ (c) $\dfrac{2x(z-x)}{z(z-1)}$

16. $\frac{3}{10}$ **17.** $\frac{9}{140}$ **18.** (a) 5   (b) $\frac{1}{64}$

19. (a) $\frac{3}{5}$ (b) $\frac{1}{3}$ (c) $\frac{2}{15}$ (d) $\frac{2}{21}$ (e) $\frac{1}{7}$ (f) $\frac{1}{35}$

20. (a) $\frac{1}{220}$ (b) $\frac{1}{22}$ (c) $\frac{3}{11}$ (d) 5   (e) 300

21. (a) $\dfrac{10 \times 9}{1000 \times 999}$ (b) $\dfrac{990 \times 989}{1000 \times 999}$ (c) $\dfrac{2 \times 10 \times 990}{1000 \times 999}$

22. (a) $\frac{3}{20}$ (b) $\frac{7}{20}$ (c) $\frac{1}{2}$

23. (a) $0 \cdot 00781$ (b) $0 \cdot 511$

24. (a) $\frac{21}{506}$ (b) $\frac{455}{2024}$ (c) $\frac{945}{2024}$

page 219   *Revision Exercise 10A*

1. $162°$ **2.** $41 \cdot 7\%$ **3.** $54°$

4. (a) 84   (b) $19 \cdot 2$ **5.** (a) $3 \cdot 4$ (b) 3   (c) 3

6. (a) $5 \cdot 45$ (b) 5   (c) 5   **7.** $1 \cdot 552$ m   **8.** 3

9. $\frac{1}{6}$ **10.** $\frac{4}{25}$ **11.** $\frac{8}{27}$ **12.** $\frac{5}{16}$

13. (a) $\frac{1}{28}$ (b) $\frac{15}{28}$ (c) $\frac{3}{7}$

14. (a) $\dfrac{x}{x+5}$ (b) $\left(\dfrac{x}{x+5}\right)^2$; $\dfrac{x}{x+5}$, $\dfrac{x(x-1)}{(x+5)(x+4)}$ **15.** $\frac{1}{19}$

16. (a) $\frac{1}{32}$ (b) $\frac{1}{256}$ **17.** (a) $\frac{1}{8}$ (b) $\frac{1}{2}$ **18.** $\frac{35}{48}$

19. (a) $\frac{1}{9}$ (b) $\frac{1}{12}$ (c) 0   **20.** $\dfrac{1}{20^3}$

page 220   *Examination Exercise 10B*

1. (a) 16   (b) 54   (c) $1 \cdot 8$

2. (a) £20   (c) £50   (d) $60°$

3. (b) (i) £114   (ii) £$3 \cdot 80$

4. (ii) (a) 190 g   (b) 70   (iii) (a) 36 kg   (b) 23 kg

5. (b) (i) 4600   (ii) 54   (iii) 23   (iv) 20%   (v) 40

6. (a) $144°$ (b) 5%

7. (a) (i) $\frac{7}{30}, \frac{23}{30}, \frac{1}{3}, \frac{2}{3}$ (ii) right-handed   (iii) left-handed

     (b) (i) $\frac{3}{5}$ (ii) $\frac{5}{12}$ (iii) $\frac{3}{20}$

8. (b) (i) $\frac{3}{10}$ (ii) $\frac{3}{5}$

9. (i) $\frac{1}{2}, \frac{1}{6}, \frac{1}{3}$ (ii) $\frac{1}{6}$ (iii) $\frac{1}{8}$

10. (a) $\frac{1}{6}$ (b) $\frac{1}{3}$ (c) $\frac{2}{5}$

11. (a) (i) $\frac{1}{3}$ (ii) $\frac{2}{3}$ (iii) $\frac{1}{3}$ (b) Yes

PART 12   Mental arithmetic

page 251   **Test 1**

1. 20, 10, 5, 2	**2.** £4·40	**3.** £2·10	**4.** £26	**5.** 8 min
6. 25	**7.** 500 ($\pm$50)	**8.** $\frac{1}{1000}$	**9.** 2·65	**10.** —
11. £15 000	**12.** £3·85	**13.** £27·50	**14.** 8	**15.** 84
16. 108	**17.** 30 litres	**18.** 7 cm	**19.** £2	**20.** 45
21. £17·70	**22.** £33	**23.** 1950	**24.** 40° ($\pm$10°)	**25.** 55 litres

page 252   **Test 2**

1. 153	**2.** 4	**3.** 7	**4.** 105 sq. yd.	**5.** 2 000 000
6. 51	**7.** 6	**8.** £6	**9.** 32	**10.** 150
11. 133	**12.** Wednesday	**13.** 34p	**14.** £9	**15.** 60 mph
16. 302	**17.** 50	**18.** 37	**19.** £62·67	**20.** 44
21. 42·4%	**22.** £157	**23.** 050° ($\pm$10°)	**24.** £10·30	**25.** £96·99

page 253   **Test 3**

1. 60°	**2.** 0·05	**3.** 80%	**4.** 8000	**5.** £16·90
6. 0·7	**7.** 5 h 20 min	**8.** $12\frac{1}{2}$%	**9.** 6 cm	**10.** 0·001
11. 7, 8	**12.** £2	**13.** 1·8	**14.** 49·2	**15.** £12·50
16. 165	**17.** 72°	**18.** 240	**19.** 105 mph	**20.** 1 h 30 min
21. £2500	**22.** 2013	**23.** £174	**24.** 37·6%	**25.** £78

page 254   **Test 4**

1. 82°	**2.** 72p	**3.** 0·25	**4.** 90 n. miles	**5.** 8
6. 11	**7.** 325	**8.** $\frac{1}{12}$	**9.** £25 000	**10.** 49 000
11. 6·3	**12.** £8·70	**13.** £2·40	**14.** 5	**15.** £37·50
16. 13·55 cm	**17.** 10 cm	**18.** £9	**19.** £12	**20.** £341
21. 10 min	**22.** 62p	**23.** 100° ($\pm$10°)	**24.** £390	**25.** 120° ($\pm$10°)

page 255   **Test 5**

1. 200	**2.** £1·20	**3.** 250	**4.** 500 m^2	**5.** £8·70
6. 0·025	**7.** 2550 g	**8.** £40 000	**9.** 150°	**10.** £150
11. 11	**12.** 9	**13.** £1·11	**14.** £15	**15.** 12
16. 8, 9	**17.** £80	**18.** £3·80	**19.** 18·3 s	**20.** 4 h 50 min
21. £5·10	**22.** £25	**23.** £2·17	**24.** £52	**25.** 48

page 256   **Test 6**

1. 50, 20, 5, 1	**2.** 86	**3.** 6	**4.** £7·50	**5.** 20 cm
6. 4	**7.** 100	**8.** $7\frac{3}{4}$	**9.** 50	**10.** 2997 m
11. £25	**12.** 120	**13.** 2	**14.** 60 p	**15.** £12
16. 39°	**17.** 1440	**18.** 10 cm (± 1 cm)	**19.** Mazda	**20.** 35 min
21. £99·85	**22.** 2107	**23.** 250° ± 10°	**24.** £83	**25.** 72° ± 5°

page 257   **Test 7**

1. (50, 10, 2, 1, 1) or (20, 20, 20, 2, 2) or (50, 5, 5, 2, 2)

2. 1 h 40 min	**3.** £10·50	**4.** 7200	**5.** 61	**6.** 146
7. 2 500 000	**8.** £18	**9.** £50	**10.** £750	**11.** 1888
12. 20	**13.** 71·6 kg	**14.** 30 p	**15.** 3	**16.** 6 cm
17. 50 litres	**18.** 6	**19.** 34 p	**20.** 66 litres	**21.** £80
22. 280° ± 10°	**23.** £54	**24.** 85 min	**25.** £9·32	

PART 12   Revision tests

---

page 258   **Test 1**

1. C	**2.** D	**3.** D	**4.** B	**5.** A
6. C	**7.** A	**8.** D	**9.** C	**10.** B
11. C	**12.** A	**13.** D	**14.** C	**15.** C
16. D	**17.** A	**18.** C	**19.** B	**20.** D
21. A	**22.** C	**23.** B	**24.** B	**25.** C

page 259   **Test 2**

1. B	**2.** C	**3.** A	**4.** B	**5.** D
6. C	**7.** A	**8.** D	**9.** B	**10.** C
11. B	**12.** D	**13.** A	**14.** D	**15.** C
16. D	**17.** B	**18.** A	**19.** B	**20.** B
21. C	**22.** D	**23.** A	**24.** A	**25.** B

page 260   **Test 3**

1. D	**2.** D	**3.** D	**4.** B	**5.** A
6. C	**7.** A	**8.** D	**9.** D	**10.** B
11. C	**12.** D	**13.** D	**14.** B	**15.** A
16. B	**17.** C	**18.** A	**19.** D	**20.** D
21. C	**22.** A	**23.** B	**24.** B	**25.** D

page 262   **Test 4**

1. B	**2.** B	**3.** A	**4.** C	**5.** D
6. A	**7.** B	**8.** C	**9.** D	**10.** B
11. C	**12.** A	**13.** B	**14.** B	**15.** D
16. A	**17.** C	**18.** B	**19.** B	**20.** D
21. A	**22.** C	**23.** C	**24.** C	**25.** A